Springer Series in Reliability Engineering

Series Editor

Professor Hoang Pham
Department of Industrial and Systems Engineering
Rutgers, The State University of New Jersey
96 Frelinghuysen Road
Piscataway, NJ 08854-8018
USA

Other titles in this series

The Universal Generating Function in Reliability Analysis and Optimization
Gregory Levitin

Warranty Management and Product Manufacture
D.N.P. Murthy and Wallace R. Blischke

Maintenance Theory of Reliability
Toshio Nakagawa

System Software Reliability
Hoang Pham

Reliability and Optimal Maintenance
Hongzhou Wang and Hoang Pham

Applied Reliability and Quality
B.S. Dhillon

Shock and Damage Models in Reliability Theory
Toshio Nakagawa

Risk Management
Terje Aven and Jan Erik Vinnem

Satisfying Safety Goals by Probabilistic Risk Assessment
Hiromitsu Kumamoto

Offshore Risk Assessment (2nd Edition)
Jan Erik Vinnem

The Maintenance Management Framework
Adolfo Crespo Márquez

Human Reliability and Error in Transportation Systems
B.S. Dhillon

Complex System Maintenance Handbook
D.N.P. Murthy and Khairy A.H. Kobbacy

Recent Advances in Reliability and Quality in Design
Hoang Pham

Product Reliability
D.N.P. Murthy, Marvin Rausand and Trond Østerås

Mining Equipment Reliability, Maintainability, and Safety
B.S. Dhillon

Advanced Reliability Models and Maintenance Policies
Toshio Nakagawa

Justifying the Dependability of Computer-based Systems
Pierre-Jacques Courtois

Reliability and Risk Issues in Large Scale Safety-critical Digital Control Systems
Poong Hyun Seong

Failure Rate Modeling for Reliability and Risk
Maxim Finkelstein

Göran Grimvall • Åke J. Holmgren
Per Jacobsson • Torbjörn Thedéen
Editors

Risks in Technological Systems

Editors
Göran Grimvall, Prof. Dr.
KTH Royal Institute of Technology
Department of Physics
Albanova Science Center
SE-106 91 Stockholm
Sweden
grimvall@kth.se

Per Jacobsson
KTH Royal Institute of Technology
SE-100 44 Stockholm
Sweden
pjn@syd.kth.se

Åke J. Holmgren, Dr.
The Swedish Civil Contingencies Agency
(MSB)
SE-651 81 Karlstad
Sweden
ake.holmgren@msbmyndigheten.se

Torbjörn Thedéen, Prof. Dr.
KTH Royal Institute of Technology
SE-100 44 Stockholm
Sweden
tort@kth.se

ISSN 1614-7839
ISBN 978-1-4471-2519-8
e-ISBN 978-1-84882-641-0
DOI 10.1007/978-1-84882-641-0

British Library Cataloguing in Publication Data
A catalogue record for this book is available from the British Library

Cover design: deblik, Berlin, Germany

Printed on acid-free paper

Springer is part of Springer Science+Business Media (www.springer.com)

Preface

The first edition of this book was published in 1998 by the Swedish Educational Broadcasting Company. The book, four radio tapes, and six TV programs were the framework of a Swedish Open University academic course. This Swedish edition was used frequently in engineering education and in courses for professionals, until the book was sold out in 2002. In 2003 an updated and restructured Swedish edition was published. It is now a course book in Swedish engineering education and it is frequently used in the education of professionals in industrial risk management.

The present English edition is based on the experience of using the two previous editions in both university and professional environments. In addition, many sections with references to specific Swedish circumstances have now been replaced by international data and outlooks.

The book provides an introduction to basic technology risks. It also aims at placing these risks in a wider interdisciplinary and global framework than is customary, and seeks to show both their social importance and their intellectually challenging character. Two of its main purposes are to stimulate critical reflection on the issue of technology risks and to provide the conceptual tools necessary for grasping them. Further, it seeks to serve the reader by clarifying key concepts, outlining alternative views, and providing both rich references and exercises for in-depth studies.

The editors, who have worked closely with the contributors, take an overall responsibility for the content of the book, but the ultimate responsibility for each chapter lies with its contributors. Finally, the editors want to thank Anthony Bristow for his linguistic corrections and improvements.

Economic support from Swedish Governmental Agency for Innovation Systems (VINNOVA) and The Royal Institute of Technology (KTH) is gratefully acknowledged.

Stockholm, September 2009

Göran Grimvall, Åke J. Holmgren, Per Jacobsson, and Torbjörn Thedéen

Contents

1

Introduction: the Global Risk Arena, Technological Systems and This Book

Per Jacobsson, Göran Grimvall, Åke J. Holmgren, and Torbjörn Thedéen

1.1 Technological Risk-driving Forces

The global citizen of the twenty-first century faces a wide spectrum of risks, such as from transportation, ionizing radiation, the propagation of contagious diseases, terrorism, natural disasters, and global climate change. Some of them, for instance the risk of a motorcycle accident, affect only individuals and have merely occasional direct impacts on society. Nevertheless the accumulation of these individual risks is leading to major long-term changes in risk patterns.

More importantly, however, large-scale global risks are emerging which affect the infrastructure systems on which society depends for the distribution of energy, water and food and for transport, telecommunication, and monetary transactions, to mention but a few. The particular risks associated with technological systems are at the heart in all these cases.

What are the main driving forces and contexts that are reshaping the risks in technological systems? One answer is to be found in the trends in *demography*, *environment*, and *socioeconomic changes* (OECD, 2003). Another answer can be found by considering issues such as *connectedness* and *mobility*, which are often characteristics of globalized societies. Finally, *technological failures* naturally have a significant effect on the risk arena. Below, the various forces driving technological risks are discussed. One must, however, emphasize that the forces mentioned are not mutually exclusive but are overlapping.

Demography

It is estimated that the world population will increase to nine billion by 2050, compared to the nearly seven billion in 2008. The additional more than two billion people will live mostly in megacities in developing countries. A third of the population in the developed countries will be over 60 years of age by 2050 (IDB, 2008). In addition, international migration is expected to intensify, particularly from the South to the North. These demographic forces may create a strain on resources and infrastructures and thus lead to the aggravation and spread of technological risks.

Environment

Human activities generating greenhouse gases are likely to cause global warming, and gradually have an immense impact on life in both developed and developing countries (IPCC, 2007). Urbanization and land use are also expected to contribute to the growing human and economic toll of weather-related disasters. Moreover, the degradation of the ecosystem and an increasing water scarcity, floods, and other water-related shocks will be major factors affecting human society all over the world. Thus, environmental changes will become an important factor increasing the known risks of technological systems and creating new ones, as well as for the mitigation and prevention of risks.

Socioeconomic Changes

Policies with regard to risks are influenced not only by governments but also by international bodies, corporations, and the media. The perception of a technological risk (Chapter 16) is nowadays shaped more by the media, than by expert opinions, in contrast to how risks were dealt with historically (Chapter 3) in the traditional society, where risks were related to popular beliefs, myths and customs. People do not behave as if they are rationally considering the probabilities and expected outcomes of alternative actions. The way in which alternative risks are presented strongly influences the real choices, as was shown by Tversky and Kahneman (1981). Media-mediated attitudes to risks are now a major part of any technological risk situation and they must be considered and integrated into policy-making. The global risk arena is becoming more and more complex and it is becoming increasingly difficult to assess risks scientifically. Models and approaches for risk and vulnerability assessment (Chapters 11–14) are limited, although fruitful approaches are being developed to cope with the complexity. As this book shows, a variety of different factors, methods, and approaches are needed for sensible risk identification, assessment, and prevention.

Connectedness and Mobility

Connectedness has been described as a distinctive feature of modern societies, linked to the crucial role of networks (Castels, 1996). Information technologies have made possible a rapid interaction between people, physically separated, to an unprecedented extent. The increased degree of connectedness not only multiplies the channels through which true information on accidents, incidents, and risks is spread, but also helps the rapid dissemination of malevolent information such as rumors relating to risks, computer viruses, and spam (Chapter 10). Connectedness makes individuals and organizations not only more accessible but also more vulnerable, even at a distance.

The increased *mobility* of people, services, goods, and information increases the frequency of interaction (Figure 1.1). These numerous interactions, which can for example involve traffic risks (Chapter 9), may generate new risks, change current risks, or mitigate risks. The large number of interactions makes the risks associated with global technology systems and their assessment more and more complex, where both the scale and the concentration of the consequences are changed. To cope with such risks, strategies, and solutions on a national level is insufficient. International solutions and cooperation as well as an exchange of best practice are essential. Technology systems must be considered in such a context.

Figure 1.1. Aker Kvaerner-H-6E, the world's largest oil rig hull, built at Dubai World dry-dock, transits the Suez Canal on its way to Norway and the North Sea. (Photo: STR/EPA/Scanpix)

Technological Failures

Technological systems have a behavior of their own that is not always fully understood by their designers. For example, a historical analysis of 800 *structural failures* showed that these were almost exclusively due to mistakes by people involved in the planning and design or in the construction procedures (Matousek, 1977). Human factors (Chapter 15), design flaws, material failures, extreme conditions or environments, or, more importantly, combinations of these were stated to be the major explanations. In Chapter 5 several examples of structural failure are described and lessons for improvement are presented. To ensure that structures do not fail or that toxic substances or radiation do not cause harm, safety factors and exposure limits (Chapters 7 and 8) are derived from data relating to failures and their consequences, such as injuries and observed health effects. The way in which technological developments and budding safety cultures have prevented technological failures is richly illustrated in Chapter 4, with the steam engine, an important element in the development of energy technology.

1.2 Technology and Three Kinds of Risk

Several *definitions of technology* have been suggested. Most definitions start with its function. Grübler (1992), for instance, proposes that technology is that which enables humans to extend their capabilities and to accomplish tasks that they could not otherwise perform. According to a more specific definition, technology is knowledge combined with the appropriate means to transform materials, energy-carriers or information from a less desirable to a more desirable form (Ayres, 1994). Some definitions focus narrowly on the most tangible parts of technology, i.e., the physical artifacts or the hardware. Others include processes, infra-structures, and software, which represent knowledge, practices, organizations, and people. The historical trend seems to be that the definition has become increasingly broader, including not only technological but also social and economic systems. This book does not suggest an all-embracing technology definition but favors a broad definition that illustrates the systemic nature of technology, embedded in the social, cultural, environmental, and economic networks.

This book, and the expertise it is based upon, deals with three basic kinds of risk in technological systems. The first is the *unwanted and unintended* consequences of otherwise beneficial technologies that were not foreseen at the time of their introduction; the health impact of chemicals is, for example, dealt with in Chapter 7. The second kind of risk arises from the known and understood adverse consequences of an otherwise beneficial technology, i.e., the *cost* of gaining a certain benefit, examples of which are energy technology risks (Chapter 6) and the risks associated with transport by air, by road, by rail, or at sea (Chapter 9). The third kind of risk involves the *intentional use* of technology to cause injury or damage, the prime example of which is antagonistic computer-related risks (Chapter 10). In addition to describing the main technological risks, the book also describes methods for analyzing and assessing risks and their consequences (Chapters 12–14), economic (Chapter 17) and others, as well as chapters reflecting on risks (Chapter 2) and man, from philosophical or literary viewpoints (Chapters 18 and 19).

1.3 The Philosophy of the Book

This book embodies two basic concepts. The first is that there is no absolutely safe technology, but that all technology can be made safer. Technology comprises much more than the devices and products that shape our environment. It also consists of the large intertwined networks and systems on which we depend for our living, nourishment, clothing, moving, and communicating with each other. Technology is created and operated by fallible and partially autonomous individuals, each with their particular limitations and intentions. Sometimes technology inadvertently leads to negative consequences such as accidents, injuries, and death. Even though the media pay most attention to the large accidents, or catastrophes, it is the ordinary everyday risks that often have the most negative consequences, such as fire risks (Chapter 14) and traffic risks (Chapter 9).

The second concept is that the unintentional but reducible technological risks are closely connected to the advantages that we associate with technology. In our daily life we can either choose to do things or refrain from doing them. Most of these everyday choices are made without any reflection as to the consequences which they may entail. Often we choose to do what we do because we think it will provide us with something desirable, e.g., safety, comfort, thrills, freedom, or enjoyment, although we may have a vague idea that what we choose to do may also lead to negative consequences for ourselves and for others. Through our choices, we run certain risks in order to favor values and needs, which we consider to be important. That which is bad, the risk, is in this case linked to what is good, the satisfaction of our desires, noted in Section 1.2 as the second kind of risk. Not to choose is also a choice, and in the same way as for an active choice this entails risks. When we try to avoid the risks associated with a certain act, we must take into account both the risks that are related to the alternative, and the possible advantages that we lose by choosing the alternative. It means that we, as decision-makers, whether at work or in private, should for each alternative action seek a sensible trade-off between risks and advantages.

1.4 Technological Risks with Which This Book Is Not Concerned

The theme of the book is risks connected to technology, but to limit the scope of the book, we have chosen to exclude economic risks that depend upon technology (e.g., computer crashes on the stock market), technology-related risks in insurance companies, political risks (e.g., the manipulation of information technology by dictatorships), risks associated with drugs and biomedical technology (e.g., gene manipulation), lifestyle risks (e.g., those related to diet, smoking, alcohol, and leisure activities) and risks associated with military technology (e.g., weapon technology development).

Chapter 7 is devoted to health risks but environmental risks have been excluded as we feel that such an important risk area could not be adequately treated in a book like this. However, since it is impossible to describe risks of energy conversion without considering the environmental risks involved, these risks have been superficially dealt with in Chapter 6.

When twenty-two researchers write about risks it is inevitable that certain phenomena recur in several chapters. We have chosen to retain these overlapping parts when they are well integrated into the whole, or are needed for the sake of the context.

1.5 The Readership

The book is addressed to four groups of readers. Firstly, it is an academic textbook for students studying to become, for example, engineers, architects, economists, lawyers, administrators, teachers, and officers. The second group of readers includes professionals who want to make themselves acquainted with risk issues relevant to their trade, company, or organization, e.g., engineers, psychologists, planners, policy-makers, politicians, or employees of insurance companies. To these students and professionals the book provides a broad knowledge of facts, perspectives, and methods for the assessment and prevention of risks, as well as rich opportunities to achieve in-depth knowledge. The third group of readers includes those whose task it is to convey information on risks, e.g., journalists and teachers. The fourth group of readers includes those in the general public who want to be able to form a well-founded opinion of what it may be reasonable to believe in media reports on risks. The book may be helpful to enable the latter two groups to acquire basic facts and to put current risk issues into their proper context.

References

Ayres R (1994) Ecotransition in the chemicals industry. CMER WP, INSEAD, Fontainebleau

Castels M (1996) The rise of the network society. Blackwell, Oxford

Grübler A (1992) Technology and global change: Land-use past and present. IIASA WP-92-69, Laxenburg

IDB (2008) International Database (IDB). US Census Bureau. http://www.census.gov/ipc/www/idb/

IPCC (2007) International Panel on Climate Change. Fourth assessment report. http://www.ipcc.ch/pdf/assessment-report/ar4/syr/ar4_syr_spm.pdf

Matousek M (1977) Outcomings of a survey on 800 structural failures. IABSE Colloquium on inspection and quality control. 5–7 July 1977, Cambridge. Wenaweser + Wolfensberger AG, Zürich

OECD (2003) Emerging risks in the 21st century: an agenda for action. http://www.unisdr.org/eng/library/Literature/7754.pdf

Tversky A, Kahneman D (1981) The framing of decisions and the psychology of choice. Science 211:453–458

2

Reflections on Risks and Technology

Lennart Sjöberg and Torbjörn Thedéen

2.1 Risks and Technology

Human activities are connected with risks. Throughout our history, managing and controlling these risks has built on the experience of generations. Industrialization meant a partly new situation. People began to work increasingly in groups, in mines and in factories. New technology was introduced, although slowly at first and in small establishments. Experience was lacking, and the technology was often difficult to understand. An accident could hit many people at the same time. In order to handle the risks, one tried to combine the few available data on accidents with engineering calculations – the start of what would become modern risk analysis.

The nineteenth century saw the advent of railways. Pressure vessels were used in steam engines and in turbines to produce energy. The large number of accidents connected with this development resulted in an increased public control (Chapter 3). The same development could be seen in other areas. In the beginning there was a patchy technology and, after some accidents – regulations and control. A characteristic feature at that time was that the management of risks came mainly as a reaction, after an accident had already happened.

Today the situation is in many respects different. Achievements such as new aircraft, roll-on-roll-off ships, nuclear reactors, pesticides, and so on represent large systems that were rapidly introduced. The number of units in such systems can be small and the economic lifetime short. Therefore, one cannot expect to gain a thorough experience of all their safety aspects. Instead one must, to an increasing degree, analyze the risks with the help of mathematical models and through a study of the interplay between man, technology, and organization (see, e.g., Chapters 13 and 15).

Media play an increasing role in modern society. News about catastrophes and accidents is rapidly spread over the globe. People find it interesting to read about accidents, and therefore risk problems are brought to public attention.

A substantial advantage of a new technology is often not sufficient for the public. They demand that one can handle, and preferably eliminate, the risks. It can be difficult to explain how the benefits can be weighed against a single possible catastrophe.

2.2 What is a Risk?

What meaning does the concept of "risk" have for individuals, for groups of people, and for the society? What research is carried out on risks? Risk is a word with many meanings. Its origin is unclear, but it probably came from classical Greek and referred to accidents at sea. Through Latin it has since been used in many modern languages. In colloquial language, a risk means a harmful event that may occur, but not with certainty. It can refer both to probability and consequences.

In research, the word risk is given a more precise meaning to describe certain concepts. One common way to use the word is to let it refer to the probability of the occurrence of a harmful event. If there is a measure of how harmful the event is (for instance the number of fatalities in a sea accident), the risk may sometimes mean the product of the probability and the amount of harm. Statisticians call this an expected value, and it has often been used to specify the concept of risk. A third use of the word has to do with the variation in the result, if a certain measure is taken. An example could be the variation in travel time going by train or car, i.e., the risk of delay. A fourth definition is the experienced risk. That is, how large an individual considers the risk to be, with the individual's own interpretation of the word risk (Slovic, 2000; Renn, 2004). Of course, the meaning one gives to a word is to some extent arbitrary. There is no "true" definition of risk. Even so, the concept of risk can be seen as a combination of a random event with negative consequences for human life, health, or environment, and the probability of that event occurring. We exclude, for instance, economic risks, even though they are often part of the basis of a decision.

Risks are usually connected with some kind of decision, for instance whether one should allow or ban the use of a certain pesticide. Often there is a choice between various safety precautions, and their cost is then a part of the problem at hand. One may identify three groups that are affected by a decision, which involves risks: the risk carriers (i.e., those who may be affected by the negative consequences), the benefit (cost) takers (i.e., those who benefit from the decision and pay for the action) and finally the decision-makers. It is characteristic of new technological systems that these three groups are not the same. That makes an analysis of risks even more important. Such an analysis contains three elements: What can occur, how likely is its occurrence, and how do we value the consequences?

Risk analysis has a long history in several technological areas, for instance in solid mechanics. A more recent aspect is the need to approach problems about risk from several different points of view. The different chapters in this book give many examples of such interdisciplinary approaches.

2.3 History of Risk Research

Research on risks took big steps forward around 1970. The background was the growing opposition to nuclear power, and concerns about the environment. Suddenly, there was a strong and increasing opinion that created a serious obstacle to certain kinds of technical development and industrial expansion. Technology and industry were no longer considered to be unambiguously good, and the concept of "quality of life" got also a dark side. Media and the public opinion together demanded a new policy that took into consideration the environment and human health – and also recognized their own anxiety, even though such anxiety was sometimes regarded as unjustified by experts.

A first approach was to estimate the magnitude of various risks, in everyday life as well as that of large but rare catastrophes in technical systems and in the environment. Smoking was already then a well-known and thoroughly researched risk; in fact one of the largest risks in the everyday life of smokers and their surroundings. One could compare the risk of smoking with that of living close to a nuclear power plant. The smoker was subject to a much larger risk, if one believed the available risk analyses. But this did not calm those who were against nuclear power, even if they were smokers. Researchers realized that it was necessary to try to understand the factors that affect how people react to risks.

The leading person in the initiation of risk research is Chauncey Starr (1969). He showed that the actual risk gives a very incomplete explanation of how society handles risks. In order to understand social reactions to risks other concepts had to be introduced. Starr suggested voluntariness. We seem to be prepared to accept much larger risks if they are voluntary than if they are forced upon us. This was a fruitful idea, and the origin of much of the subsequent research on risk perception. However, it was only a start. Why is leisure sailing more dangerous than traveling on a car ferry? The latter is not really "involuntary." Many examples can be given where voluntariness seems to be an unlikely explanation of risk acceptance. Other dimensions were suggested, leading to the psychometric model of risk perception (Fischhoff et al., 1978) where dread and novelty of risk emerged as important factors.

Risk appears to be very important in policy discussions. Why is it so? It is not obvious that risk is an important aspect of our actions. We can choose between avoiding risks or taking the chance to get what we strive for. It is not unusual that the most risky alternatives are also those that may give the largest benefit. To bet money on a horse that few believe is a winner means a large risk. It is probable that we shall lose what we have bet. But if, on the other hand, the horse comes in as the winner, we can collect a nice profit. Risk thus has two faces, and we may take risks to achieve what we want; power, money, joy, excitement, etc.

Nevertheless, the public image of risk is mainly negative. The mountaineer takes a risk and he does so voluntarily – he seeks the risk. But does he really believe that there is a significant risk of dying? Hardly so. He believes that, through his skill, the danger can be kept under control. Perhaps he yearns for exactly that feeling of control, and to show courage and skill. One theory, called the theory of risk homeostasis, says that people want a certain level of risk, and therefore adjusts their actions to reach that level. If you use winter tires, the safety on a snowy road increases, but most car drivers who use winter tires drive faster and thereby restore the risk level to some extent, even if they do not do so fully. Is that because they want to drive at a certain risk? The example shows merely a willingness to accept a certain risk level. What the driver really wants is to save time.

2.4 What Is an Acceptable Risk?

We have seen that risk research first started from the notion that people found that certain risks were "too large," and researchers tried to explain this attitude. Later, researchers became interested in what can be called risk denial. For instance, it turns out that most people think that they are exposed to much lower risks than other people. This could be true for many, but not for all. Furthermore, almost all investigations of risk perception show that many people find almost all risks to be utterly small: 3–4 times as many as those who find risks to be alarmingly large. But not all risks are non-existent or extremely small. Society must be prepared to anticipate and reduce risks before serious accidents happen.

Who is to decide whether a certain risk is to be accepted or not? Legislation and regulations can prevent people from taking risks that could hurt themselves, and from exposing other people to risks. The latter aspect is not uncontroversial, but is it reasonable to prevent people from taking risks that would hurt only themselves? Consider, for example, the use of seatbelts in cars. It was relatively rare until it was made mandatory through legislation. Of course, such a law was introduced for the benefit of people, and it involves only a minor limitation of the individual's freedom. Furthermore, society had a strong interest in reducing the high costs for hospital care and other medical consequences – particularly when most of those costs are borne by a social welfare system. But still – is it reasonable that society decides which risks an individual may take?

There is much to be said about ethics and risks. A certain activity can be beneficial for society, but perhaps not equally so for each individual (Figure 2.1). A nuclear power plant is not of much direct use in a sparsely populated part of the country, where hydroelectric power is abundant. Who should bear the risk associated with the nuclear waste? Where should the waste be deposited? Should it be in the densely populated region where the nuclear plants are located, and where most of those live who benefit from the nuclear power? Or should it be deposited in remote areas where few people live, in spite of the fact that the people in this region could meet their energy needs with a small hydroelectric plant?

In the theory of economics, it is assumed that people try to maximize their expected benefit, and that this means that a decision is based on the simultaneous consideration of the possible cost (damage) and the benefit. But risk research has shown that the possible damage is in most cases much more important. The benefit often plays a less prominent role. This is something familiar to decision-makers. In almost all countries it has been difficult to find a site where the local inhabitants are willing to accept a nuclear waste deposit, regardless of assurances that the technical problems have been solved and that there is no risk, now or after thousands of years, to those living nearby. People have not been willing to accept economic benefits in return for what they consider to be health hazards to themselves, their children and their grandchildren. The dominant factor by far is the risk aspect.

Figure 2.1. A crane lifts a nearly 200-ton nuclear reactor safety vessel at the Indira Gandhi Centre for Atomic Research at Kalpakkam, India, June 24, 2008. The reactor is a 500 MW prototype fast breeder and it is planned to begin commercial production by 2011. (Photo: Babu/Reuters/Scanpix)

A common idea, often ascribed to the American psychologist Maslow (Maslow, 1970), is that people's needs can be arranged in a hierarchy. At the lowest level are certain primary needs of food and drink, etc., which must be satisfied before one engages in higher levels of needs. Thus, for example, poor people would not be expected to be as concerned with long-term environmental risks. But research has shown that Maslow was mistaken. Even people living under very difficult conditions retain their ability to engage also in that which lies beyond the toils and threats of their daily lives. This is very important for politicians and businessmen. Many of them have thought that exporting the waste to developing countries could solve part of the environmental problems of the rich world, such as hazardous waste.

But is risk something that really "exists"? Of course, an accident or a disease can hit people, no one denies that, but that is not the same as saying that risks exist (Breakwell, 2007). A risk is an expectation that something negative may happen. Thus it is a subjective phenomenon. Of course, this does not mean that the actual thing, which is expected, is also a subjective phenomenon. On the contrary, accidents and damage are very real. And our expectations contain some valid knowledge, otherwise we would not survive. Having said that, we now must step back and take a new look. Anticipations and worldviews are culturally dependent. In that sense, risk is a culturally dependent phenomenon, and not entirely a function of reality. Many other aspects come into play, but nevertheless – people's anticipations usually have a basis in something real. It would be completely wrong to draw the conclusion that risk perception is entirely a question of subjectivity and culture and that it lacks connection to reality. This question is further considered in Chapter 16.

2.5 Society's Reaction to Risks

In industrially developed countries, people are often concerned with risks associated with technological progress. An increasing part of the issues discussed in political circles deals with risks. It is interesting to ask why this is so. One explanation is that society cannot launch new and expensive welfare reforms to the extent that was previously possible. The politicians therefore turn their attention elsewhere. Another explanation is that we know much more about risks, or that risk levels have actually increased. It is true that we know more, but the risks themselves have hardly increased, as is corroborated by a higher standard of living and a significant improvement in the public health.

The introduction of new technology often leads to fierce debates. Research shows, however, that it is hardly the "novelty" per se that is negative. On the contrary, "novelty" is in many cases a positive argument. Instead, the reaction seems to depend on whether or not the new technology is conceived as being necessary, or perhaps even irreplaceable (Sjöberg, 2002). The Internet is a new technology, as is e-mail, but people do not seem eager to abolish it. On the contrary, the attitude is very positive. Railways represented a new technology in the nineteenth century, and in spite of accidents, few advocated that the building of railways should be stopped. Why not? The reason was probably that the new technology had many positive aspects offering faster and cheaper, and even safer, travel. At that time it was an irreplaceable technology.

Why does it seem to be a characteristic of man to worry about risks and to react to them? We can only speculate about this question, but it seems reasonable to view risk avoidance as being evolutionarily important. Of course that holds particularly for a being that does not have big muscles and sharp teeth and claws, and is not particularly fit to run fast and for long time to escape an enemy. Such a being must be prudent and anticipating and avoid the danger before it becomes a reality, i.e., to engage in matters of risk.

In modern society, we are constantly being confronted with media, and the media present a lot of information about risks. Some psychologists suggested that it is the intensity of the media coverage that creates the risk perception (Combs and Slovic, 1979). That is a common view, which strongly appeals to common sense, but it has been difficult to verify the hypothesis. Of course it is often media that inform us about the existence of a certain risk, for instance when it became clear in 1996 that the "mad cow" disease can be transferred to humans through food. Since then media have been quick to note anything that is related to this new disease. But is it the media per se that create the risk perception? There are other possible explanations, for instance the associations and ideas evoked by the concept of the "mad cow" disease. We have a system of concepts ready to help us understand new situations. One would not like to eat beef from a "mad cow." "Nuclear waste" sounds threatening, dirty, and dangerous through the word "nuclear" and its connection to nuclear weapons, and the generally unpleasant word "waste," adds to this feeling. It becomes even worse if it is called "nuclear garbage," as realized by opponents to nuclear energy.

There are risks of many different kinds. Technological risks are just one form, another being natural catastrophes. People react differently to nature and its risks. Technology that is perceived as disturbing the order of nature, and this holds particularly for nuclear power and gene technology, is considered to be particularly dangerous for this reason (Sjöberg, 2000). It is assumed that man is biologically adapted to a life in harmony with nature, in nature's pristine condition before man has changed it. For some, this is a religious belief. What is in fact then meant by "nature" is no simple question, see Chapter 16.

Perhaps people are sometimes fatalistic, and do not consider it possible to avoid nature's risks. Further, there is no human responsibility for these risks – or at least there is perceived to be none. "Accident" is a concept that partly implies the idea that there is nobody to be held responsible for what has happened. One can never be sure that accidents will not occur.

Of course, all this is a simplification. Nature creates risks. But they can often be forecasted and we can protect ourselves. There is an interplay between the forces of nature and the technological systems that could sometimes have been avoided. For example, strong winds combined with a deficient construction of ships may lead to catastrophes; compare with Chapter 9.

Until now, we have discussed only risks to which we are involuntarily subjected. Such risks are strongly disliked. One of the first debates about risks dealt with fluoridating drinking water in order to prevent caries. It was an involuntary risk that was forced upon people, or at least it was probably viewed as such by many (Martin, 1989). Yet another type of risk is completely voluntary, for instance when we smoke or drink alcohol. It is characteristic of these cases that the activity in itself is pleasant, and that the risk each time one smokes or drinks is perceived as extremely small or completely negligible. We also think, in these cases, that we are subject to a much smaller risk than other smokers or drinkers (Sjöberg, 1998).

There are risks in our everyday life that we almost never think about, and simply discard. What if our neighbor smokes in bed and the house we live in is set on fire one night? What do we actually know about the habits of our neighbors, and how could we protect us against such a risk? Or the next time we fly – can we be absolutely sure that "the captain is sober"? (Of course, we cannot ask.) It is simply unreasonable to worry about all the risks in daily life, and even if we did worry about them, we could usually not protect ourselves without large costs – and at the same time take other risks.

We would prefer to completely avoid any environmental or technological risks. We want an "absolutely safe" technology. In the 1970s a law was introduced in Sweden decreeing that there must be an absolutely safe method to take care of the nuclear waste from Sweden's nuclear power plants before they were allowed to operate. But taken literally, is impossible to live up to such a requirement. This holds for any technological system. Some risks always remain, even if we claim to have done everything possible to avoid them. Moreover, it is rare that we want to do everything possible to avoid the risks irrespective of how small they may be. Other risks can arise, replacing those that were eliminated. The cost of decreasing the risk level usually becomes larger, the smaller is the risk. We get to a point where we are not willing to pay for further risk reduction.

This may seem obvious, but political examples show that this is not always the case. As another example from Sweden, the authorities have adopted the official policy that no person shall die in traffic accidents – "zero tolerance" policy. If this goal is taken literally, one may ask what the costs would be and what changes are required in the society in order to achieve it. Of course one can view it as a goal to strive for, but it is unrealistic to think that it can ever be reached. Is it good policy to have goals that can never be achieved?

Taking risks also has another face. Although it may be irrational to require absolute safety, this does not necessarily mean that we should always accept very small risks. Very great damage is something we may want to avoid, "at all costs." Perhaps we prefer a solution that has a higher risk, but where the consequences are smaller and easier to handle. An insurance company does not normally accept to insure something that could force the company into bankruptcy if the worst were to come to the worst, even if the probability is extremely small. In this connection, it is important to realize that small probabilities can almost never be determined with a high degree of accuracy. They rest on assumptions that may turn out to be incorrect.

One may also question our tendency to wait to act against risks until something happens. Risk policy is largely reactive rather than proactive and the actions that follow from a catastrophe may turn out to be very expensive and demanding (Renn et al., 1992). The handling of risks in modern society is full of examples of this. The economic analysis is given low weight after a catastrophe, when public opinion demands, "it must never happen again." Industry has a strong interest in restoring the confidence of consumers and the public, perhaps at a high cost. But there are also good examples of how safety can be increased without overly high costs, as exemplified by the air transport sector.

2.6 Risks in the Public Debate

What is rational when one has to decide about a risk such as a core meltdown in a nuclear power reactor? Of course one factor of importance is the knowledge one has about the field. In some sense, specialists and experts can be said to arrive at the most rational judgment, based on known and proven theories and models. But even they do not have, and never will have, completely certain knowledge about the issue under consideration. It is always necessary to make assumptions, and the expert's judgment can sometimes be questioned.

Often, other experts have objections and a fierce debate may start in the media. The public must base its risk assessment on its impression of the experts' reliability and on other information they have about them. We judge all people, also experts, based on psychological and social dimensions, some of which can be rational and very reasonable while others may be merely prejudices. Common to all these dimensions is that they give incomplete information. A Nobel Laureate often appears to be more reliable than an "ordinary engineer," but the Nobel Laureate may have worked on problems quite different from the risk issue, while the engineer has spent many years of qualified practical work in the field. We should therefore be careful not to let status be the only decisive factor in our judgment of reliability. A person with a strong personal and emotional engagement, and who shows that in a debate, can appear to be partial. But it may also be that he or she, on impartial grounds, has arrived at a strong conviction in an issue that is very demanding, emotionally and in terms of value. Problems of risk are often of such a character, and we should not be too quick to dismiss such a person. Perhaps the others are just better at disguising their emotions.

In risk debates, it is fairly common that people are talking at cross-purposes (Sjöberg, 1980). One of the reasons is that the word "risk" in itself is so ambiguous. One person can think about the probability of damage and claim that it is small, while another person may think of the size of the injury or consequences and, if that is very large, he or she can keep arguing that the risk is too large, no matter how much the first person claims that the probability is small. This is a type of argument that rarely works, since probability is a difficult and theoretical concept that few understand very well. Moreover, it is difficult to get an intuitive idea of small probabilities. Statements about small probabilities are also built on models whose assumptions can be questioned.

It is not uncommon that the debate strays off the subject. The parties accuse each other of being ignorant or "bought," or at least suspect that this is the case. Risk communication has emerged as a field of study to find ways out of the dilemma. Sometimes it is about attempts from industry to achieve public "acceptance," sometimes with the help of PR consultants (Stauber and Rampton, 1995). The latter have experience of advertising and may know consumers as being fairly uninterested but who can be influenced by irrelevant tricks. But when it is an issue about risks involving the life and health of an individual and of his or her family, people are not uninterested or credulous or particularly easy to sway with easy tricks. At the same time it should not be denied that the area of risk communication is very difficult and that research still has a long way to go. In the end it seems to be a question of democracy. People must be given real influence, either directly or through their representatives. Compare with Figure 2.2.

Figure 2.2. Environmentalists camp in the shadow of Drax power station, Britain's biggest coal-fired power station. (Photo: John Giles/PA-EMPICS/Scanpix)

A special case that clearly illustrates the need for democratic control arises in cases where risky structures are built close to borders between countries (Löfstedt, 1996). In some countries one may have confidence in experts and authorities, in contrast to the situation in other countries. There can be historical reasons behind these different attitudes. In the USA and in Russia, there is a history of badly managed risks, for example in relation to radioactive radiation. In Eastern and Central Europe, mass media were for a long time in the hands of the rulers of the state, and critical information about such events as environmental pollution as not made public (Sjöberg et al., 2000). As a result, environmental pollution went a long way before the public realized what was happening and demanded improvements. At the same time, public trust in authorities and experts was undermined. Data indicate that people in these countries nowadays perceive great risks, and that they don't trust their own authorities and experts. Trust is easier to lose than to acquire. When lost, it is a demanding task to restore it.

Social trust is important for understanding reactions to risk (Slovic et al., 1991; Frewer at al., 1996). Often it is assumed that risk debates can and should lead to the building up of public trust in experts and authorities. But some skepticism is always healthy, since experts can be wrong (Sjöberg, 2006). Furthermore, many members of the public are skeptical as to whether science really has such completely correct answers that risk analyses on a scientific basis can be considered to be completely trustworthy (Sjöberg, 2001). In fact, all researchers know that science does not hold the final answers about anything, science always asks new questions. Uncertainty must therefore be debated. People want to know about it (Frewer et al., 2002). Risk is a theme that involves many psychological and social factors, in addition to the purely technical complex set of problems that has to do with measuring and controlling risks. The technical part is of course indispensable, but so is the human part. We cannot hope to be able to manage risks in a better way in society if we do not understand how people react to them.

References

Breakwell G (2007) The psychology of risk. Cambridge University Press, Cambridge

Combs B, Slovic P (1979) Newspaper coverage of causes of death. Journalism Quarterly 56:837–843, 849

Fischhoff B, Slovic P, Lichtenstein S, Read S, Combs B (1978) How safe is safe enough? A psychometric study of attitudes towards technological risks and benefits. Policy Sciences 9:127–152

Frewer LJ, Howard C, Hedderley DR (1996) What determines trust in information about food-related risk?. Risk Analysis 16:473–486

Frewer LJ, Miles S, Brennan M, Kuznesof S, Ness M, Ritson C (2002) Public preferences for informed choice under conditions of risk uncertainty. Public Understanding of Science 11:363–372

Löfstedt R (1996) Risk communication: the Barsebäck nuclear plant case. Energy Policy 24:689–696

Martin B (1989) The sociology of the fluoridation controversy: a reexamination. The Sociology Quarterly 30:59–76

Maslow AH (1970) Motivation and personality. 2nd edn., Harper & Row, New York

Renn O, Burns W, Kasperson RE, Kasperson JX, Slovic P (1992) The social amplification of risk: theoretical foundations and empirical application. Journal of Social Issues 48:137160

Renn O (2004) Perception of risks. Toxicology Letters 149:405–413

Sjöberg L (1980) The risks of risk analysis. Acta Psychologica 45:301–321

Sjöberg L (1998) Risk perception of alcohol consumption. Alcoholism: Clinical and Experimental Research 22:277S–284S

Sjöberg L (2000) Perceived risk and tampering with nature. Journal of Risk Research 3:353–367

Sjöberg L (2001) Limits of knowledge and the limited importance of trust. Risk Analysis 21:189–198

Sjöberg L (2002) Attitudes toward technology and risk: Going beyond what is immediately given. Policy Sciences 35:379–400

Sjöberg L (2006) Rational risk perception: utopia or dystopia?. Journal of Risk Research 9:683–696

Sjöberg L, Kolarova D, Rucai AA., Bernström ML (2000) Risk perception in Bulgaria and Romania. In: Renn O, Rohrmann B (eds) Cross-cultural risk perception. A survey of empirical studies. Kluwer, Dordrecht

Slovic P (ed.) (2000) The perception of risk. Earthscan, London

Slovic P, Flynn JH, Layman M (1991) Perceived risk, trust, and the politics of nuclear waste. Science 254:1603–1607

Starr C (1969) Social benefit versus technological risk. Science 165:1232–1238

Stauber J, Rampton S (1995) Toxic sludge is good for you! Lies, damn lies and the public relations industry. Common Courage Press, Monroe, Maine

3

Risks in the Past and Present

Birgitta Odén

3.1 Background

Our present society is generally thought of as a risk-society. Individuals die prematurely, cities and villages are destroyed, the surface of the earth is altered and areas of cultivation are lost (Beck, 1986; Nordin, 2005). It can be argued, however, that mankind, throughout its long history, has always lived in a "risk-society." The surface of the earth has repeatedly been the victim of natural catastrophes, ice ages have destroyed our means of living, bacteria and viruses have periodically exterminated large portions of the population, and lack of nourishment has threatened reproduction (Diamond, 2005).

For a very long time, people believed that catastrophes and accidents belonged to the supernatural sphere. Rock-carvings and paintings suggest this (Dunbar, 2004). Magic was employed to minimize risks. When the great religions evolved, God was held responsible. Accidents and catastrophes were believed to be punishments inflicted on humans by God, and prayers were a means of averting his wrath.

A major shift in European thinking about risks is believed to have occurred after the Lisbon earthquake in 1755. The explanation for that disaster was sought not from theologians but from scientists (Broberg, 2005).

People have also developed counter-measures and used their creative brains and communicative ability to diminish risks (Gärdenfors, 2000). Plans were made for how to provide security in the face of catastrophes.

One of the most important preventive measures has been to increase and spread information about risks. During the development of national states, the crown, the state or the society was made responsible for the safety of its citizens. The state needed information about the size of its population, how it was provided for, births and deaths – in order to be able to carry out this duty. Statistical data began to be collected and analyzed. The measuring of society became a method in the development of systems for increased safety (Johannisson, 1988).

In 1756, Sweden set up a nationwide Bureau of Statistics and took on an internationally leading role in that field. The existence of long series of statistics on fatal accidents in Sweden makes that country a good subject for the study of risks in changing society – past and present. The information that these statistics provided about risks, prompted the state, entrepreneurs, and individuals to take actions to prevent accidents (Odén, 1972; Molander and Thorseth, 1997; Beck, 1986). Such information is available from the nineteenth century and has been published annually. National statistics were also developed in other European countries.

A great deal of modern research is focused on studies of the risk-society. A project with the working title "Risk generation and risk evaluation in a social perspective" was initiated in 1975, and the task of exploring the historical perspective was given to me, a historian interested in social and interdisciplinary research (Zitomersky, 1982; Sjöberg, 1982).

In 2000, Gunnar Broberg and Svante Nordin, both historians of ideas and science, initiated a new project with a historical perspective: "Risk and History." The aim of this project was to study "in what way has our society been shaped by fear of the threatening perils from wars, revolutions, fire, plague and famine" (Broberg and Nordin, 2000). The project is carried out at Lund University. Only a smaller number of essays have so far been published, including one about the earthquake in Lisbon in 1755 (Broberg, 2005). The project considers how the spread of information through mass media has influenced how people have perceived risks.

This chapter is a follow-up of the 1975 project. See the acknowledgements for further information about the project.

3.2 The Problem

One of the initial assumptions of the risk project of 1975 was that modern development had created "new, ever more complicated and large-scale technical systems." Another was that "large systems are economically efficient, but also vulnerable and thereby create risks." It was also assumed that "the ever increasing standard of living is thus obtained at the price of certain future risks."

It was therefore from the start of the project assumed that modern large-scale society has exposed people to new risks; risks which individuals are unable to assess for themselves and which are triggered as a result of decisions made in organizations or social institutions. This assumption will here be examined in a historical perspective. How large were the risks in small-scale society? How did people in traditional societies assess risks? And how did they try to defend themselves against threatening risks?

The lack of appropriate sources makes it difficult to examine how individuals in a historical perspective evaluated threatening perils. Using a number of different sources, an attempt will be made to identify what risks individuals were aware of, and then relate these risks to the factual evidence of loss of life and health which people actually suffered. Statistics on fatal accidents will be used as an indicator of the degree of risk. By making use of these two types of source, changes in the risk panorama from a small-scale, low-technological to a large-scale high-technological society can be traced.

3.3 The Historical Perspective

A historical perspective on the present and the future can serve a number of different functions. It can for example be used to explain the development which has led to contemporary phenomena – how we have reached our present position. The aim may also be more ambitiously oriented towards the future: the history of a specific development can be traced in the form of time-series which can be used to study trends and to make predictions about the future. Other aims might be to establish similarities or bases for comparisons between older and newer situations. In such cases, the aim is rather to indicate how the studied phenomena are connected to other factors in the community, thereby making it possible, through comparable conclusions, to demonstrate which factors are crucial for the changes.

The idea that processes of change conform to certain laws linking the past, present, and future is implicit in most "historical perspectives" that researchers more or less clearly acknowledge. For this reason, it is worthwhile to look at the theories of change which have governed the thoughts of researchers.

There are four main types of theories of development:

1. Theories based on a concept of progress, regarding historical change as development. Change occurs by gradual stages, and is often perceived as linear, becoming increasingly complex and adaptable.
2. Theories which look at historical change as taking place within a theoretical framework based on balance, change being seen as a restoration of homeostasis, stability.
3. Theories which look at historical change from a perspective based on conflict, as a class struggle which regularly leads to instability and social conflict.
4. Theories about the rise and fall of civilizations, which look at historical processes from an organizational perspective, as culturally determined processes of expansion and decline.

There is no generally accepted theory of social change among historians, and for this reason each prognosis has to be very carefully analyzed and critically examined (Appelbaum, 1971; Wehler, 1975; Österud, 1978). This study is based on a theory of change in which the scale of society is seen as increasing from "small" to "large."

3.4 Changing Society

The transition from a traditional to a modern society in Western Europe is usually described in terms of certain inter-related processes of change:

1. *Economically*, the process is determined by industrialization and the commercialization of agriculture, leading to economic development, occupational differentiation, occupational specialization, the use of new technology, and the exploitation of new sources of energy. Older patterns of decision-making are

replaced by economic calculations and future-oriented planning. The change of scale in this process can be assessed by a greatly accelerated increase in the consumption of energy, an increase in individual consumption, and in a trend towards consolidation and growth in industrial units.

2. *Culturally*, the transition is determined by modernization in the broad sense of the word, entailing changes in the educational system, longer education within institutional frameworks, more efficient, widespread information, the "democratization" of traveling, increases in political participation, and the development of public welfare. The role that an individual plays in society is no longer determined by inherited social position but by personal merit and the class structure of society. The aim of politics is to eliminate differences and to assert equality before the law. The change in scale in this process can most readily be measured by the spread of cultural values and social benefits to larger portions of the population, which may be seen in the increasing numbers of "participants" in elections, schools, universities, hospitals, the expansion of mass communication, tourism, and the decentralization of culture.

3. *Socially*, the transition is determined by urbanization and centralization (in the broad sense), which involves domestic migration, concentrated housing, the depopulation of sparsely populated areas, the separation of residence from place of work, and social restructuring. The changes in scale involve, in particular, an increase in the geographic size of cities, attempts at megalopolis formation, increases in the size of the administrative units in sparsely populated areas, increases in the radius of transport networks, and increased commuting. More and more time is spent on traveling.

4. *Technologically,* the transition is determined by the development of technology making possible increases in the scale of buildings, transport systems, and the consumption of energy. The elaboration of electronics and information technology (IT) has changed the control systems in society.

5. *Politically*, the transition is determined by a shift of emphasis in political power and responsibility from groups whose decisions concern a small area (parish, county) to groups whose decisions concern a large area (nation, supra-national units). Problems concerning size and democracy are fundamental and of great interest, in general depicted in terms of the centralization as opposed to decentralization of decision-making functions, of growing bureaucracies as opposed to a few elected representatives.

These changes from the small to the large scale in society having been identified, the assessment of risks in the two different ideal types of society will be discussed – with the explicit reservation that all generalizations of this kind serve to conceal essential differences in the actual course of events.

3.5 Assessment and Prevention of Risks

3.5.1 Risks in Traditional Society

In traditional society, the assessment of risks and perils was intimately related to popular beliefs, myths, and customs (Nordin, 2005).

A summary of the official view of the problems of risks and perils in society can be obtained from the prayers said in church services, especially the Litany. The Litany was medieval and Catholic in origin, but it retained its position after the Reformation and was still of importance in the pre-industrial Swedish society discussed here. The following risks are especially stressed in the Litany printed in the 1811 church service book:

From plague and famine, from war and strife,
riot and discord, from hail and storm,
fire and peril, from a violent death –
preserve us, dear Lord.

The Litany's assessment of risks thus first and foremost listed contagious diseases, war, bad weather, poor harvests, and fires, all factors which can also objectively be associated with risks.

Certain groups were considered as facing special risks with regard to life and health, namely pregnant women, women in confinement, and all travelers on land and at sea. The church service book also contained special prayers for particular situations that were fraught with risk: times of war, famine and scarcity, epidemics, storms and tempests. It is clear that these "abnormal" periods of risk were thought of as punishments, as acts of God, which were legitimately inflicted on mankind, and should therefore be suffered meekly. They were to be accepted as punishments to be endured in order to escape a yet harsher fate. There were also special prayers for persons whose occupations exposed them to risks: seafarers, fishermen, and miners.

Even if the official Christian view was that societal risks were natural and that periods of heightened risk constituted a legitimate punishment, the prolific literature of prayers for private use indicates two totally different attitudes towards risks and perils. The pessimistic view of the world and of human life regarded disease, privation, and death as a natural state of worldly affairs, e.g., "the vale of tears" brought about by Satan and his followers of evil spirits. As a result of this belief, people tried to reduce risks by doing penance. At the same time, there existed a more optimistic faith concerning conditions on earth, wherein attempts to reduce risks instead were focused on the mercy of the Lord, i.e., on prayer.

In the medieval legislation of small-scale, traditional society, the same risk panorama as that of the Church is to be found with regard to the kinds of accidents that could be expected. These were linked to the following risk phenomena: sea voyages, sleigh journeys, fires, wars, and contagious diseases.

Popular notions about risks and ways of preventing them were to a great extent based on what we now refer to as blind faith and superstition. The threat came from invisible forces which had to be placated, or from evil persons whose influence had to be curtailed by means of ritual practices. This was especially the case with regard to popular notions concerning risks to life and health (Oja, 1999).

In spite of the belief in traditional society that risks and perils were part of a divine scheme for salvation, people did develop various strategies to reduce dangers threatening the individual and to spread risks. We can see examples of this in the following review of the various risk phenomena and the changes that these have undergone.

3.5.2 War

War constituted a recurrent risk to life and health in traditional society, and people tried to seek safety from it through the feudal system of hierarchically based allegiances and defenses.

Peasants in the traditional small-scale Swedish society negotiated local treaties, known as peasant peace-treaties, along the borders of the country, to reduce the risks of death, ill-health, and economic ruin caused by conflicts at the national level. But it is also clear that increased national integration during the sixteenth and seventeenth centuries made such local treaties more difficult to achieve. The loss of human life during the wars in the period of Swedish expansion in areas around the Baltic region was significant, and involved both troops and civilians along the borders. In modern large-scale society, an international organization (United Nations) and the European Union have tried to eliminate risk of total war as well as local border conflicts.

Sweden has not been actively involved in any war since 1812 and this has reduced the number of deaths caused by war to an insignificant figure in the mortality statistics (Östergård, 2006). In an international perspective, however, it is obvious that the twentieth century, through science and technology and the large scale of society, has experienced an extremely high number of deaths during war. The risks to life and health in the event of a nuclear war are enormous and well known – the "over-kill capacity" of new technology has given rise to an entirely new situation seen from a historical perspective.

3.5.3 Homicide and Manslaughter

The collective violence of war has its counterpart at the individual level in homicide and manslaughter. Contrary to what is generally believed, the number of deaths by homicide and manslaughter in Sweden has steadily fallen from 1860 to 1950 (Figure 3.1).

During recent decades, a slight increase can be noted, and it is to this increase that people refer when attempts are made to prove that crime is on the increase in modern society. This increase has not, however, reached the level of individual violence that existed in the pre-industrial society. Historical research into the social history of crimes of violence has demonstrated a falling trend from medieval to modern times. The same pattern can be seen in Western Europe, in which the rate of intentional and completed homicide in most countries at the end of the twentieth century was less than 2 per 100 000 inhabitants and often less than 1 per 100 000 (Österberg, 1996; ESCCJS, 2003).

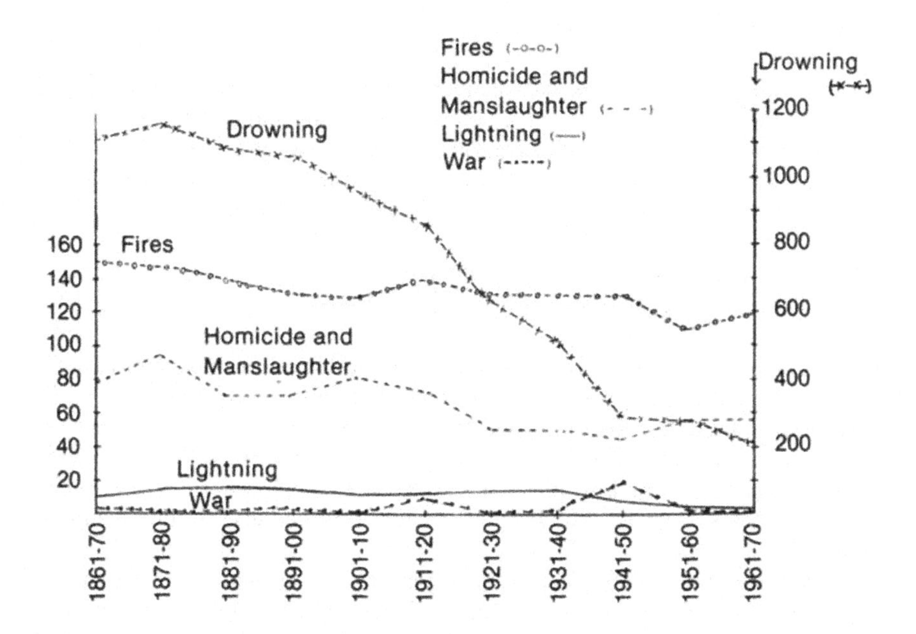

Figure 3.1. The number of deaths in Sweden caused by drowning, fires, homicide and manslaughter, lightning, and war [annual averages] (Odén, 1982)

3.5.4 Contagious Disease

The risk of contagious disease was recurrent and a dreadful threat in traditional society. The small scale provided no protection for individual life and health, in spite of the fact that people did not travel much. It was not until more was known about the medical aspects, antibiotics and vaccinations, and until the nutritional condition of the majority of the inhabitants had been improved, that the situation was radically changed. When the cholera epidemics of the 1830s and the 1850s raged, knowledge about the way the disease spread was so deficient that totally inadequate measures were adopted to fight and gain control over it (Zacke, 1971).

What effect did the transition to a large-scale society have on the risks associated with contagious diseases? On the whole, no negative effects have been noted. Effective rules for quarantine and improved methods of locating the carriers of contagion at an early stage have for the time being minimized the risks of epidemics in modern communities. The large scale in medical services has definitely increased the potential for effectively combating infectious contagion as well as ameliorating the consequences for its victims. The statistics on mortality demonstrate a rapidly falling trend in epidemics. Increased global communication, however, poses a threat, e.g., the spread of HIV.

3.5.5 Transport and Communications

It is interesting to note how conspicuous the position of the transport and communication system was in the panorama of risks to life and health.

In Sweden, as it was organized during the medieval and the early modern period until about 1800, the most important means of transportation was by water; boats with varying degrees of safety operated in all kinds of waters.

Behold, we float between the sea and the sky,
a feeble plank being all that separates us from death,

according to the words of a Swedish prayer from the seventeenth century (Rodhe, 1923). Death by drowning was frequently registered in the early death and burial records and in the oldest mortality statistics, especially in the coastal regions, of course, where the sea provided a living and was not merely a means of communication. Death by drowning was seen to be a common event when official mortality statistics from the middle of the eighteenth century became available (Imhof, 1976).

The increase in population and the struggle for a living in some of the coastal areas in Sweden created a class of landless fishermen using open or semi-open boats. Through improved technology, especially the introduction of iron hulls and steam turbines, the risk of drowning at sea was reduced. This development was particularly notable in the history of passenger transport to the United States during the emigration era. The risks of disaster and individual ill-health were much greater during the 1850s, when the voyage took place in sailing vessels, than 50 years later when steamships were used and social legislation was passed to protect poor passengers from bad conditions at sea (Bonsor, 1955; Hvidt, 1971; Fleisher and Weibull,1953; Olsson, 1967).

It is clear that scale increased with the changes in technology. An increasing number of passengers could be transported on each voyage. This development towards an ever-increasing scale continued until 1912, when the sinking of the Titanic broke the trend. The international classification system of the insurance companies aimed at diminishing the risks, but efficiency was not enough. The sinking of the ferry Estonia in the Baltic in 1994 exemplifies the technical deficiencies and the large scale which led to the deaths by drowning of 852 passengers (Chapter 9).

In the development of air traffic, the same pattern was repeated: improved technology, growing numbers of passengers, increased speed – but in this case linked to aircraft. The serious accidents that happened to airships during the 1930s had a deterring effect, however. The statistics of fatal accidents also show a rising trend in Sweden. The aircraft accidents that have occurred in Sweden during the most recent decades have, however, hit private operators and miscellaneous commercial flights (taxi and aerial work) but not charter flights or regular passenger flights. The increase must also be seen in relation to the great increase in air travel. Although occasional aircraft disasters have had serious consequences, travel by air is considered as "natural" and as having great advantages for the individual (Chapter 9).

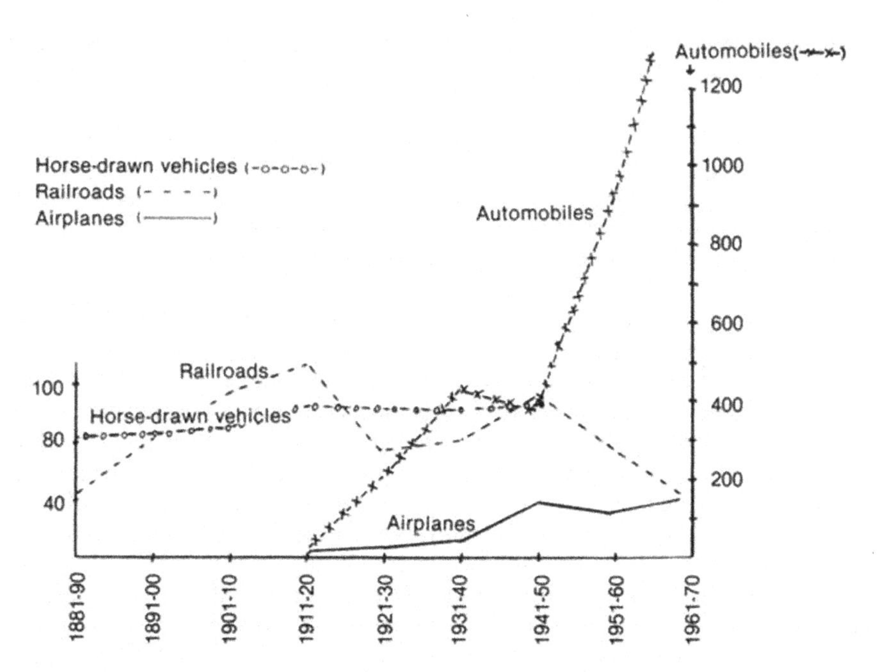

Figure 3.2. The number of deaths in Sweden through accidents involving horse-drawn vehicles, railroads, airplanes, and automobiles [annual averages] (Odén, 1982)

The horse was the means of transport used on roads in traditional society. Originally used to carry riders and loads on difficult routes between settlements isolated by forests, the horse provided a means of transport that represents communication on a very small scale. The transition to transport by carriages and stagecoaches brought about an increase in scale – the volume of travel increased and greater distances could be covered. Of course there were serious accidents associated with transport by horse and carriage, even though we have little or no information about the accident rate during earlier periods. From 1881, however, statistics are available concerning deaths caused by bolting horses, tilting carriages, or by being run over (Figure 3.2). The numbers are very low and show little variation. The increase to 1911–1920 is presumably due to the general increase in road traffic at the end of the nineteenth century and also, possibly, to the introduction of motorcars on the roads, an innovation which frightened horses, making them bolt.

Technical developments revolutionized road transport in two ways and at two different periods: the railroads during the nineteenth century, and motorcars and buses during the twentieth century.

The railway system in Sweden was constructed during the period 1856–1928, when, generally speaking, it reached its present size. When the transition to railway traffic was made, it was preceded by a heated public debate, but the risks that were discussed were not the threats to human life and health, but the economic risks and the risk of theft of rails and carriages. Expressions of fear can, however, be found in private sources, fears which often concerned the high speeds. A diary written by a young woman in 1857 contains the following account of popular beliefs in Sweden (Peyron, unknown year):

> *You are locked up in a carriage or a kind of cupboard, in which, while surrounded by smoke, dust and deadly peril, you roll on at an incredible speed from one station to the next. It is not possible to see anything during this dangerous journey, so that it is not until you have arrived that it is possible to discover what part of the world you have come to.*

The first rail carriages were, as is well known, constructed like extended stagecoaches and created the impression that it was the old familiar kind of carriage that had developed a new and dangerous speed and on a much larger scale.

The frequency of railway accidents was low. There was a period with a rise in the number of deaths, coinciding with a rise in the volume of traveling, after which there followed a period with a steep rise in traveling without any corresponding increase in the death rate. The risks diminished in spite of the increase in scale.

The fear of railway journeys was short-lived and in the source material it is not possible to observe any changes in attitude towards railway traffic even after serious accidents. In part this is due to the fact that some of the worst accidents – the Getå disaster in Sweden in 1918 for instance – occurred at a time when there was no real alternative to railway travel for ordinary people. It should also be noted that most of the victims of fatal railway accidents were railway employees. The risks that the general public faced when using this means of communication were insignificant (Chapter 9).

The transition from the use of horses to motorcars in road transport was experienced as a considerable improvement of the environment in big cities – droppings from horses had begun to become a large problem in cities trafficked by horse-drawn trams. But there was also a debate about the risks that the new technology involved. In Sweden, the number of cars increased from 3036 in 1916 to ten times as many by 1921. In 1950, the number of registered motorcars had reached 345 000. In 1925 there was one motorcar per 100 inhabitants; in 2005 there were 47 cars per 100 inhabitants. This means that in theory there are today two persons for each motorcar (SCB, 2007).

Statistics relating to car accidents exist from 1911 onwards. These show an enormous increase in the number of deaths caused by motorcars. This increase must of course be seen in relation to the increase in the number of motorcars and in the number of journeys made by car. The decrease in risks, which was evident occurred in railway traffic, does not seem to have had any parallel with regard to motorcar traffic until recently. It is not until the last few years that the compulsory use of safety belts has reduced the risks run by drivers. At the same time, however, both drivers and pedestrians have become less cautious; the number of people who have been run over has increased.

The increase in the number of deaths associated with accidents on roads must be seen first and foremost as an indirect result of improvements in the economy of large groups in the community, improvements which permit investments in motorcars as a private means of transport. The motorcar represents a definite breakthrough in road transport. The personal advantages are apparently experienced as so great that the obvious risks are considered to be acceptable.

A special type of increase in scale may be studied in the development of bus traffic, i.e., of public road transport. Long-distance bus traffic made its breakthrough during the 1920s and to begin with mainly served densely populated urban areas. As from the end of the 1930s there was a gradual change – in part as a consequence of measures dictated by traffic policy – which provided the sparsely populated regions with most of the bus services.

As a result of the availability of the public road transport systems, low-income groups in the community also had access to better communication. It seems that the techno-social increase in scale in this case has led to a decline in risks – public motor traffic is in general safer than private motorcar traffic. On the other hand, there now appears to be an increase in risks, when people walk between bus stops and their homes, compared to the corresponding move between the garage and the front door. There is no way of historically analyzing this, however, since the source material dealing with bus traffic in the earlier period is very sparse.

It is within the transport sector that we find the first systematic attempts to spread risks through the use of risk assessment. This took place within shareholding systems and ship-owners' associations, which later developed into marine insurance pools. It is characteristic that the earliest measures concerning the spread of risks aimed at the distribution of risks threatening economic values. Life insurance and insurances against accidents for travelers and those employed in the transport sector constitute a later attempt to diminish the consequences of risks to life and health.

Attempts have been made to minimize the risks in the traffic sector by various road-safety campaigns and measures. The increasing scale in traffic intensity, caused by the greater geographical integration and the democratization of travel, has also led to socio-economic demands on public welfare which are familiar in the current public debate. A number of social utopias have been projected, a starting-point being the necessity to reduce the demand for communication by deliberately trying to diminish the scale of the geographic area in which people live and work. In a future perspective it will also be necessary to consider a gradual reduction in private automobile traffic, since the costs for motor fuel will increase as a consequence of a shortage of oil and the price policy of the oil-producing nations (Chapter 9).

3.5.6 Fire

The risk of fire has also undergone a great change over time. Wooden housing in both town and country meant that the risk of fire was acute. Rules and regulations regarding fires and the kindling of fires were set up at an early stage in laws, statutes and village by-laws. Fires in towns were especially devastating since the housing there was so congested (Figure 3.3). Accounts of fires in traditional society seldom refer to human losses or to disablement caused by these violent conflagrations, but these events must obviously have exacted human victims. Mortality statistics, however, demonstrate a falling trend, the explanation of which is to be found in extensive public measures. Town planning was implemented with fire breaks and regulations concerning the construction of buildings, fire-wardens and fire brigades. In the countryside, land-parceling reforms meant that densely populated villages were broken up, thus reducing the risks of fire. The development of fire-resistant materials was also important.

Figure 3.3. The Great Fire of London, 1666. (Image: Topham Picturepoint/Scanpix)

Already during the Middle Ages, the disastrous consequences of fires in rural areas led to legislation concerning "risk-pooling" in the event of fires. This meant that those whose buildings were destroyed by fire could turn to every other household in the district to obtain contributions in the form of grain as well as money for reconstruction. The towns did not have this system of contributions from its inhabitants. Instead funds for fire insurance and commercial insurance companies were established at a fairly early stage. These public solutions also attracted the increasing number of land-less persons in the rural areas. The contribution system was linked to landownership and thus excluded all those who did not own land.

It is clear that the modern large-scale and technologically advanced communities have rapidly and efficiently reduced the risks to life and health from the hazards of fire. A combination of fire-proof construction materials and increased measures for local fire protection has reduced the scope of disasters. So far, the continually increasing size of buildings in cities has not led to any increase in the risks to life and health through fire – although this same development is perceived as a threat to social values and self-fulfillment. The increase in risks through the introduction of certain plastic substances has been checked through various prohibitive measures (Chapter 14).

3.5.7 Natural Catastrophes

The number of deaths due to common accidents, such as lightning strikes, has on the whole also undergone a downward trend. In this case too, technological development has facilitated an increase in the scale of buildings without any corresponding increase in the risk of disaster caused by lightning. Other natural catastrophes, common in countries with different environmental conditions, such as floods, landslides, earthquakes, and tornadoes, have been rare and limited in scope in Sweden.

Harvest disasters – due to climatic influences such as frost, hail, and storms – which in earlier periods had serious consequences and created high mortality, were brought under control through improved harvests on the one hand and more efficient communications on the other. The last great famine in 1867–1868, which hit northern Sweden particularly severely, was aggravated because the railway system had not yet been fully extended. But since the scale in agricultural production and the transport system has increased, the risks of famine have diminished in modern Swedish society.

3.6 Fatal Accidents in the Past and the Present

Changes at the national level in the risk panorama can be summed up by relating the total number of fatal accidents to the number of deaths and – via the death ratio – to the population trend. The death ratio is the number of deaths per year in relation to the average size of the population in the same year. The number of deaths per year is remarkably stable during this particular period. If instead, the death ratio is considered, this results in a slightly falling trend. There is a figure illustrating the main demographic factors 1860–1970 in "The Biography of a People" published in 1974 (UD, 1974).

Figure 3.4 shows a slight downward trend up to the 1920s, after which there is a certain increase. It was the inter-war generation that mainly profited from the diminished risks, while the generation after the Second World War faced a slight increase in the risk of fatal accidents.

What is more remarkable, however, is that the differences are so small. During the whole period, the number of fatal accidents remains at a level of approximately 40 deaths per year per 100 000 inhabitants. The conclusion to be drawn from this, when working at a macro-historical level, is that fundamental structural changes in society, on the whole, have not had any great influence on the frequency of accidents.

It is also of interest to compare the risk development in different countries. Modern Western countries all have a level close to 40–50 fatal accidents per year per 100 000 inhabitants – if the effects of the Second World War, which have been included in the statistics of the Netherlands and Italy, are excluded. Italy and the USA were worse at the beginning of the period but gradually became increasingly safe societies, if a reservation for the war years is made here too. All countries, except the USA, show a slight increase in the indicators during the 1960s. It is also puzzling to find a concentration of the indicators around 40–50 per year after the

Second World War. Does this mean a convergence in the risk rate as the structure of the welfare-state is established?

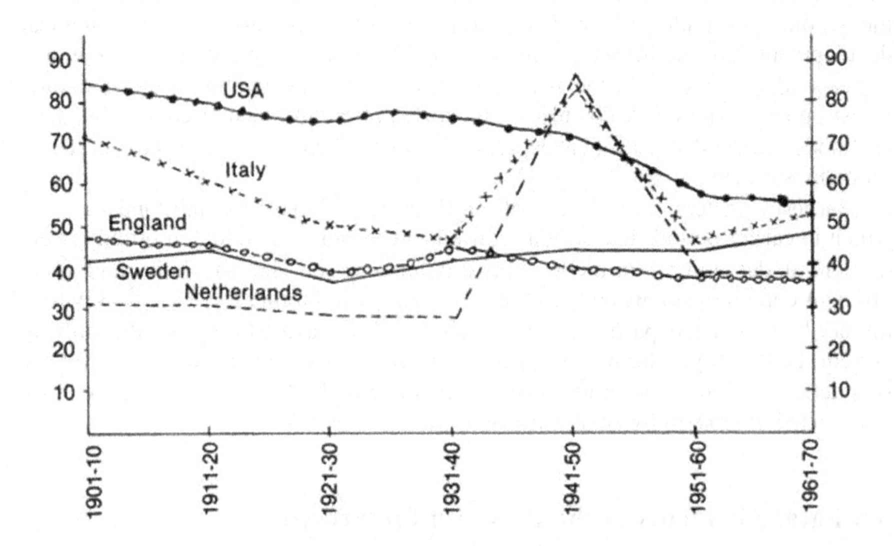

Figure 3.4. The number of fatal accidents per 100 000 individuals in the USA, Italy, England, Sweden, and the Netherlands [annual averages] (Odén, 1982)

To sum up, it can be argued that, generally speaking, the risks of fatal accidents are constant and constitute an inevitable part of the human condition in modern society (Odén, 1982). The explanation for the fact that the rate of fatal accidents has been relatively stable from 1860 to 1970 is probably that disasters have always attracted the attention of every community. People have been prepared to face risks because they have sought advantages which could be gained only at the cost of risks. But they did that because they felt sure that attempts were being made to minimize the risks through legislation and technological improvements. Even if the risk of one type of accident was diminished or eliminated thanks to improved technology, new problems arose which in their turn had to be dealt with. Although it has never been possible to eliminate accidents from the death statistics, an optimistic view of the future still seems possible.

During the latest 25 years, risks have taken on a new dimension which can also be related to a change in scale. The borders of national states do not shut out radio-active pollution, greenhouse effects, or acid rain. It is difficult to agree on international legislation and even more difficult to control its compliance. The traditional methods of dealing with risks have become obsolete. Human activity, whether individual or in organizations, threatens to violate the limits of nature and we have not yet found any effective method of reducing the risks. As a consequence of the vast scale of international communication in modern society, individuals living in countries with a relatively low risk have become much more aware of natural disasters taking place in other parts of the world. All these factors have created the pessimistic view of the future that sociologists and philosophers have expressed and which every day is spread through mass media (von Wright, 1986 and 1993).

A few lines from the Nobel laureate Harry Martinson's poem *Aniara* (1956) reflect this new pessimism:

There is protection from almost everything,
from fire and damage due to storms and frost,
add whatever blows may come to mind –
but there is no protection from mankind.

Acknowledgments

The first version of this study was written 1976 and published as a separate report in 1977. A somewhat enlarged English version has been published in Zitomersky (1982). A version in English, French, and later also in Spanish was published in the same year in the journal *Diogène*. A Swedish version was published in Sjöberg (1982).

References

Appelbaum RP (1971) Theories of social change. Markham Publishing Company, Chicago
Beck U (1986) Risikogesellschaft. Auf dem weg in eine andere moderne. Suhrkampf, Frankfurt am Main
Bonsor NRP (1955) North atlantic seaway: an illustrated history of the passenger services linking the old world with the new. Prescott: T Stephenson, Lancashire
Broberg G (2005) Tsunamin i Lissabon (The Tsunami in Lisbon) (in Swedish). Bokförlaget Atlantis AB, Stockholm
Broberg G, Nordin S (eds.) (2000) Risk och historia. Sex uppsatser om katastrofer och livets vanskligheter (Risk and history. Six essays about catastrophes and the risks in life) (in Swedish). Ugglan 14, Avdelningen för idé- och lärdomshistoria, Lund university, Lund
Diamond J (2005) Collapse: how societies choose to fail or succeed. Viking Penguin/Allen Lane, New York
Dunbar R (2004) The human story – a new history of mankind's evolution. Faber and Faber, London
ESCCJS (2003) European sourcebook of crime and criminal justice statistics. 2nd edn. The Federation Press, http://www.europeansourcebook.org/
Fleisher EW, Weibull J (1953) Viking times to modern: the story of Swedish exploring & settlement in America, and the development of trade & shipping from the vikings to our time. University of Minnesota Press and Almqvist & Wiksell, Göteborg
Gärdenfors P (2000) Hur homo blev sapiens. Om tänkandets evolution (How homo became sapiens. On the evolution of thought) (in Swedish). Nya Doxa, Nora
Hvidt K (1971) Flugten til Amerika eller drivkraefter i masseudvandringen fra Danmark 1868-1914 (The escape to America) (in Danish). Universitetsforlaget, Aarhus
Imhof A (1976) Aspekte der bevölkerungsentwicklung in den nordischen ländern 1720–1750. Francke Verlag, Bern

Johannisson K (1988) Det mätbara samhället. Statistik och samhällsdröm i 1700-talets Europa (Measurable Society. Statistics and social dreams in 18th century Europe) (in Swedish). Norstedts, Stockholm

Martinson H (1956) Aniara. Bonniers, Stockholm

Molander B, Thorseth M (eds.) (1997) Framsteg, myt och rationalitet (Progress, myth and rationality) (in Swedish). Daidalos, Göteborg

Nordin S (2005) Risktänkandets rötter (The origins of thoughts about risk). In: Brink I (ed) Risk och det levande mänskliga (Risk and the living human) (in Swedish). Nya Doxa, Nora

Odén B (1972) Historical statistics in the nordic countries. In: Lorwin VR, Price JM (eds) The dimensions of the past. Yale University Press, New Haven

Odén B (1982) An historical perspective on the risk panorama in a changing society. In: Zitomersky J (ed.) On making use of history: research and reflections from Lund. Lund Studies in International History 15, Esselte Studium, Lund

Oja L (1999) Varken gud eller natur. Synen på magi i 1600-talets och 1700-talets Sverige (Neither God nor nature. Ways of thinking about magic in 17th and 18th century Sweden) (in Swedish). Brutus Östlings bokförlag Symposion, Stockholm

Olsson NW (1967) Swedish passenger arrivals in New York 1820–1850. Swedish Pioneer Historical Society Press, Chicago

Österberg E (1996) Våld och ära, böter och skam. Kriminalitet och straff i Sverige från medeltid till nutid (Violence and honour, fines and shame. Crime and punishment in Sweden from medieval times until the present). In: Åkerström M. (ed) Kriminalitet, kultur, kontroll (Crime, culture, control) (in Swedish). Carlssons förlag, Stockholm

Östergård U (2006) The history of Europe seen from the North. European Review 4:282–297

Österud Ö (1978) Utviklingsteori og historisk endring (Theories about development and historical change) (in Norwegian). Gyldendal, Oslo

Peyron A (unknown year) Några anspråkslösa minnesblad från fremmande land (Some humble pages of memory from foreign countries) (in Swedish). Manuscript in the Archive of Mrs. A Lindencrona, Stockholm. Remark: No year of the manuscript is noted. However, the year the railway journey took place was 1857

Rodhe E (1923) Svenskt gudstjänstliv (Swedish divine service) (in Swedish) Uppsala. Litanian (The litany) in 1811 års kyrkohandbok (The church service-book of 1811). Svenska kyrkans diakonistyrelse, Stockholm

SCB (2007) Statistical yearbook of Sweden 2007. Statistiska Centralbyrån, SCB (Statistics Sweden), Stockholm, http://www.scb.se/statistik/_publikationer/OV0904_2007A01_BR_A01SA0701.pdf

Sjöberg L (ed) (1982) Risk och beslut. Individen inför samhällsriskerna (Risks and decisions. The individual in the face of risks in society) (in Swedish). Liber, Stockholm

UD (1974) The biography of a people. Past and future population changes in Sweden. Conditions and consequences. Royal Ministry of Foreign Affairs (Utrikesdepartementet), Allmänna Förlaget, Stockholm

von Wright GH (1986) Vetenskapen och förnuftet (Science and common sense) (in Swedish). Söderström & Co Förlags AB, Borgå

von Wright GH (1993) Myten om framsteget (The myth of progress) (in Swedish). Bonniers Förlag, Stockholm

Wehler HU (1975) Modernisierungstheorie und geschichte. Vandenhoeck & Ruprecht, Göttingen

Zacke B (1971) Koleraepidemien i Stockholm 1834. En socialhistorisk studie (Cholera in Stockholm) (in Swedish). Stockholms Kommunalförvaltning, Stockholm

Zitomersky J (ed.) (1982) On making use of history: research and reflections from Lund. Lund Studies in International History 15. Esselte Studium, Lund

4

The Dangerous Steam Engine

Jan Hult

4.1 Safety Devices

In large parts of our modern world, electric power is produced by heating water to produce steam, which is then fed to a steam turbine driving a generator. Every such energy plant contains a pressure vessel – a boiler.

Uncontrolled heating may raise the steam pressure to a level high enough to cause the boiler to burst, and this may result in great damage to people and plant. Such explosions were not uncommon in the early years of high-pressure steam technology.

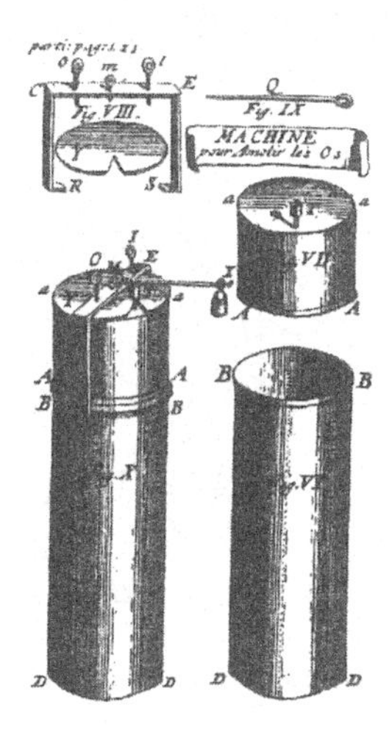

Figure 4.1. Papin's safety valve. Source: Höjer (1910)

A safety valve, first conceived by the Frenchman Denis Papin, found early use as a standard device to limit the steam pressure (Figure 4.1). By adjusting the position of the weight on the lever, the pressure was kept below a chosen limit. If it rose above that level, the valve opened and let out steam, thus reducing the pressure. Three additional safety devices were later taken into use, permitting continuous observation of water level, steam pressure, and temperature.

The water level could be read directly on a vertical glass tube connected to the boiler (Figure 4.2). A "steam gauge," or manometer (Figure 4.3), showed the pressure, and the temperature could be read on a thermometer. As an ultimate means to prevent excessive heating, a safety fuse was sometimes mounted at the bottom of the boiler (Figure 4.4). A central hole in the plug was sealed by a metal alloy with a low melting temperature. If the temperature rose above that point, the seal melted and the plug opened, letting water and steam out of the boiler into the hearth to quench the fire.

Before these devices had found common use, knowledge about the state of the boiler was often meager, and the risk of a blow-up could be felt as a menace.

The most serious accidents occurred in steamships, where many unprotected people were often gathered in the vicinity of the engine room. Steamships had their first breakthroughs in coastal transport and on lakes, rivers, and canals. In those cases, there was no need for the ship to transport large amounts of coal and water on board, since bunkering could take place at various depots along the route as soon as the need arose. Long before trans-ocean steamship traffic had started, regular steamship routes had been established – an important innovation in commerce.

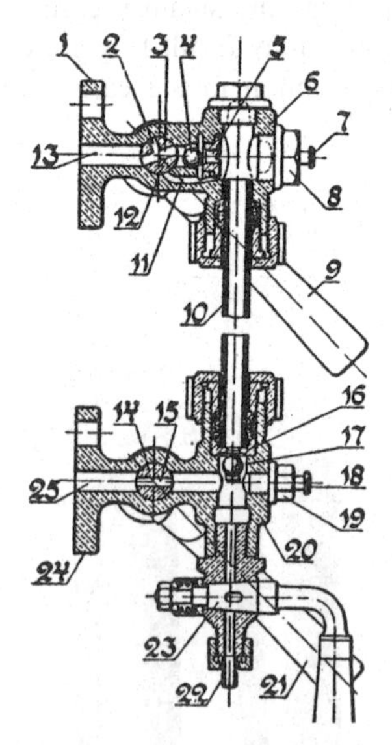

Figure 4.2. Device to monitor the water level. Source: Höjer (1910)

Examples of general references treating the issues that are discussed in this chapter are: Gordon (1978), Hammond (1956), Petroski (1985 and 1994), and Scanning (1976). More specific references to works dealing with pressure vessels can, for example, be found in: Bednar (1981), Gill (1970), Nichols (1979–83 and 1980), Spence and Tooth (1980), and Steele and Stahlkopf (1980).

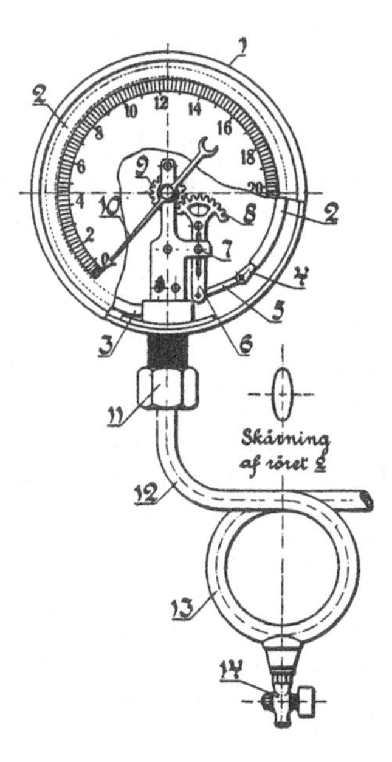

Figure 4.3. Manometer, to measure the steam pressure. Source: Höjer (1910)

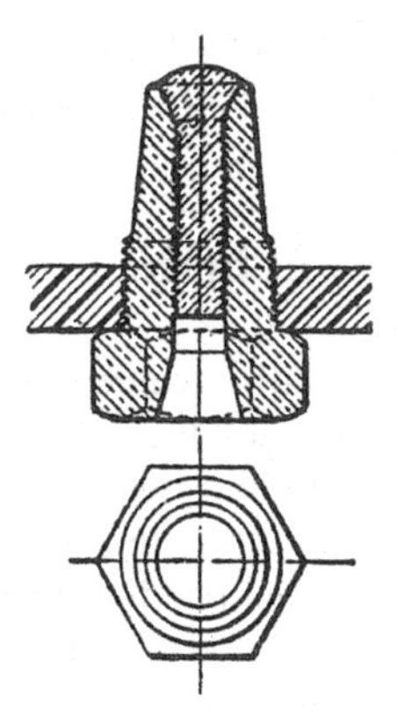

Figure 4.4. Safety fuse. Source: Höjer (1910)

4.2 Early Accidents in the Home Country of the Steam Engine

The cheapest and most comfortable way of traveling from London to Newcastle or to Edinburgh or Aberdeen in the 1820s was by steamboat. But the risks involved were not negligible. Between 1817 and 1839, 23 boiler blow-ups were registered in British steamships. After a much publicized accident in 1838, a parliamentary commission was created, which submitted its report after a year. Public safety had become a concern for the expanding technological society.

The British Institution of Mechanical Engineers (IMechE) was created in Birmingham in 1847. An occupation, which had until then been looked upon as just a trade, had begun to gain a prestige on a par with that of the builders of canals, roads, and bridges, the Civil Engineers. Thirty years later, the Mechanicals moved to London, where their headquarters are still located, adjacent to those of the Civils just off Parliament Square.

An urgent task for the Mechanicals was to come to grips with problems of bursting steam boilers, but the reasons for such disasters often evaded the investigators. One inexplicable boiler blow-up was finally – after much deliberation – attributed to an assumed decomposition of water into hydrogen and oxygen, which then ignited! The ignorance was formidable, and boiler problems continued to cause worries for a long time. A report submitted to the IMechE in 1866 presented an account of 1046 blow-ups, which had caused a total of 4076 deaths.

IMechE then decided to attack the problems – unprejudiced and on a broad front. A statistical survey showed that 43 % of the failures were due to erroneous design or less professional repair work, whereas 28 % could have been warded off during a prescribed inspection. About 25 % were due to outright mismanagement, and 4 % were attributed to external circumstances, impracticable to control.

This survey resulted in a proposal for the compulsory regular inspection of steam boilers and for requirements that every boiler should be equipped with dual safety valves, and dual water-level gauges, in addition to a steam gauge. A national British pressure vessel code had begun to be created.

4.3 Simultaneous Development in the USA

By means of large flat-bottomed paddle steamers, the Mississippi had become an important transport route for both goods and passengers. Lucrative business possibilities opened up for owners of rapid steamships, and several ways were tried out to enhance the ships' performance, e.g., by placing a heavier counterweight on the safety-valve lever. Contests between paddle steamers, such as the one between Natchez and Eclipse (Figure 4.5) created excitement, but disaster was lurking, when the engines were taxed to the extreme.

Several events with bursting boilers on American river steamers in the early nineteenth century caused severe damage. Many passengers were killed, and cries for restrictions began to be heard, but the Constitution of the United States provided no legal means to act in such matters.

Between 1818 and 1824, the number of people killed by bursting steamship boilers rose to a total of 47 due to 15 blow-ups. When yet another such disaster occurred 1824 in New York harbor, killing 13 people, Congress finally acted and embarked upon a program to reduce the risks for the general public.

In that same year, the Franklin Institute was founded in Philadelphia, with the task to perform research into "the mechanical sciences." The scientific *Journal of the Franklin Institute*, from its start, devoted much attention to the aggravating steam boiler problems. The need for legislation was discussed, but agreement could not be reached, so a committee was set up to investigate the reasons for recorded blow-ups and also to perform tests with boilers under increasing internal pressure.

In the meantime, boiler disasters continued at an appalling rate. From 1825 to 1830, a total of 42 explosions were recorded in the US, killing 273 persons. After "50 or 60 persons" had been killed by a disaster in 1830, onboard SS Helen McGregor near Memphis in Tennessee, the federal government finally yielded, and gave a substantial research grant to the Franklin Institute. This was the first occasion on which the US government gave financial support to an engineering research project.

Figure 4.5. The race between Natchez and Eclipse in 1855. (Published with permission from Science Museum, London, UK)

The ensuing technical report recommended legislation prescribing compulsory boiler tests every third month, with a pressure up to three times the normal working pressure. But, again, the resistance against legislation was so strong that the report was rejected. Boiler explosions continued to occur for years ahead, on the Mississippi and other major rivers.

In 1880, The American Society of Mechanical Engineers (ASME) was founded, and steam boiler safety soon took an important place on its agenda. The ASME Boiler Codes, first published in 1911, have since become a very influential basis for the development of the safe design of all kinds of pressure vessels, in the US as well as in many other countries.

4.4 Nuclear Engineering – Steam Technology at the Crossroads

The steam age may be stated to have been born in 1712, when Thomas Newcomen's first pumping machine was erected at a coal mine near Dudley Castle in England (Figure 4.6). During the ensuing almost 300 years, the technology has passed through a series of stages, from atmospheric to high pressure, from a pumping machine to a producer of rotary motion, and from a stationary to a mobile energy transformer.

Common to all these incarnations is the boiler, where heating of water produces steam. Different heat sources have been used: wood, coal, oil, gas, or nuclear fuel.

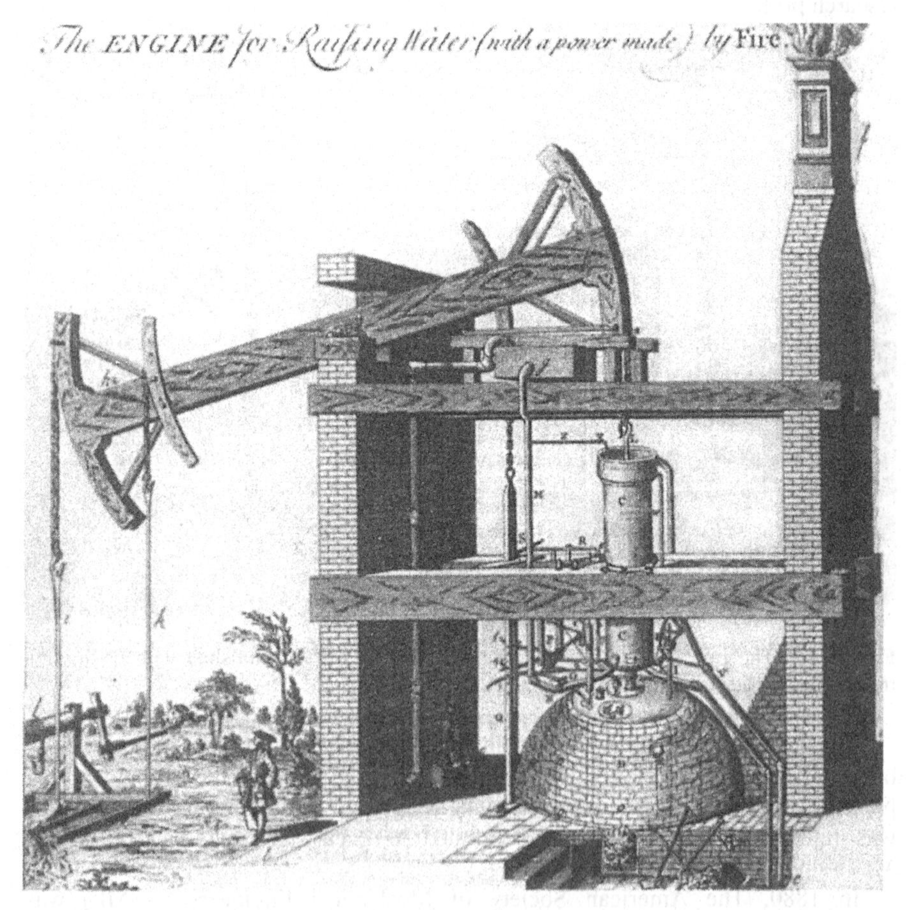

Figure 4.6. The oldest known illustration of Newcomen's pumping machine (1771). (Published with permission from Worcester College, Oxford, UK)

In the early stages of the steam age – as shown earlier – the uncontrolled rise of the steam pressure was an ever present menace. In addition, corrosion (rust) in the boiler walls created a hazard, growing with the age of the boiler. A manhole became a compulsory requirement for every boiler, making regular inspection of its inside possible.

Taken together, all such safety arrangements – with firm regulations both as to handling and as to inspection intervals – have resulted in a technology where the risks of boiler blow-ups have become negligibly small.

In addition, theoretical analyses of stresses and strains in pressure vessels have been developed to include the effects of various small deviations from the intended design.

The types of damage caused by bursting boilers during the first centuries of the steam age are not experienced as serious threats anymore – with one exception, namely boilers in nuclear power stations. Here, a blow-up of a main pressure vessel would probably mean the end of nuclear engineering as we know it today. The state of the boiler itself is, therefore, always analyzed with particular respect to the deterioration of the boiler material, due to long-term irradiation. In addition, material samples which have been immersed in the boiler ever since the plant became operative are examined during annual closure periods.

Special attention has also been given to the consequences of certain other events, such as bearings in machinery running hot, or suddenly excited vibrations, both of which might require an emergency stop of the entire plant. An extremely dangerous situation, termed "LOCA" (loss of coolant accident in nuclear engineering parlance), could cause a meltdown in the reactor hearth, in the event that all safety systems became inactive at the same time. For several reasons, such a situation is, however, extremely unlikely:

- Pressure vessels and pipelines in nuclear power stations are designed by the same experts who have successfully designed safe boilers as well as pipework for conventional power stations.
- Before a nuclear power station is commissioned to operate, rigorous inspection programs with the testing of controls are carried out.
- These tests are repeated at regular predetermined intervals, throughout the lifetime of the plant.

Reactor safety studies have also included hypothetical events such as a sudden rupture of the main steam pipe from the reactor to the turbine, causing steam under high pressure to gush out. The reactive force exerted on the reactor vessel would cause severe damage to the plant, but all previous experience indicates that such a situation is extremely unlikely.

Nuclear power stations are normally not located in earthquake-prone regions. Disaster could of course be inflicted upon a nuclear power plant due to sabotage or acts of war, but such kinds of risk will not be further discussed here.

In conclusion, lengthy experience of severe disasters in steam technology has come to foster a deep sense of societal responsibility among engineers and management. Few branches of engineering are as strictly ruled by codes and regulations.

4.5 Other Examples of Historical Safety Measures

4.5.1 Dramatic Accidents and Slowly Acting Dangers

Dangerous events occur in many walks of life, events which have inspired important safety improvements. A few examples, not relating to high-pressure steam technology, will be shown next.

A fire in a department store in Boston, Massachusetts, in the late nineteenth century had frightful consequences. The main doors could open only inwards, and therefore a great number of visitors could not escape, and were burned to death. Since then, it is prohibited by US federal law to use such doors in buildings open to the general public.

Day-to-day safety improvements are continuously being carried through, often based on recordings from insurance companies. TV supervision in car tunnels, regular tests of alert systems, and flying inspections of car drivers have assisted in diminishing accidents in road traffic.

Dramatic disasters often dominate in discussions of risks in modern life, whereas slowly acting disturbances on humans in our technological society tend to be disregarded. The danger of high sound levels in industrial workplaces – or in music arenas – may often become manifest only after an extended time. Monotonous work at computer terminals may induce pain in muscles or joints. Long-term exposure to certain chemicals – even in a very diluted form – may eventually lead to detrimental effects.

Much effort and penetrating research have been devoted to minimizing risks in our technological society. I shall conclude this brief account with some examples of technical solutions to safety problems.

4.5.2 The Falun Mine

One of the main problems in deep underground mining has always been to keep the mine free from water penetrating from roofs and walls. The Newcomen machines, precursors of all later steam engines, were constructed for this purpose. If the pump failed, it was important to warn the miners so that they could climb the ladders and exit the mine in time.

The automatic warning system at Falun consisted of a small chiming bell, driven by a water wheel, which was in turn fed by water from the mine pumps. If the pumps failed, the chiming ended abruptly, signaling imminent danger.

This "reversed" system is an example of a totally reliable solution to the warning problem: Danger is indicated not by a sounding signal but by a sudden silence. Everyone then knows that the pumps must have stopped – or that the bell has got stuck. In any case – for safety's sake – it is best to get out immediately!

4.5.3 The Otis Elevator

When the first person elevators were installed in high-rise buildings – in particular in major cities in the United States – insistence on safety design was soon brought forward to prevent a disaster, should the rope break.

The designer of one elevator, Mr. Elisha Graves Otis (1811–61), created quite a sensation during a demonstration at the Crystal Palace Exhibition in New York 1854 (Figure 4.7). Ratchets had been installed on each side of the shaft, and pawls on the cage were held clear of the ratchets as long as the rope remained stretched. If the rope broke, springs would immediately force the pawls out to engage the ratchets – and stop the fall of the cage.

During the demonstration, Mr. Otis stood calmly in the open cage and, when an assistant used an axe to cut the rope, the cage fell a few inches before it halted – and the spectators applauded.

Figure 4.7. Elisha Graves Otis' demonstration in New York 1854. (Source: TAILPIECE. Elisha Otis. After Steelways 9, no. 5, back cover, 1953. American Iron and Steel Institute, New York. D.E. Wodall)

4.5.4 Sundry Safety Equipment

Safety equipment, simple or more advanced, may serve the protection of human health in traffic (seat belts, crash helmets, crash cushions), or at the work place (protective eye glasses, ear plugs) and also the limiting of property damage (electric fuses, fire alarms).

Hospitals are vulnerable institutions, where a sudden power failure may have perilous consequences, e.g., in an operating theatre. An emergency system to protect vital systems may consist of the following: a heavy flywheel, driven by an electric motor, is kept in steady rotation. If a power failure occurs, the flywheel is immediately clutched to a diesel engine, which then starts and takes over to drive a replacement generator.

Modern industrial production often uses hydraulically operating machinery, e.g., presses for plate-forming in car manufacture. The operator places the work piece in the press, and must then use both hands to press two start buttons simultaneously, so as to avoid getting one of them crushed. The press cannot be started by pressing only one of the two buttons.

Already in their first generation, electric locomotives were equipped with a "dead man's handle," which automatically actuated the brake, should the driver release his hold. This type of safety arrangement is now commonly used in much other equipment, such as power lawnmowers.

The railways have always been highly safety minded for the protection of the passengers. Thus, the doors of a railway carriage *cannot* normally be opened when the train is moving.

In contrast to the small risks involved in train travel, automobile traffic – for obvious reasons – leaves the passengers less protected. It is a challenge for society to develop much safer conditions for travel on the highways.

4.6 Concluding Remarks

In our technological world, national characteristics have become less dominant than before. International cooperation relating to safety in modern societies has resulted in almost universally accepted rules and regulations.

Nearly three hundred years ago, the steam engine led the way in research into technological safety problems which required scientific study before they could be mastered. It has become a common interest between actors in this field to show openness, a hopeful development for the future.

References

Bednar HH (1981) Pressure vessel design handbook. Van Nostrand Reinhold Company, New York

Gill SS (ed.) (1970) The stress analysis of pressure vessels and pressure vessel components. Pergamon Press, Oxford

Gordon JE (1978) Structures, or why things don't fall down. Penguin Books, London

Hammond R (1956) Engineering structural failures. Odhams Press, London

Höjer EB (1910) Lokomotivlära, Statens Järnvägar, Stockholm, Sweden

Nichols RW (ed.) (1979–83) Developments in pressure vessel technology, 1–4. Applied Science Publishers, London

Nichols RW (ed) (1980) Trends in reactor pressure vessel and circuit development. Applied Science Publishers, London

Petroski H (1985) To engineer is human. St Martin's Press, London

Petroski H (1994) Design paradigms: case histories of error and judgment in engineering. Cambridge University Press, Cambridge

Scanning J (ed) (1976) Great disasters: catastrophes of the twentieth century. Treasure Press, London

Spence J, Tooth AS (eds) (1980) Pressure vessel design: concepts and principles. Chapman & Hall, London

Steele LE, Stahlkopf KE (eds) (1980) Assuring structural integrity of steel reactor pressure vessels. Applied Science Publishers, London

5

Risks and Safety in Building Structures

Håkan Sundquist

5.1 Introduction

All structures and materials are degraded and destroyed with time. The only thing which varies is the time-scale of the degradation. A sandcastle on the beach can perhaps survive for a few hours, whereas the breaking down of a mountain can take billions of years. In the case of a building structure, it is important that it is designed so that it has the lifetime which has been intended. Throughout its envisaged lifetime, the structure should also have a high level of safety to resist the forces and loads to which it is subjected and the environment to which it is exposed. In this chapter, we shall discuss risks and safety aspects in relation to different types of structure and in particular the influences of nature and of human beings. In this context building structures include everything which humans build, e.g., houses and civil engineering structures of various kinds, such as bridges, locks, masts, tunnels etc.

5.2 Structures Are Degraded by Use and by the Environment

There are many factors seeking to degrade and destroy building structures. These degrading factors are usually divided into three categories, namely: *loads*, *environment*, and *use*. Use often involves loads, so that this division is not completely logical, but it has practical advantages in a building structure context.

5.2.1 Commonly Occurring Loads

The loads are usually those which the structure has been designed to support or withstand. The walls and roof of a house are made to provide protection against, e.g., snow and wind, and the structures are therefore designed so that these loads can with great probability be borne. That the snow load varies with the geographical location is self-evident. Walls are designed so that the greatest wind force which occurs within a period of 50 years can with a high degree of probability be withstood. Loads of this type are called natural loads. Snow and wind are commonly occurring natural loads. There are also more rare natural loads, to which we shall return later.

A bridge or a system of structural elements has been designed to carry the loads for which the bridge or the house has been designed. We may call these loads functional loads, since it is precisely these loads for which the structure has been made. To find the correct dimensioning load for a bridge, it is possible to proceed in two ways. One way is to create rules for the maximum size and weight of the vehicles which are to be allowed to use a certain system of roads, and the second way is to measure the real loads through weighing stations or in some other way. As in all human activities, however, there are those who do not follow the rules and regulations and it happens that vehicles which are much heavier than the permitted limit are in fact driven over bridges. Since such behavior can lead to an accident, bridges are always dimensioned so that they can indeed withstand certain overloads. However, it is theoretically possible that a series of unfortunate events occur at the same time, and that the safety margins are thereby exceeded. Nevertheless, building structures belong to those systems in society which have the greatest margin of safety, and accidents involving finished structures are very rare.

Injuries and accidents are in fact much more common during the constructional stages. Every year building workers are injured or lose their lives because, for example, the formwork shuttering is unable to support the loads from a concrete casting. It may seem illogical that different margins of safety apply during the building period and during the useful lifetime of the structure, but tradition and economy have led to different safety levels for these two situations. Similar illogical divisions can also be found within other spheres.

5.2.2 The Environment

The influence of the environment on a building structure is unavoidable. The most common environmental load is probably corrosion in steel, but all materials are degraded in the environment in which they act, even though their lifetimes can be extremely long. Other environmental loads include various forms of chemical action, and erosion by wind and water. Those human beings who use the structures also contribute to a kind of "environmental load" as another term for "use."

5.2.3 Use

A structure is of course intended to be used, and this means that, in addition to the various loads mentioned, it is also subject to wear and tear. The activities which take place in the buildings wear on the structure. Protective paint is scraped away, blows and impacts wear away material, and small but repeated loads cause small cracks which in the long run mean that the strength of the structure is reduced.

5.2.4 Accidental Loads Due to Human Activity

In a structural context, the concept of *accidental load* is used to indicate an unexpected and rare influence on a structure. There are a great number of possible accidental loads which are due to human activities. We can imagine cars which drive into buildings and break pillars, explosions from gas stored in houses, an airplane which crashes onto a building, or terrorists who seek to blow up a monumental building. A few cases of this type are discussed in Section 5.5. Buildings are to various extents designed and built so that they can also withstand this type of load up to a certain level. Nuclear energy plants situated close to airports are of course dimensioned so that nothing serious would occur if an aircraft were to crash on them. All great buildings are designed and built so that a single load-bearing part, e.g., a column, can be broken without any serious consequences. Very important buildings are specially designed so that they can withstand terrorist activity, etc. Nevertheless, there are influences which cannot be coped with within a reasonable economic framework. The probability of the size and effect of the accidental load must be assessed by those responsible for the design and construction of the building.

The most important and most common accidental load is fire, although this is so common that it is often given a special heading. Practically all buildings are dimensioned so that they resist fire for a certain time. It should be possible for a large and important building to be exposed to fire for a considerable time without collapsing, so that it is possible to evacuate people from the building. Consideration is also given to the safety of the fire-fighters in attempts to ensure that the structural framework does not collapse while the fire is being extinguished. There are nevertheless certain cases that cannot be coped with, since it is uncertain what is stored in the building. Here, there are also different risk levels for the people who are expected to be in the buildings and for the firemen whose task it is to extinguish the fire (see Chapter 14).

5.2.5 Nature Reclaims

The earth is not a dead body. The surface of the earth is continually changing, e.g., through tectonic movements in the earth's crust which give rise to the violent natural loads described in the next section. Inland ice also gives rise to movements in the earth's crust, and rivers erode loose soils. When these movements have occurred, the earth's surface wants to restore the state of equilibrium. This gives rise to earth slips on slopes, i.e., landslides and avalanches. Geotechnicians and engineering geologists work to assess and reinforce earth and rock slopes and to create stable rock and earth mineshafts.

5.2.6 Violent Natural Loads

Our earth is subject to a large number of unusual natural loads of the accidental type. A few examples are listed here. The first that one thinks of are earthquakes, but nature also affords many other surprises such as volcanic eruptions, typhoons, floods, and meteors. Whether and to what degree a structure is to be designed to withstand such loads is dependent upon the probability with which one judges that such an event can occur. In areas where it is known from experience that earthquakes occur, the structure is dimensioned for earthquake loads. Maps are drawn up which show the expected probabilities and magnitudes of such loads and the dislocations in the ground which can be expected.

The serious earthquake in Kobe in Japan in 1995 is, however, an example of the fact that one did not have sufficiently good knowledge. The area in which the earthquake occurred had not been classified as one of the areas at most risk, and the strength and spread of the earthquake surprised the experts. The next section describes how earthquakes are classified.

We read about serious damage to buildings caused by typhoons, tornadoes, and floods. The fact that damage occurs is often due not to a lack of knowledge of the risks but rather to a lack of economic resources to safeguard buildings and other installations against these rare and large natural loads. Some people live under a constant awareness of the fact that natural loads can cause great damage. Large areas of Holland can be seriously damaged by floods. However, security systems have been built up over a long period of time and extensive resources are devoted to ensuring that the security level is high. The Dutch are perhaps the most skillful engineers in the world with respect to building in water, and the security considerations which are adopted there have been an example for many other countries.

There are of course natural loads against which one cannot provide protection. The population of Iceland, for example, lives all the time with the threat that a volcanic eruption may take place in the central part of the island where it is crossed by the Mid-Atlantic ridge.

The earth is constantly being bombarded by meteors. Large such meteors can reach as far as the surface of the earth. The world's largest explosion in the modern era was probably when a meteor exploded at a height of between 5 and 10 km close to the Podkamennaya Tunguska River in Siberia in 1908. The explosion was so powerful that it is considered to have corresponded to the explosive strength of 700 atomic bombs of the size of that dropped on Hiroshima.

There is no absolute safety, and the risk level which is provided for damage due to meteorites is usually the level which is used as the lower limit of the acceptable risk. It is pointless to build so that the risk level is lower than this, since a meteorite can strike at any time and cause enormous damage. The risk that this may take place is, however, very small.

5.2.7 The Richter Scale for Earthquakes – an Empirical Scale

The strength of an earthquake is usually indicated by its magnitude on the Richter scale. Strong earthquakes have a magnitude M greater than 6 and earthquakes with $M > 8.5$ are very unusual. Originally, the magnitude was related to the reading on a particular American seismometer. The Richter scale was created amongst other things to provide journalists with a simple measure of the strength of an earthquake. Later, attempts were made to relate the magnitude M to the energy E liberated. Figure 5.1 shows graphically an approximate relationship, where it is evident that an increase of one unit on the Richter scale corresponds to a 30-fold increase in the energy liberated.

The relationship in the figure is very uncertain, but the energy conditions in an earthquake are so complex and varying that one must be satisfied with a rough empirical relationship between the magnitude M and the energy liberated E.

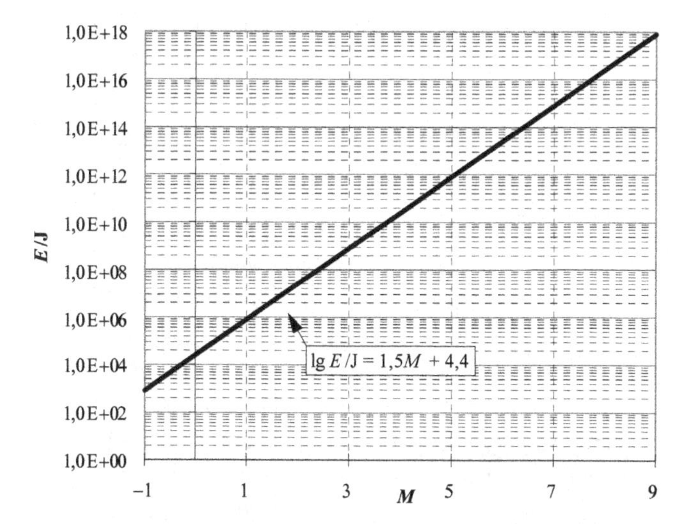

Figure 5.1. The energy E liberated as a function of the magnitude M on the Richter scale

5.3 How Does a Design Engineer Think?

Those who work with structures within the sphere of buildings and civil engineering structures have built up their own safety thinking, often with the help of regulations developed by society. The regulations which apply in the building sphere are probably society's most extensive safety regulations.

The provision of safety and security in building structures, i.e., in everything which is built by human beings, is one of the most important tasks of the design engineer and constructor. Risks and safety act as two mutually dependent opposites in all design work. In many cases, the building standards contain regulations which help the designer, but the design engineer nevertheless can and should often reflect on the concepts discussed in this chapter. Matters relating to the quality work and the quality control of the finished product are often regulated by the designer. In this case, the risks and safety must also be evaluated as a part of the control.

Issues of these kinds have occupied engineers for a long time and it is plain that in ancient times the interaction between safety, risk and quality and also how to regulate this was well understood. We quote freely from Hammurabi's building code from ca. 2000 BC, shown in Figure 5.2.

- If a master-builder builds a house for a person and makes the structure so weak that the house collapses and causes the death of the house owner, the master-builder shall be put to death.

- If the collapse is such that the property of the house owner is destroyed or damaged, the master-builder shall provide compensation for this property and rebuild the house at his own expense.

Figure 5.2. Hammurabi's building code from circa 2000 BC was written with cuneiform script on clay tablets. The regulations in the code placed strict responsibility on the master-builder for that which was built

The engineer is of course aware that absolute safety can never be guaranteed. If overly high levels of security were applied for the strength of, e.g., an aircraft wing, it would be so heavy that the plane would not be able to take off. In the case of aircraft, one relies not only on a very qualified design work, but also on constant checks and revisions of the aircraft.

In the case of bridges, the national road or railway administration can have a system whereby the bridges are inspected regularly on an annual or biennial basis, where the inspection intervals are dependent on the results of previous inspections.

Thus, the work of the design engineer includes the task of achieving a suitable balance between risk and safety. In this analysis, the factors to be considered will depend on whether the task involves:

- economic risks, or
- risks to human beings

and the results may be entirely different. In both cases, it is a question of trying to identify and study the risks, and this can be fairly difficult. The risks can be classified in different categories:

- Risks which are known about in advance and for which there is statistical documentation based on earlier studies.

- Risks indicated by previous accidents, as a result of which it is possible, through some form of check list, to evaluate the possible risks which exist.
- Risks which are shown to be possible on the basis of separate and independent studies. When working with very dangerous systems (e.g., nuclear energy), it is not possible to rely solely on the experience gained from previous accidents. Here, it is essential to try to take into account all the potential deficiencies and faults in the systems. For obvious reasons, this type of risk assessment is extremely difficult and requires very qualified experts.

The various measures available for handling risks can be divided into the following categories:

- Self-evident measures
- Standards and norms
- The application of previous experience
- The design engineer's own assessments
- Risk analysis.

Within the building sphere, the regulations in Building Codes and Standards are primarily directed towards the risks for the health and safety of human beings, where the risk is the probability that the structure will collapse. Modern Standards refer to the consideration of the "ultimate limit state" under normal loads, of which there is knowledge and experience. For instance, measurements are made of how much snow falls within a certain area, and the structure is then normally designed for a snow load which statistically is not likely to occur more often than once in 50 years.

For the design of dams and structures in watercourses, accurate measurements of the water flow are made over as long a time period as possible. On the basis of these measurements, a statistical calculation is then carried out to assess how great the probability is that the flow rate is of a given magnitude once in, e.g., 100 years. Usually, the data are available for much shorter measurement periods than would be needed for making a certain statement about the probability, and the weather can have periodic variations with a repeat period of, e.g., 100 years.

Certain controls of the failure safety with respect to so-called *accidental loads* are also carried out. These should perhaps be referred to as "disaster loads" in accordance with the terminology introduced above. Such loads are often very schematic and are intended, at a certain (often not unambiguously determined) level, to take into consideration the risk that, e.g., a truck is driven into a house or that an explosion occurs because LP-gas has been kept in apartments.

However, there are many risks that are ignored, for example, the risk that an airplane may crash or a meteorite strikes a house. The reason why this type of risk is ignored is, of course, that the probability of such an occurrence is deemed to be extremely small. However, the designer is free, through a suitable design and other measures, to build in safety also against this type of risk. In the case of a nuclear energy plant, some of these are in fact also taken into consideration. Naturally, there remain risks which cannot be eliminated, e.g., that a meteorite falls and hits the plant, but consideration is, e.g., given to the risk that a small airplane might crash into it.

The Building Standards also contain regulations which take economic and hygienic risks into consideration, e.g., that a building structure develops deformations so great that it is unpleasant to live in, or that cracks arise which mean that there is a risk of heat and moisture leakage. Such a consideration is usually referred to as a check on the *serviceability limit state.*

5.4 Human Errors

In all design and construction work, one must include “the human factor” in the calculations. Human beings are all liable to make errors, and a structure must to a certain extent be designed so that such errors do not lead to too great risks. See Chapter 15 for more about “the human factor.”

5.4.1 We Are All Human

An error can lead to serious accidents and of course economic losses. Methods for eliminating errors have therefore attracted great interest. A certain variation in the design and quality of the products must always be expected, and the Codes and Standards recognize this by demanding, e.g., that the material properties shall with a certain probability fulfil the requirements strived for. However, the material manufacturers are usually so anxious to obtain good test results that a result below the lowest acceptable level is extremely rare. If such a result is obtained, it is usually because a gross error has occurred, e.g., there has been a mix-up of materials. Thus, it is important to create systems so that gross errors do not occur.

5.4.2 Quality

By quality we mean that the products shall have the properties which are required and expected by those who buy and use the product. To deliver higher quality than is required and expected is poor quality, since this can cause additional costs for which the supplier will not be paid. Systems for achieving quality have always attracted considerable interest.

Practically all quality systems seek to delegate the responsibility for quality down to those who work with the products or systems. In this way all those concerned feel such a responsibility that no errors arise. However, the human factor cannot be entirely eliminated and the available knowledge is not always sufficient to meet all the problems which can arise in the processes.

5.4.3 Control

An alternative way to that described above for supervising that the products and structures do indeed have the quality aimed at is through control and checks. The old systems where special supervisors were employed to oversee, measure, and test the products have practically disappeared from the construction industry, since the method was not sufficiently effective. The efficient way is to delegate responsibility and quality awareness to as many as possible. All errors can still not, however, be avoided.

5.5 Learning Through Accidents!

The skilled designer starts his or her work by finding out not only what others have done correctly, but also what mistakes have previously been made. History is the principal source from which to be able to learn and gain understanding for the future.

5.5.1 The Leaning Tower of Pisa

The world's perhaps most well-known faulty structure is the leaning tower of Pisa. Already during the building process, the ground below the tower started to settle, and to settle unevenly. Nevertheless, the construction work continued, while attempts were continually made to correct the error. The result is that the tower is "banana-shaped." To date, the tower has sunk ca. 2 meters, and the inclination is now ca. 5.3 %, as shown in Figure 5.3. The faulty structure has become a tourist attraction, but experts have been working persistently to seek methods to reduce the increase in inclination. These efforts have earlier not been completely successful, but recently applied counteraction methods have probably stopped the increase in inclination and the safety of the tower is considered to be satisfactory.

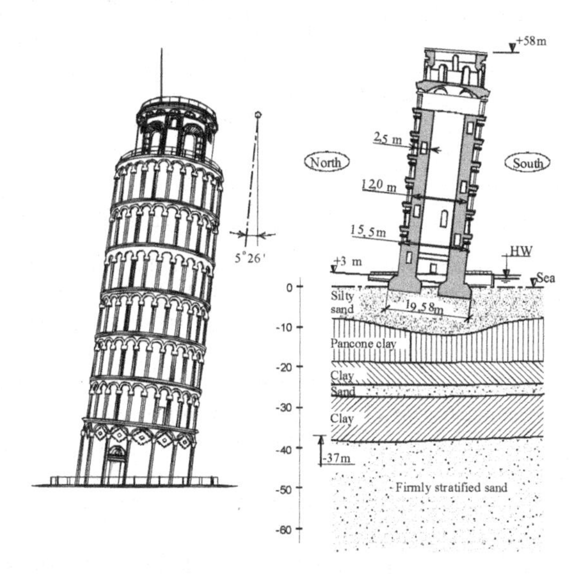

Figure 5.3. The leaning tower of Pisa. The tower has sunk more than 2 meters into the clay. Redrawn by the author, essentially from information in Leonhardt (1997)

5.5.2 The Getå Accident – the Worst Train Accident in Sweden

Roads and railway embankments are structures which human beings designed and constructed so that they will bear the loads associated with traffic. They are designed by soil mechanics engineers who, on the basis of geological information and the results of drillings, either accept that the existing ground can carry the loads from the embankment and traffic or design various reinforcing structures. Recently, a large number of effective methods have been developed to reinforce soils and rocks which in themselves have a poor ultimate bearing strength. As in other disciplines, it has also been necessary to learn from history within this sphere.

In the afternoon of 1 October 1918, the train between Norrköping and Stockholm with an engine and seven carriages fell down a slope by Getå on the Bråviken shore. The three last carriages in the train remained on the embankment. Of the approximately 125 passengers and personnel on the train, 41 persons died.

Just before the train passed the stretch in question, a landslide had taken place. As shown in Figure 5.4, the embankment and the main road parallel to the embankment had been built up on sloping ground.

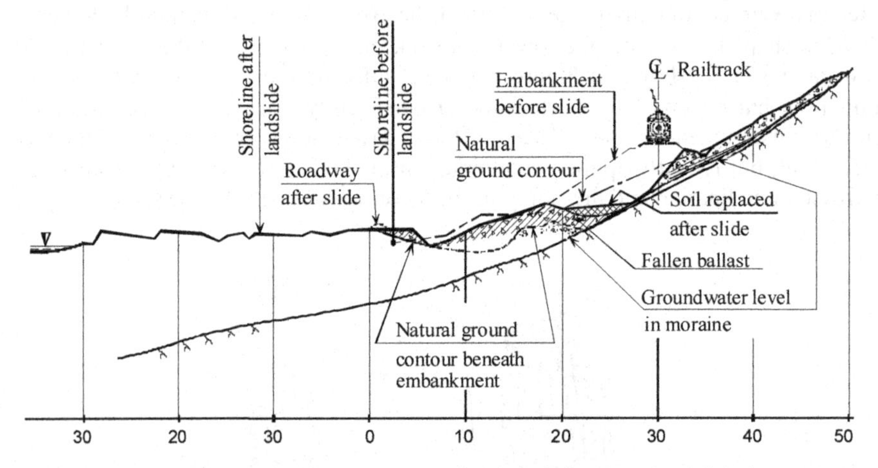

Figure 5.4. A section through the ground at the site of the Getå landslide in 1918. The surface contour is shown before and after the landslide, with the approximate earth-layer sequence. Redrawn by the author, essentially from information in SJ (1918)

The embankments themselves gave rise to an increased load on the relatively loose earth which consisted of clay and silt, on top of which material, which had slid down from the rock above, had settled in layers. Owing to a lot of rain in the period before the accident, it was stated that the weight of the embankment had increased and that an overpressure in the soil layers beneath the embankment had led to deterioration in the load-carrying strength.

The geotechnical commission of the Swedish State Railways had already been formed to investigate the safety of embankments, etc. The Getå accident speeded up research and development work within the geotechnical field and led to good contributions to the understanding of how to provide protection against landslides.

One can never be completely certain, and every year landslides of a greater or lesser magnitude occur, often along rivers and streams which flow through clay areas. It is very difficult to provide complete protection against this type of phenomenon, unless one can choose to build only on soils with good bearing-strength properties. As emphasized in the introduction to this chapter, nature will sooner or later also break down that which we consider to be solid ground.

5.5.3 The Worst Building Accident Due to a Structural Fault in Modern Times – a Combination of Errors

On Friday July 17, 1981, the atrium in Kansas City's newest hotel, the Hyatt Regency, was filled with festivelydressed people who were celebrating the weekend with dancing and listening to a big band. The premises had been built as a high glass-roofed well with a ceiling height of 18 meters. From the ceiling, walkways were suspended at the third and fifth floor levels, to provide convenient passage between different parts of the building. These walkways were also filled with people beating time to the music with their feet.

Suddenly, the disaster occurred. Two of the walkways collapsed, casting all those who were on the walkways down onto those who were on the floor; 114 people were killed and almost 200 were injured. This accident is the worst which has happened to an American building structure due to structural mistake. No such large accident in peacetime caused by a building collapse has occurred in Europe in recent times.

An investigation into the cause of the accident rapidly ascertained that the original drawings of the suspension of the walkways had not been adhered to, and that a change had been made during the actual building process. The reader is encouraged to study Figure 5.5 carefully.

The walkways were supported on welded box girders (A) which, with a nut and washer (B in the original design and C after the change), were suspended in long threaded vertical rods (D) from the ceiling. The figure shows the suspension of the upper of two walkways, one above the other, which were suspended in the same rods. At the time of the accident, the box girders gave way so that the nuts with their washers were pulled through the girders which were badly deformed.

What was the reason for this design change? In the original design, the vertical stays were 14 meters long and they had to be threaded all the way up. (Alternatively, the rods could have had a greater diameter at the threaded sections, but this would have been an expensive solution.) It was totally impossible to handle such long objects and to get the bridges in place. Two walkways were suspended above each other, and the upper was to be held by nuts which were to be rotated 9 meters along the thread on the stays. In other words, the design was a complete impossibility. Somebody had the idea that one could use several short stays which would then only need to be threaded closest to the ends, and the modified design shown in Figure 5b was chosen.

The investigation showed that the original design in itself was under-dimensioned and had only 60 % of the strength prescribed in the Building Standards. The Standards presuppose a certain safety factor, and it may be assumed that the Standards include a dynamic addition to take into account resonance effects caused by dancing people, like the well-known problem of a marching army. In this case, however, the safety factor was only about 1, i.e., there was no margin whatsoever. This meant that a fracture in one place led immediately to an overload in other places which could not even support the load until all the people had been able to reach safety. In other words, the structure was not *fail safe* and contained no *redundancies*, i.e., it was not robust and tolerant of partial failure in the event an overload. The result was an immediate total collapse.

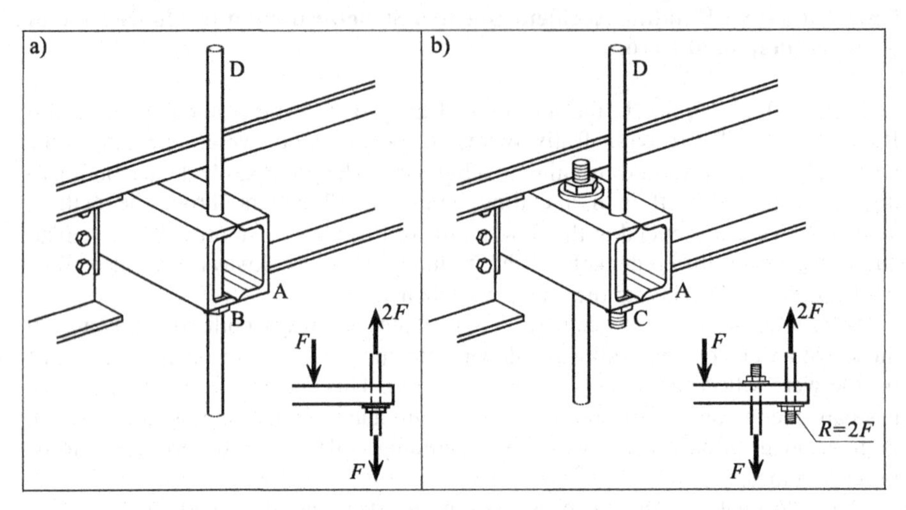

Figure 5.5. A detail of the building design for the steel bars that supported hanging walkways which caused the greatest accident due to a structural fault in the West in recent years. **a** This is how it was conceived and drawn. The force acting on the nut and washer (B) is here F, where F is the load coming from one walkway. **b** This is how the part was built. The force acting on the nut and washer is here $2F$. Redrawn by the author, essentially from information in Marshall et al. (1982)

This was nevertheless not the principal reason for the accident. The reader is encouraged to look again at the figure. In the original design, the nut (B) takes up the load from a single footbridge. In the modified design, the nut (C) takes up the load both from the upper footbridge and from that which is suspended below it. This nut is thus overloaded by a factor of 2. The result was that a poor structure was replaced by a highly dangerous one.

One may ask why nobody questioned the design. There was good reason to do so, when it was found necessary to make a change. Warning signals were also present. The building workers noticed that the walkways swayed and they therefore made detours with their wheelbarrows to avoid crossing them. With the obvious safety risks, the structural engineer should have reconsidered the design of the structure.

There were other alternative solutions. If the threaded rods were too long, it would have been possible to join several shorter ones together with threaded sleeves, and thus have retained their load-bearing ability.

When the walkways were rebuilt, they were supported by columns from the floor. In this way, the impression of the free volume of the hall was of course lost, but the walkways were much more stable.

This example has been taken from a thought-provoking book: *To Engineer is Human* by the American professor Henry Petroski (1985). His theme in the book is that one must take risks in order to be able to make progress. It is not strange that one sometimes fails, but one must learn from experience.

5.5.4 What Has Been Learned About the Risk of Gas Explosions

Another case which has been very important for the development of Standards and Codes for buildings is the collapse in May 1968 of a part of a residential block of apartments at Ronan Point, East London, as shown in Figure 5.6.

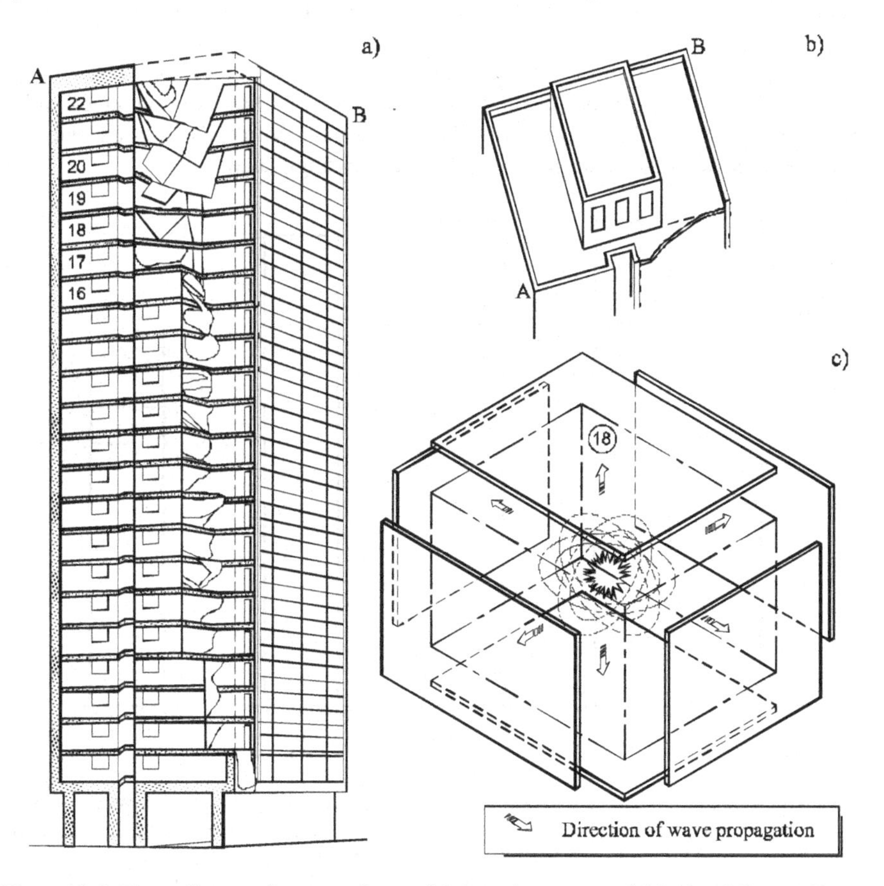

Figure 5.6. The collapse of a part of a prefabricated constructed block of flats at Ronan Point, East London. The collapse began as the result of a gas explosion in a flat on the 18th floor and then spread through almost all the building. (**a** Building after explosion, **b** top-view of buildning, and **c** initial explosion). Redrawn by the author, essentially from information in Griffiths et al. (1968)

The collapse was initiated by a gas explosion in one of the corner flats on the 18th floor. The house was a structure based on prefabricated elements and several walls fell out. Owing to a poor connection between the elements, more elements both above and below the accident floor collapsed like a house of cards, and many people were killed or injured.

In most countries, this accident has led to the requirement that a building shall have a certain protection against so-called progressive structural failure. In most cases, the protection means a requirement regarding various connecting devices between the pre fabricated elements.

5.5.5 Loads Due to Terrorism

A disaster which will influence the design and building of all buildings for a long time is the flying of aircraft into the World Trade Center (WTC) and the Pentagon in the USA in 2001.

Already in 1945, a small bomber flew into the then highest building in the USA, the Empire State Building in New York. The building was damaged only locally and the number of people who died was small. A fire also started, but the amount of fuel was small and the fire was rapidly extinguished.

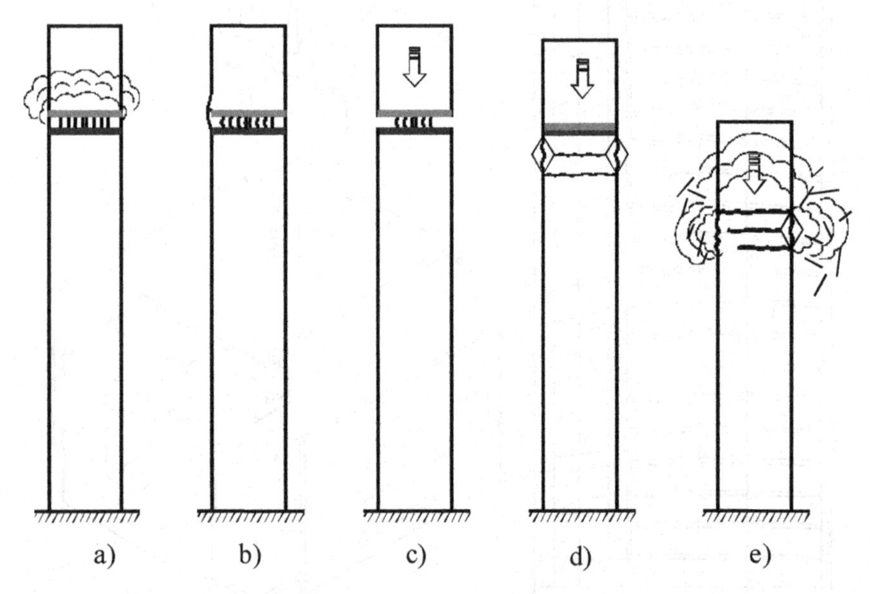

Figure 5.7. Schematic sequence of the collapse of one of the towers of the World Trade Center in New York on September 11, 2001. **a** and **b**, the collision between the aircraft and the building caused an explosion and seriously damaged the outer load-bearing shell; **c** the fire softened the internal supporting columns, **d** the remaining framework in the damaged part was unable to support the weight of the whole upper part of the building, and **e** when the collapse had begun, nothing could prevent it from continuing down through the whole building

Designers thus became aware of the risk of aircraft flying into tall buildings and, in the design of special buildings, the risk of an aircraft flying into the building has to a certain extent been taken into consideration. To build so that no damage arises is probably almost impossible, but to construct so that the possible damage would not be much worse than in the accident in 1945 has been the target in some cases. The intention was thus to design the World Trade Center towers so that they would resist being flown into by a large air liner which had got off course by mistake, without the building collapsing. In the event, the combination of the local damage caused by the airplane crashing into the tower and the subsequent extremely strong fire led to the total collapse of the towers, as indicated in Figure 5.7:

- The collision between the aircraft and the building seriously damaged the outer load-bearing shell, Figures 5.7a) and 5.7b)
- The fire softened the internal supporting columns, since a large part of the fire protection had been damaged by the explosion which accompanied the actual collision, see Figure 5.7c)
- The framework remaining in the damaged region was no longer able to support the load due to the weight of the whole upper part of the building, Figure 5.7d)
- When collapse had actually started, the kinetic energy of the collapse was so great that nothing could prevent it from continuing down through the whole building, Figure 5.7e)

At the present time, there is an intensive debate among designers as to what can be done to reduce the risk of total collapse of tall buildings in a manner similar to that which took place in the World Trade Center. It is unavoidable that there is great damage and that many people are killed if an aircraft flies into a building, but specialists are agreed that measures can be taken to improve the safety of important buildings in the face of terrorist attacks. These measures involve chiefly better fire protection and designs which ensure that different structural parts are held together better if some parts fail.

5.5.6 Do Not Always Trust Beautiful Computer Calculations

Another very interesting example is the collapse of the Sleipner A oil rig in 1991. This type of platform is a Norwegian speciality and many have been built, and several intended for even greater depths of water are being planned. These gigantic structures have been built in a special way, utilizing the conditions existing along the Norwegian coast with great depths of water. The design of the platform support is indicated in Figure 5.8.

The structure shown in Figure 5.8 was towed out and partially filled with water, the intention being that it would sink to a depth which would make it possible to float out the superstructure (the actual platform) and place it on the towers. While being submersed, the whole structure suddenly collapsed because a wall between the internal water-filled triangular spaces was pressed in. This caused the platform to start to sink and several walls collapsed. In a few minutes, the structure lost its buoyancy and the whole structure, worth ca. USD 150 million at the time of the collapse, sank to the bottom with such a force that the impact was sufficient for readings to be recorded by seismologic stations. All the people on the structure were able to get to safety, so the whole event was only an economic disaster.

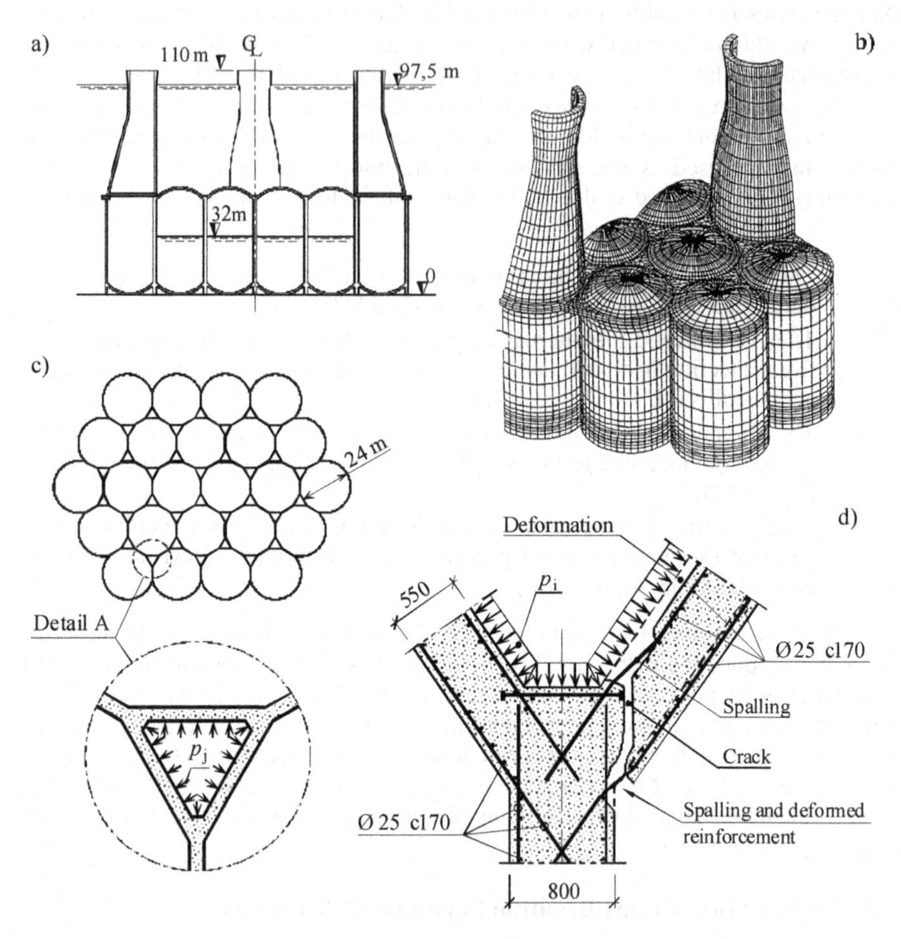

Figure 5.8. The Sleipner A drilling platform: **a** the dimensions of the platform, **b** the FEM-model created to calculate the stresses and deformation in the structure, **c** the probable fracture surface in the structure, and **d** the approximate fracture surface found in the tests subsequently carried out by the Swedish Government Testing Institute. Redrawn by the author, essentially from information in Schlaich and Reineck (1993)

What then was the cause of this disaster? Two reasons have been put forward:

- Incorrect calculations caused by too coarse and incorrectly evaluated finite element method (FEM) analyses. An FEM analysis is today the most widely used method for carrying out an approximate analysis of forces in a complicated structure.
- Unsuitable reinforcement design (see Figure 5.8).

How can this occur? For this type of structure, very sophisticated control systems are applied, and it should not be possible for any fault to arise. One type of check – which had been made – involved, for example, the execution of several different independent FEM analyses. These were, however, based on the same theoretical documentation and a similar coarse mesh division, and no real independent check was therefore involved. The calculations also showed that it should be possible to use the unsuitable reinforcement design shown in Figure 5.8. With a better understanding of the mode of action of reinforcement in a concrete structure, the reinforcement should have been given a much better anchoring. The problem is actually so elementary that the whole situation could have been assessed with calculations carried out on a paper napkin, if one had built up a simple model of the type taught at every Civil Engineering University.

5.5.7 Bridges Which Collapse – Learning from History

Nearly always when one speaks of the development of the study of the strength of materials, examples are taken from the sphere of bridges. This is natural, since bridges are spectacular structures which if they collapse arouse interest and give rise to considerable anxiety. Today, we experience this type of debate in the case of present-day accidents with car ferries. In the earliest development of all types of structures, there were no theories to follow. One had to use the "trial and error method," i.e., one had to feel one's way forward. If the building failed, the master-builder, who at that time was often the same person as the designer, was replaced and a new concept was tested. Thus, a bridge functioned as a kind of full-scale experiment in the field of the strength of materials.

Today, bridge structures are among the technical systems where the safety is the greatest. Even though accidents occasionally occur, the number of accidents should be compared with the enormous number of bridges in the world, and the large amounts of traffic which pass over these bridges. Compared with the risk of meeting with a traffic accident, the risk that the bridge will collapse beneath the car is extremely small.

5.5.8 Bridges Which Shake Down. The Engineer Must Also Learn from History

The case of the bridge at Tacoma Narrows, which collapsed in 1940 after it had been set in oscillation by a relatively moderate wind, is very well known. Many people have seen the film of the collapse and have been surprised at the force of the wind, but those who designed the bridge had not studied their history books properly, for this type of collapse had in fact occurred several times before. Almost exactly 100 years earlier, for example, the Brighton Chain Pier structure had collapsed in the same way. At that time, there were no film cameras, but the wreck was documented by artists who described the course of events in simple pictures.

Other suspension bridges which had been wrecked by the influence of the wind were the first suspension bridge over the Menai Straits in Wales, which was damaged in 1854, and the Clifton Bridge at Niagara, which was totally wrecked in 1889. One learned early that dynamic processes can cause a bridge to collapse. The most serious known bridge accident took place when a marching troop caused a suspension bridge, the Basse-Chaine in Angers in France, to collapse and 226 soldiers drowned, as depicted in Figure 5.9. The accident is considered to have been due to a combination of unfortunate circumstances, where both the wind and the rhythm of the troop marching in step made the bridge vibrate to such an extent that the rust-damaged structure broke.

Figure 5.9. The collapse of the Basse-Chaine bridge in Angers in France in 1850

5.5.9 Fatigue Failure and Other Phenomena in Metallic Materials

Less known than the cases referred to above but equally instructive are the large number of steel bridges that collapsed during the nineteenth century and some way into the twentieth century. When steel became increasingly less expensive, one started to build, e.g., railway bridges in this material in the middle of the nineteenth century. Steel was a new material which opened the way to new construction principles, but the adoption of a new and untried technology led to serious consequences. Approximately one bridge in four suffered serious damage or collapsed. It was realized that it was vibrations from the trains which led to the collapse of the bridge, but the concept of fatigue was unknown at that time.

A typical example of fatigue fracture is the collapse of the Point Pleasant Bridge in Ohio, USA, in 1967. The case is interesting, since it teaches us that it is inappropriate to design a structure so that the load-bearing ability is dependent on a single connection. The collapse, which was the worst bridge disaster in modern times, led to the loss of 46 human lives.

The accident occurred in the rush-hour traffic on a cold day in 1967, after the bridge had been in service without any problem for 40 years. The failure, which began in the suspension rod shown in Figure 5.10, meant that the whole bridge (445 meters) including approach spans, fell into the river. This was a typical fatigue fracture which can largely be referred to the steel grade used when the bridge was built. An inappropriate structural design, with no extra margin of safety, i.e., no redundancies, led to the total collapse.

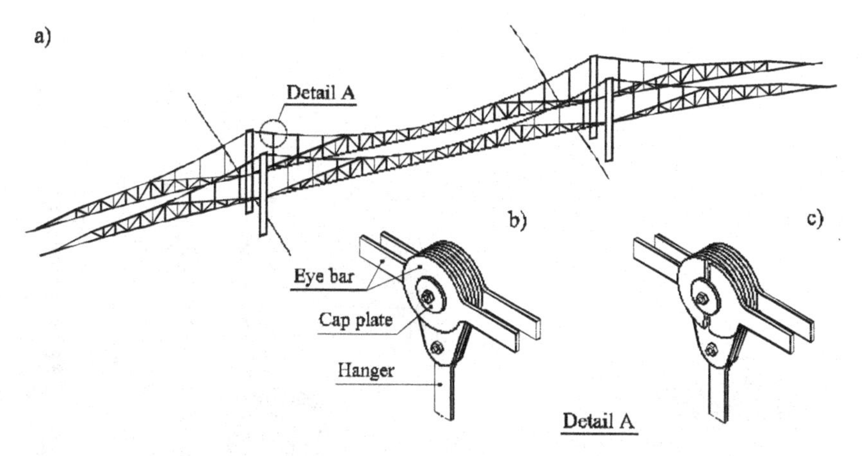

Figure 5.10. **a** The structure and bridge type for the Point Pleasant Bridge in Ohio, USA, which collapsed in 1967. **b** Close-up view of detail before faiulure. **c** Close-up view of detail after failure. Redrawn by the author, essentially from information in Levy and Salvadori (1992)

In a similar case, the I-35W Bridge over Mississippi River in the USA totally collapsed due to one poorly designed joint between the steel members; 13 people were killed and many injured.

5.5.10 The Designer's Nightmare – the Unforeseen Load

An interesting example of an unforeseen load is the collapse of the first Tjörn Bridge in Sweden. Its load-carrying structure consisted of two tubular arches which were anchored in the rock a distance up on the land. In January 1980, the superstructure of a ship collided with one of the arches, whereby the plate material in the arch buckled so that the arch immediately lost almost all its load-bearing ability, and the bridge collapsed, as shown in Figure 5.11. There were no people on the bridge and nobody was injured on the ship, but there was no time to stop the traffic approaching the bridge, and seven cars drove straight out into the air and eight motorists died.

The piers for bridges are normally dimensioned for being driven into by cars or by ships, depending on whether the bridge goes over land or water. In this case, the piers were on the land and there was thus no risk that a ship could hit a bridge support. Obviously, no one had considered the risk that the superstructure of a ship might hit one of the arches. Concern for this type of unforeseen accidental event has increased considerably in recent years. The new Tjörn Bridge is a modern cable-stayed bridge with supports so far up on the land that the risk of being hit by a ship is non-existent. Modern bridges which are built over traffic routes, such as for example the Öresund Bridge between Sweden and Denmark, are designed so that they can withstand impacts from even very large ships.

Figure 5.11. The first Tjörn Bridge after having been hit by the ship Star Clipper on 18 January, 1980. (Photo: Pelle Lund/Scanpix)

5.6 Safety Regulations Within the Building Sector

There are many regulations dealing with questions of safety within the building sector. There is probably no sector in the society that is so regulated as the building sector.

Within the European Union (EU), work is in progress to create joint regulations, so-called Eurocodes, for safety within the construction sector. The intention is that they shall apply for all building work in Europe. The reason for establishing joint regulations is of course the desire to increase the opportunities for trade between the countries and thereby give lower building costs for the consumers. The work on the regulations has been rather complex, since the different countries have very different traditions and experiences. It will probably take a long time before the same regulations apply over the whole of Europe.

5.7 Concluding Remarks

What can we learn from these examples from the history of technology? Obviously, it is enormously important to check all the details in a design again and again, and constantly to ask oneself whether anything will break and what the consequences then will be.

Reliability technology is needed to guarantee that the system has a safe function and a sufficient lifetime. Here, we can only mention a couple of methods that would have functioned excellently, for example, in the hotel building in Kansas City. A *design examination with a check list* would have directed the necessary attention to the different details in the design. *Failure mode and effects analysis* (FMEA) is another method that is particularly suitable for identifying the weak points in a design. The adoption of this latter method would probably have been able to prevent the disaster with the Sleipner A oilrig. Both methods build ultimately on the concept of ensuring that there is time for extra reflection and for respecting common sense, and they also presuppose that staff with experience and competence are engaged. For more thorough discussion on the subject of risk analysis, see Chapter 13.

In this chapter, we have described different approaches to the question of safety within the building sector. The factors which distinguish this field from many other fields of technology are the long lifetimes for which one projects, and also the close interplay with nature and with the forces of nature which influence everything which we build. In spite of the difficulties in predicting the influence of nature, the safety level within the building sector is very high compared with that in many other spheres. Some people consider that it is perhaps unnecessarily high, since the costs are also high. On the other hand, it has become customary to have a very high safety level, since everybody wishes to live and reside in comfort and in safety. For further reading, see, e.g., Ferguson (1993), Gordon (1978), Griffiths et al. (1968), Leonhardt (1997), Levy and Salvadori (1992), Petrovski (1985), Roddis (1993), Schlaich and Reineck (1993), and Timoshenko (1953).

References

Ferguson ES (1993) How engineers lose touch. Invention & Technology 8(3):16–24

Gordon JE (1978) Structures or why things don't fall down. Da Capo Press, New York

Griffiths H, Pugsley A, Saunders OA (1968). Report of the inquiry into the collapse of flats at Ronan Point, Canning Town. Her Majesty's Stationery Office, London

Leonhardt F (1997). The Committee to save the tower of Pisa. A personal report. Structural Engineering International 7:201–212

Levy M, Salvadori M (1992) Why buildings fall down. WW Norton & Co, New York

Marshall RD, Pfrang EO, Leyendecker EV, Woodward KA, Reed RP, Kasen MB, Shives TR (1982). Investigation of the Kansas City Hyatt Regency walkways collapse. Building science series. National Technical Information Service (NTIS), Technology Adminstration, US Department of Commerce, Springfield

Petrovski H (1985) To engineer is human, the role of failure in successful design. St Martin's Press, New York

Roddis WMK (1993) Structural failures and engineering ethics. Journal of Structural Engineering 119:1539–1555

Schlaich J, Reineck K-H (1993) Die Ursache für den Totalverlust der Betonplattform Sleipner A. Beton- und Stahlbetonbau 88:1–4

SJ (1918) Årsberättelse för 1918. Statens Järnvägar (SJ), Stockholm

Timoshenko SP (1953) History of strength of materials. McGraw-Hill, London

6

Risks in the Conversion of Energy

Monica Gullberg and Torbjörn Thedéen

6.1 Introduction

Modern society is dependent upon the availability of an abundant supply of cheap energy. The use of energy is basically a positive factor necessary for our prosperity, but like all other industrial activities, the production of energy can have negative effects on human beings and on the environment. Today, energy is obtained and converted to a great extent in large systems. The chemical energy stored in fossil fuels is, for example, converted by combustion into heat energy, which can be used directly for heating purposes. The heat can also be converted into kinetic energy via steam in turbines and via generators into electricity. When this is subsequently delivered to the consumer, some of the electrical energy is lost.

At the consumer level, the electrical energy is again converted into kinetic energy or heat. Each stage in the progressive conversion of energy is to some extent associated with risks and negative effects; and this is also true when energy systems are built up, maintained, or discontinued. Environmental effects are certainly very important, but that is such a large field in itself that we only briefly consider it in this chapter and focus on other risks related to the conversion of energy.

6.2 Energy Concepts

It is a basic law of physics that energy can be neither created nor destroyed. It can only be transformed from one form to another. In a closed system, the sum of all the energy is therefore constant. Our atmosphere is, however, an open system. It receives its energy primarily from solar radiation and it loses a large part of this energy by radiating heat into space and by reflecting the solar radiation. In between, the energy flows on the earth and adopts different more or less permanent forms. Some of these are used by human beings for heating, cooling, industrial processes, and transport.

Amongst other sources, energy is obtained from fossil fuels, in other words from solar energy stored in plants millions of years ago. Living plants are also used, after felling or harvesting, as fuels and sources of energy. Hydropower stations utilize the energy in the hydrological cycle (i.e., water's evaporation, condensation and return to earth as precipitation) to convert potential energy into electrical energy. Wind-power stations utilize the kinetic energy in the wind in order to generate electricity. Incoming radiation energy is a prerequisite for both the wind and the hydrological cycle. Examples of types of energy that are not derived from the sun are nuclear energy, which utilises atomic fission energy, geothermal energy, which utilizes the earth's internal heat, and energy based on the flow of tidal waters.

The energy in fuel (chemically bound energy) or potential energy can be stored for a long time. Many types of fuel can also be transported over a long distance, which is a practical advantage. After conversion to heat or electricity, energy is more transient. Heat can be stored by insulation; electricity can be stored in batteries. In large electrical power systems, however, most of the electricity is used directly and is never stored. The power, i.e., energy per unit time, varies continuously in an electrical power system in accordance with the electricity consumers' requirements.

Some facts about energy: for power given in kW, energy is given in kWh. The energy content in oil is about 41 MJ per kg (MJ = megajoule, 1 kWh = 3.6 MJ).

6.3 The World's Energy Supply

At the present time, more than 80 % of the world's total commercial energy (i.e., energy which is bought and sold on the market) is based on the combustion of fossil fuels. Oil is the dominant energy source, followed by coal and natural gas. The term "commercial energy" does not include traditional fuels such as wood for direct use in private households. Traditional fuels are an important source of energy in most of the developing countries, but they represent only about 10 % of mankind's total energy supply (UNSD, 2004).

Trade with fossil fuels, particularly oil, is international. The largest net exporters on the oil market are Saudi Arabia, Russia, United Arab Emirates, Norway, Iran, Kuwait, and Venezuela. The largest importers are USA, Japan, China, Germany, South Korea, France, and India [figures from 2006 (DoE, 2008)].

Trade in coal is dominated by Australia's export to Japan and South Korea. These countries also purchase a lot of coal from China and Canada. Gas is transported in large quantities from Russia, primarily to the Ukraine and Germany but also to many East European countries, and from Canada to the USA. In addition to Canada and Russia, Norway Algeria, Holland, and Indonesia are also exporters of gas.

Hydropower and nuclear power are responsible in almost equal amounts for the rest of the world's power supply. Apart from fossil fuels, hydropower, and nuclear power, other types of energy are negligible in international statistics. Most hydropower is produced in China, Canada, Brazil, USA, Russia, and Norway. In the case of nuclear power, USA, France, and Japan are the world-leaders followed by Germany, Russia, and South Korea (UNSD, 2004). Uranium is also bought and sold on the international market. Australia is currently the dominant net exporter, while USA and Spain are the largest importers.

Fossil fuels can be converted to electricity in power stations, to mechanical energy in, e.g., motor vehicles, or to heat. A large and increasing proportion of the fossil energy, and practically all hydropower and nuclear power is used to generate electricity. The conversion and distribution of energy for different purposes requires both technology and organization.

6.4 Energy Systems and Energy Use

An energy system can be limited in different ways. A system is commonly considered at a national level, even though smaller independent systems of course exist. In addition, national systems are integrated by, for example, high-voltage cables or gas pipes. The design of such energy systems is the result of national and international economics, politics, and technical development. When the system limits have been chosen and defined, it is possible to describe the energy balance in the system.

Here the world's current energy usage is briefly described with a few examples of national energy balances, i.e., how energy is supplied to, or generated and used in a country. The energy usage in the transport sector is included as an important factor in a country's energy statistics. With regard to risks, we shall in this chapter follow accepted practice and consider the energy system as the high-voltage network with its power stations (the energy producers) and the loads on this network (the energy consumers). In those cases where heat energy is also delivered in a distance-heating network, this is included in the energy system. On the other hand, risks related to transport are dealt with in Chapter 9, which is concerned with transport systems.

The nations that are currently dominating with regard to energy consumption are USA and China, even though the per capita consumption in China is not very high. The most energy-intensive countries in the world, calculated as per capita consumption, are Norway and Iceland. Less than one-fifth of the world's population live in OECD countries, but they are responsible for more than half of the world's energy consumption (IEA, 2007).

In the industrialized countries, virtually all the inhabitants have access to electricity from more or less large distribution networks. In the poorest countries, however, electrical energy consumption is concentrated to the larger cities. In the rest of the country people meet their energy requirements by collecting wood and by purchasing, e.g., paraffin, diesel oil, LP gas, candles, and torch batteries.

In the middle of the twentieth century, oil was the most important energy-carrier in almost all the industrialized countries. In connection with the energy crisis in the early 1970s, when there was a several-fold increase in the price of fossil fuel on the world market, changes were made in many places to reduce the dependence on imported oil, and internal domestic resources and non-fossil alternatives became important. Many countries began to extend nuclear power as a complement to the fossil-burning power stations. Hydropower was given priority in many regions. Other energy-producing alternatives such as wind-power also increased, although their use has not yet become widespread. In the transport sector, petroleum products are still the only real alternative in most countries, whether they are industrialized or not, so that oil is still often a large part of the total energy consumption.

The total energy consumption is increasing both globally and in most countries. Energy consumption within the OECD countries currently consists mainly of three equally large parts: transport 34 %, the home and service sector 33 %, and industry 30 %. During the last three years, energy consumption, particularly in the transport sector but also in the home and service sector, has increased more than the total energy consumption, while industrial energy consumption has decreased somewhat. In the non-OECD countries, the transport sector is increasing even more rapidly, and the industrial energy consumption has more than doubled since 1973 (IEA, 2007).

6.5 Examples of National Energy Balances

We now illustrate national energy balances with data taken from IEA (2007). In the USA, fossil fuels are of central importance, partly because of the large petrol-consuming transport sector and partly because domestically mined coal has been chosen as the raw material for electricity production. In USA, oil is now responsible for 40 % of the energy consumption, natural gas, 25 % and coal, 22 %. More than half of the oil and almost all the gas are imported, but very little coal. Nuclear power is responsible for 8 % and hydropower for only 1 %. Industry uses most energy today, almost 40 %, while transport is responsible for almost 35 % of the consumption. The remaining energy is used in the housing and service sectors.

Since 1970 the industrial consumption of energy in the USA has increased by almost 30 %. Energy consumption in the transport sector has increased by more than 65 %. In the USA, roughly 15 % of the total energy consumption is in the form of electricity. More than half of the electricity is currently generated in coal power stations. Nuclear power is responsible for one-fifth of the electricity production. Oil and gas, which were important for electricity production in the early 1970s, are now responsible for less than one-fifth. Electricity is used in almost equal proportions in the industrial and the housing and service sectors, but to a negligible extent in the transport sector.

France is one of the countries that has invested heavily in nuclear power. As much as 33 % of the total energy need is met by domestic nuclear power, which is responsible for 75 % of the electricity production. Nuclear power was established in France during the 1980s. In the early 1970s, less than 10 % of the country's electricity production was based on nuclear power. Electricity is used primarily in the housing and service sectors (more than 60 %) and in industry (more than 35 %). The transport sector in France also uses a certain amount of electricity, but this is nevertheless less than 5 %.

Among other energy-carriers in the present French energy system, the most important are imported oil (from Saudi Arabia, Norway, and England) and natural gas (from Norway, Russia, and Algeria). As in 1970, the total energy consumption is still dominated by the housing and service sector, but the transport sector is growing most rapidly and has increased from less than 20 % to more than 30 % since the 1970s. Industry is using 25 % of the energy in France today, compared with over 35 % in the early 1970s.

6.6 Risks in Energy Systems

An energy system is often a complicated technical system integrated into society's total infrastructure. This leads to the risk of accidents of various types, from disasters to long-term environmental consequences. The risks depend partly on how the energy system is built up and partly on the competence with which the system is maintained, and on the design of the communication system and the remaining infrastructure.

Decisions concerning the design of the energy system are taken at many different levels. A private individual can, for example, decide to use electrical equipment in the home, but the manner in which the electricity is generated is usually decided at a national level. International aspects and national guidelines govern the national decision. In addition to the choice of equipment and energy-generating technology, decisions are made concerning the methods of obtaining fuel, the transport of fuel, the technical design of power stations, transmission systems, electricity prices and electricity meters, etc. A number of factors are considered whenever a decision is taken, including the possible risks.

When a decision is taken concerning the choice of an energy system, three groups can be considered: risk-bearers, beneficiaries, and cost-bearers (see Chapter 13). These are not always the same persons. A modern energy system is so complex that it is impossible for decision-makers to take all the risk-bearers into consideration. The area of responsibility is limited in both time and place. In addition, the understanding of possible consequences is often limited. As a result of this uncertainty and of the priorities of different parties, there will always be more than one opinion regarding the benefits and risks associated with the different parts of any energy system. Here, we have chosen to concentrate on two generally important risk categories for energy systems, namely disasters and environmental hazards.

A disaster can be defined as a large-scale accident concentrated in time and place, which occurs with low probability. Some environmental hazards also have the character of a disaster, e.g., oil discharges at sea, but in many cases, environmental pollution is not a sudden disaster but a continual environmental effect through relatively undramatic processes, e.g., a change in the ecological system when this is exposed to acid precipitation over a long period of time.

6.7 Methods for the Evaluation of Risks in the Conversion of Energy

The assessment of disaster risks is based on statistics, expert opinions, and models (see Chapters 12 and 13), often in combination. A method developed at an early stage for the assessment of environmental and health risks in industrial activities is the so-called Environmental Impact Assessment (EIA). This shows how a system for energy conversion affects its surroundings and the assessment is often carried out at the projecting stage.

During the 1970s, a method was also developed to describe material flows related to the production of any given product, a so-called Life Cycle Analysis (LCA). It was then of special interest to illustrate environmental effects. These analyses were ambitious studies of what substances were used in different products, where these substances came from, what environmental effects they had, and to what they could possibly be recycled when the product had served its purpose. This way of looking at a product has also been adopted in the assessment of energy conversion processes.

In LCA, the system limits for an environmental analysis were extended to include the projecting and dismantling of an installation, and thus also the environmental consequences of reducing material to its technical components. During the 1990s, the emphasis has been on the quantification and evaluation of risks rather than on a qualitative description.

6.8 Examples of Risks in Energy Conversion

6.8.1 Disaster Risks

Here follow examples of accidents, which can be considered as likely causes of large-scale damage. The estimated probability of such disasters can sometimes be calculated on the basis of historic incidents and knowledge of the currently prevailing conditions.

1. Petroleum fuels:

- Fire or structural damage to an oil platform at sea
- Blow-out in sea-based oil pumping (oil in the water)
- Wreck of a super-tanker
- Transport accident on land with light petroleum fuel
- Explosion of a storage tank
- Rupture of a tank for the storage of lead additives
- Large refiner fire

2. Natural gas:

- Large discharge of gas, ignition of gas cloud
- Fracture of a natural gas pipeline, ignition of gas cloud

3. Coal:

- Explosion in a coal mine
- Landslide (slag heaps)

4. Hydropower:

- Uncontrolled outflow from a water reservoir

5. Nuclear power:

- Large discharge of radioactive substances from a nuclear power station or nuclear reprocessing unit

Petroleum Fuels

The pumping of oil at sea involves a risk of large accidents with many deaths, as shown in Table 6.1.

Table 6.1. Examples of large oil industry accidents at sea (CNN, 2001)

Year	Description	Deaths
1980	The oil platform Alexander Kjelland in the North Sea capsized as the result of a fatigue failure	123
1981	A USA-flagged oil-drilling ship sank in the South China Sea	81
1982	The oil platform Ocean Ranger in the North Atlantic overturned	84
1984	Explosion and fire on a Petrobras oil platform in the Campos Basin, Brazil	36
1988	The world's greatest accident on an oil platform so far occurred when the Piper Alpha exploded in the North Sea, due to a gas leak	167
1992	A Super Puma helicopter left the Cormorant Alpha oil platform and crashed into the sea	11
1995	An explosion on a mobile oil platform off the Nigerian coast	13
2001	An explosion occurred on the world's largest sea-based oil platform in Brazil, and the platform sank five days later	10

An American compilation of international oil discharges shows that there were 11 discharges of raw oil in quantities exceeding 50 000 ton during the period from 1967 to 1991 (NOAA, 1992).

The handling and transport of light petroleum products (e.g., LP gas and petrol) on land involves risks to humans (Figure 6.1 and Table 6.2).

Figure 6.1. The Bahamas-flagged oil tanker Prestige split in half and sank on November 19, 2002, about 233 kilometers off the northwestern Spanish coast. The oil spill is considered to be the largest environmental disaster in Spain's history, and polluted thousands of kilometers of Spanish and French coastline, as well as causing great damage to the local fishing industry. (Photo: AFP PHOTO/EPA/EFE/Spanish Navy Press/Scanpix)

Table 6.2. Examples of incidents in connection with the transport of petroleum products on land (UNEP, 2001)

Year	Description	Deaths (injured)
1974	LP-gas leakage during road transport in Eagle Pass, USA	17 (34)
1983	Explosion of LP-gas transport on the Nile, Egypt	317 (44)
1990	Explosion when an LP-gas transport collided in Bangkok, Thailand	63 (90)
1994	LP-gas leakage during transport by road in Onitscha, Nigeria	60
1998	Explosion and fire as a result of LP-gas leakage in a pipeline, Nigeria	500
2001	Explosion in petrol on a ship under repair in Lagos, Nigeria	10

Natural gas

The use of natural gas can lead to explosion and suffocation accidents. The risks vary depending on the technology used and on the distribution of responsibility in the distribution and installation of gas-fired equipment.

During the 1970s in France, 15–20 persons died each year in connection with explosions and a further 50 through suffocation. The large number of suffocation deaths was due to the fact that hot-water boilers in apartments were heated by direct combustion of natural gas. During 1984–1993 in Holland, an average of 10 persons died each year in accidents with natural gas. Most of these accidents were due to faults in the end-use installations.

Coal

Coal mining is one of the most dangerous activities in the energy sector. The probability of a major coal mining accident varies according to the local conditions and the technology used. Underground mining is more dangerous than opencast mining. In the year 2000 alone, several thousands of miners died in Chinese mining accidents.

Hydropower

Water reservoirs all over the world are often nearly 100 years old, even though most of the dams in use today were built during the 1970s. A dam may have several functions. In many cases, they have been built to regulate the masses of water and to protect nearby towns and villages from natural floods. The water restrained by the dams is usually utilized for irrigation or for power generation.

Figure 6.2. The drained upper portion of the Taum Sauk reservoir near Lesterville, Missouri, USA. On the morning of 14 December, 2005, a triangular section on the northwest side of the reservoir failed, releasing about 4 million m^3 of water in 12 minutes and sending a 7-meter surge wave of water down the Black River. The federal inquiry that followed showed that a combination of control system failures caused the reservoir to continue filling even though it was already at its normal level. (Photo: Julie Smith/AP Photo/Scanpix)

Most large modern dams are rock-fill or earth-fill dams (ICOLD, 1998). Other dams are of concrete built directly on the rock. The structure is exposed to pressure both from its own dead weight and from the water masses. Cracks may develop in such structures, but they are usually discovered at an early stage, and failure of the dam can usually be prevented. Movements in the ground or erosion are examples of possible causes of crack formation. Abnormal quantities of rain can also lead to a pressure greater than expected. It happens, however, that dams break. The risks of disaster are of course greatest in the case of high dams, especially if there are nearby towns or villages downstream. Table 6.3 lists a number of large dam accidents of historical interest (see also Figure 6.2).

One of the twentieth century's greatest flooding disasters occurred in China in 1998 when the flood-banks of the Yangtze River broke and it was estimated that 3 000 persons may have lost their lives. Controlled, but nevertheless disastrous, discharges have also taken place from, e.g., dams in Kaniji, Jebba, and Shiroro in Nigeria (1998 and 1999) and from Zambia's, Zimbabwe's, and Mozambique's joint Kariba dam (2000 and 2001). In these cases, heavy rains meant that it was necessary to open the power station dams for safety reasons. As a consequence, villages and crops were destroyed. The number of deaths directly due to these water discharges was less than 50 persons, but the events led to starvation and illness for estimated tens of thousands of people.

Using statistics from ICOLD one can estimate the risk for failure per year in a dam higher than 15 meters to be 10^{-4} in an international context.. The probability of a dam failure leading to a large loss of life is judged to be several orders of magnitude lower than the probability of a dam failure itself.

Table 6.3. Dam accidents with a large number of deaths (ICOLD, 1998)

Year	Place	Deaths
1929	Saint Francis, USA	450
1959	Malpasset, France	421
1959	Vega de Tera, Spain	144
1961	Bab-i-yar, Soviet Union	145
1963	Vajont, Italy	1 989
1967	Koyona, India	180
1972	Canyon Lake, USA	240
1979	Macca 2, India	2 500
1979	Gujarati, India	15 000
1980	Orissa, India	1 000
1983	Candinamarca, Colombia	150
1985	Val di Stava, Italy	270

Nuclear Power

In 2008, 439 nuclear power reactors were in operation in the world and an additional 36 were being built (WNA, 2008).

In 1979 a serious accident, which might have led to a nuclear meltdown, occurred in the nuclear power plant on Three Mile Island in the USA.

The meltdown in Chernobyl in 1986 was an extremely serious nuclear power plant accident (Figure 6.3). It has led to improvements in the safety at nuclear power plants in the former Soviet Union. The level of safety is, however, still in dispute in many of the nuclear power plants in Europe. The Chernobyl disaster is also dealt with in Chapters 8 and 13.

Figure 6.3. General view of the destroyed fourth power block of Chernobyl's nuclear power plant, from a helicopter in April 1986, a few days after the catastrophe. (Photo: Vladimir Repik/AFP PHOTO/Scanpix)

Wind, Bio and Solar Energy

It is generally agreed that no accident associated with installations for the utilization of wind energy, bio-energy, or solar energy could have catastrophic consequences. The storage of gas from biomass gasification could lead to a local accident, as could the possible storage of, e.g., solar-energy-based hydrogen. There are risks involved in the extraction of bio fuel and the building of suitable installations, but they are not of a catastrophic character.

6.8.2 Risks During Normal Operation

The risks that are of concern in connection with a process or system, vary with time. In the 1950s, the dominant problem of combustion was a local health issue, whereas it is the international environmental problem that has attracted attention more recently. The nuclear power debate was at first concerned primarily with the risk of a nuclear meltdown with subsequent large radioactive fallout, and the risk for an increased proliferation of nuclear weapons. In recent years, the risks associated with the disposal of nuclear waste and the final storage of the fuel have dominated in the debate.

For the politician or layman who has to weigh the advantages against the risks in an energy system, disaster risks thus are only a part of the basis for a decision. There are also risks, such as environmental and health hazards, associated with normal operation.

There are relatively small risks associated with the normal operation of a nuclear power plant. The storage of nuclear waste involves certain small risks in the foreseeable future. The risk assessment referring to much longer time spans is very uncertain. In the case of hydropower, the normal regulation of the water flow has a certain influence on the ecological system in and around the river. The magnitude and importance of these effects are poorly documented.

In the case of combustion plants, there is a greater understanding of the risks associated with normal operation. As a result of the combustion, the surroundings are exposed to smoke and gases. The recipients can be purely local (for most particles, etc.), regional (for, e.g., sulfurous and nitrous gases which can be transported over long distances), or global (for greenhouse gases). The environment can also be harmed if substances from a landfill are leached out and are carried to the groundwater. If particles from the combustion are breathed in with the air, they can be a health problem for persons in the vicinity of a combustion plant. Not only are these health risks a direct problem for human beings, but the environmental impacts lead to a progressive deterioration in the conditions for our own existence and for that of other living organisms. In the western world, one has succeeded in reducing the sulfur emissions but the achievements are not as good with regard to the emission of nitrogen compounds.

Acidification that can lead to permanent damage to the ecological system is a regional problem, while the source of the precipitation may be in another country.

In recent years the dominant discussion about risks connected with energy has been the greenhouse effect. This influences the whole of our atmosphere regardless of where the effect originates. There are strong reasons for believing that the increase in temperature of the earth's atmosphere during the last 50 years is due to human activities and the associated discharge of carbon dioxide, ammonia, methane, nitrous oxide, and fluorocarbons (freons). Carbon dioxide is the most important gas in this context (IPCC, 2007). The most important causes of an increase in the carbon dioxide concentration in the atmosphere are the combustion of fossil fuels and deforestation in tropical and sub-tropical regions. Combustion always leads to a release of carbon dioxide. The amount of carbon dioxide formed is directly related to the amount fuel consumed, and carbon dioxide cannot today be removed from the fuel gases. (See, however, the final discussion in this chapter.)

One argument in favor of using bio-fuels as a replacement for fossil fuels is that one would then not increase the amount of carbon dioxide in the atmosphere, since the plants use carbon dioxide in their photosynthesis processes. The argument thus assumes that there is a good regrowth of bio-fuel. The quantity of carbon dioxide, which is discharged to the atmosphere from a bio-energy system, after subtraction of the plant's consumption of carbon dioxide is therefore only equal to that due to the harvesting and transport, etc., of the fuel, if these processes consume fossil fuel.

6.9 Evaluation and Comparison of the Risks Associated with Energy Conversion

A classical dilemma associated with risk evaluation is that we have a tendency to see risks which affect us here and now as more serious than risks which lie far ahead in the future or at a distance from us. It is also often true that the more we know about a risk, the smaller it appears to be. One way of evaluating a risk is to choose several different types of negative effects and then to assess whether a given process has small, medium, or large consequences of this type.

In the early 1990s, methods were developed for assessing the external costs of energy conversion processes. An external cost arises when the social and economic activities of one group affect those of another group, and when this influence is not fully taken into consideration by the first group. For example, a power station can emit sulfur dioxide, which damages the façades of buildings, and nitrogen oxides, which lead to respiratory problems in asthmatic persons, which finally give rise to external costs, because the consequences for those who own the buildings and those who suffer from asthma are not normally accepted by the power plant. The definition of what is external is dependent on local conditions. The health effect of soot particles may have become an internal matter as a result of some form of health insurance in one community but not in another.

Since only some costs can be estimated, the method should be used with extreme caution. In the project ExternE within the EU, an attempt has been made to determine the external costs of generating electricity in European countries. The energy conversion processes, which have been most completely analyzed so far, are those, which are based on the combustion of fossil fuels. There is still considerable uncertainty as to the financial cost of the effect of acidification on the ecological system and of the influence of the greenhouse effect on our climate. The analysis shows, however, that the greenhouse effect can lead to large costs. It is also clear that injuries to human beings, both those working in some stage of the energy conversion process and those who live close to combustion plants, lead to high external costs. Damage to the built environment is also considerable. Thereafter, of somewhat less importance, comes damage to agriculture and forestry (EUR, 1995).

In addition to the difficulties in identifying and valuing the different risks, it is also important to define the limits of the system to be analyzed. A common weakness is that the energy conversion process and parallel activities are included in the analyses to different extents. This means that the comparisons between different types of energy are inadequate. In the analysis of wind power, for example, the manufacture of the turbine and the installation of the power unit may be included, whereas the building of the power plant is often ignored in the analysis of an oil-fired power station. Another example can be that the fuel transport is neglected for bio-energy in a comparison between bio-fuel and fossil fuel power plants.

The effort put into the assessment of risks in energy conversion is fortunately not in vain, even if the results do not always enable comparisons to be made. The assessment of risks has been very important in the development of the various energy conversion systems, thus progressively reducing the risks. Risk assessments have also motivated the development of completely new technologies.

Technical and scientific risk assessments have provided reference material for the general energy policy debates, even though they often differ from the general opinion of the public. People often feel that the risks are much greater than is indicated by the calculations (e.g., in the case of nuclear power), see also Chapter 16.

6.10 Energy Today and in the Future

Among other things, the energy sector is characterized by the fact that any decision relating to the choice of energy system is a long-term measure, which must often be considered at a national level. The risks – both for disasters and for more long-term environmental effects – are essentially different for different types of energy, and the risks of, e.g., an economic or other character must also be taken into consideration. The installations and distribution networks are usually large and expensive, and they have a long depreciation time. For example, the choice of natural gas means not only expensive pipelines but also contractual links to a few suppliers, which may mean that there is risk of interrupted supplies in the event of a crisis. In the case of national power sources such as hydropower, the dams are to be considered as an irreversible effect on the environment, but once built they are associated with only small risks.

The long-term aspects of energy systems, and their complexity and degree of irreversibility, mean that it is difficult, if not impossible, to reach an optimal decision. Technical advances can sometimes radically change the conditions for the total risk evaluation. This favors the choice of a flexible reversible system. In the case of fossil fuels, the availability in the long term (a few decades or more) is limited, but this type of fuel will be needed for a long time in the future. It may be, however, that the environmental risks will set a limit to how much of these resources are in fact used rather than the magnitude of the resources themselves (which are large if all the sites are considered). Techniques are available for greatly reducing most emissions, with the exception of carbon dioxide, which is the gas that contributes to an increased greenhouse effect. Solutions are being discussed, such as blowing the gas into the sea or into old gas fields, or conversion with hydrogen to new hydrocarbons, etc., but it is difficult today to say whether such ideas are technically or economically feasible.

In the nuclear power sector, pilot plants are being projected with a new type of "inherently safe" reactor. Accelerator-driven reactors, which cannot lead to a nuclear meltdown, are technically feasible. Several aspects of this theme are now being investigated in many places, even though the pace is slow. Similar technology – so-called transmutation – is being studied to reduce the amount of radioactive nuclear waste with a long half-life, and to increase the amount of energy obtained. If there is sufficient financing for the required research, we should know within a decade or so how realistic these ideas are.

During recent years, there have been rapid developments in the fuel cell sector, and there is talk of small power plants in which hydrogen can be utilized to produce energy in fuel cells. The renewable types of energy based on solar or wind power plants are also becoming increasingly more efficient.

Environmental and safety aspects play a leading role when alternative methods of providing energy are being considered, and the environmental aspects are decisive in almost all developments of future energy technology. Technical developments have also meant that it has been possible to greatly reduce or totally eliminate previous large technical risks and environmental hazards, e.g., by purification measures, new combustion technology or improved regulatory and surveillance systems. The development of new technology has been a powerful and efficient means of improving safety and reducing environmental and other risks associated with the energy system.

In summary, we can say that attention to risks and to the environment is essential in any choice of an energy system. Every energy system embodies some type of risk. In some cases, it is possible to identify risks of disaster, which should be eliminated or reduced to an acceptable level. Our increasing understanding of disaster risks (nuclear power disasters, climate disasters) has had a powerful effect on the technical developments in the energy sector. Risk analyses and risk assessments thus play an essential role in the design of the energy systems of the future.

Attention must be given to the energy system as a whole, including the extraction of fuel, transport, and the handling of waste. But, since energy investments must often be long-term, consideration must also be given to possible technical developments and other uncertain factors, such as possible environmental effects. From a risk point of view it can thus be politically wise to avoid becoming locked into an irreversible system, and to avoid placing all the energy-eggs in one basket.

References

CNN (2001) Major oil industry accidents. http://edition.cnn.com/2001/WORLD/americas/03/20/oil.accidents/index.html

DoE (2008) Energy information administration. US Department of Energy (DoE), Washington DC. http://tonto.eia.doe.gov/country/index.cfm

EUR (1995) ExternE, externalities of energy. Vol 1, Summary. European Commission, Energy Technology Support Unit (ETSU). Office for Official Publications of the European Communities (EUR), Luxembourg

ICOLD (1998) World register of dams. International Commission On Large Dams (ICOLD), Paris

IEA (2007) Key world energy statistics. International Energy Agency (IEA), Paris

IPCC (2007) Climate change 2007: Synthesis Report – Summary for policymakers. Intergovernmental Panel on Climate Change, United Nations. http://www.ipcc.ch/pdf/assessment-report/ar4/syr/ar4_syr_spm.pdf

NOAA (1992) Oil spill, case histories, Significant US and international spills, 1967–1991. Hazardous materials response and assessment division, National Oceanic and Atmospheric Administration, Washington DC

UNEP (2001) Transport disasters. Production and consumption unit (APELL), United Nations Environment Program, Washington DC

UNSD (2004) Energy statistics yearbook. The United Nations Statistics Division, New York

WNA (2008) World nuclear power reactors 2007–08 and uranium requirements. World Nuclear Association. http://www.world-nuclear.org/info/reactors.html

7

Chemical and Physical Health Risks – Chemicals, Radiation, and Noise

Ulf Ulfvarson

7.1 Environment, Technology, and Health

Imagine yourself standing in early morning at the best viewpoint of a big city. The sum of all noises from the city forms a tremendous roar, which increases during the morning hours. Separate sounds may be discerned; the sound from an accelerating truck, a siren, a pile driver, a passing aircraft. In the city, the traffic noise is more obvious. Sometimes it is strong enough to make conversation impossible, a sign that the noise may damage one's hearing. As the traffic becomes denser, it stirs up dust from the streets. The exhausts emit particles and gases. The smell itself is annoying – for some people very annoying. Sensitive people may get respiratory problems. Other sources of air pollution are emissions from refuse dumps and the incineration of garbage. Air pollution comes from chimneys of houses or factories. The sunlight hits air pollutants and transforms them to compounds that can be even more harmful.

In the city, the temperature is several degrees warmer than at the viewpoint. Almost all energy used is eventually converted into heat, and this increases the outdoor temperature in the city. The majority of people stay indoors most of the time, at home and at work. The indoor environment is not always healthy, since the houses emit and enclose air pollutants. Countless sources generate electric and magnetic fields. Fluorescent lamps and monitors flicker in work sites.

The city is a result of the endeavor to satisfy human needs with its houses and flats, its traffic (Figure 7.1), and its work sites. However, not all attempts to satisfy human needs promote health. Technical solutions may have unwanted side-effects. The question is how all these environmental factors influence our health. Although of great interest as such the possible influence of man's activities on global warming is in this context disregarded. This chapter is limited to health effects of chemical compounds, noise, and radiation (including electromagnetic fields) in the environment created by man.

Figure 7.1. Traffic in Bangkok. (Photo: Ulf Ulfvarson)

7.2 How Do We Know What is Harmful?

An important aim is to protect people as far as possible from being injured or becoming ill because of exposure to risk factors in our environment. Therefore, mechanisms causing injuries and diseases should be investigated. We should also know how to prevent injuries and diseases. So far, most knowledge has been acquired by analyzing injuries and diseases that have occurred. We have learnt much of what we know about illness related to the environment from work environments. Long before industrialization, very hazardous work environments existed. However, industrialization increased both the risks and the size of the exposed groups. Lack of will to invest in the necessary resources was in the early days of Western industrialization a more important obstacle to limiting the most obvious risks in work environments than any lack of knowledge about the risks. Gradually the task of reducing risks in the work environment changed. When obvious risks have been eliminated, risks that are more complex remain. It has become more and more difficult to understand and reduce the risks. The efforts of generations of scientists were required to accumulate today's knowledge. The same applies to the process of transforming knowledge into effective risk reduction. Over a long period, the work environment in industrialized countries has been radically improved, but new risks emerge with the development of technology and old risks rise again because of a lack of maintenance of knowledge and a deficient preparedness. The need to describe, evaluate, and control risks in the environment will probably remain.

The main sources of information about health risks in the environment are the following: (1) tests of functions and other tests, (2) epidemiological investigations, (3) statistics of injuries, (4) data about exposure to various risk factors, (5) investigations of material and equipment. Each of these will now be discussed.

7.2.1 Tests of Functions and Other Tests

There are many ways to investigate how the function of the body is affected. People exposed to irritating or lung-damaging substances may be examined with respect to how their lungs perform. Those who may suffer from allergic contact eczema caused by some substance can be examined by a patch test, where small amounts of sensitizing substances are applied to the skin and the skin reaction is observed. The hearing of a person is examined by audiometry.

By experiment, with animals, plants, or microorganisms, it may be possible to detect risks that would otherwise not be found until injuries or deaths among human beings began to appear. Such experiments on other species can only indicate expected health effects in human beings.

7.2.2 Epidemiology

Epidemiology is the science of occurrence and distribution of diseases in the population. In epidemiological investigations of environmental factors, exposed and unexposed groups are identified. These groups are compared in order to find out whether the occurrence of diseases or deaths may be associated with the exposure. The epidemiologist is interested in the fraction of a group affected by a certain disease at a given point in time (*prevalence*) or the frequency of new cases of a disease occurring during a certain period of time (*incidence*). In this way, the observed number of sick or dead in the exposed group is related to the expected number in a non-exposed group made up in the same way. Since the occurrence of diseases and mortality depends on age, it is necessary to consider differences in the age distribution between the groups being compared. After age standardization, a quotient is formed between the observed and the expected numbers of sick or dead. The studied environmental factor may be said to increase morbidity or mortality if the quotient exceeds 1.

A single or only a few epidemiological investigations of a certain factor must be interpreted with great care. Many circumstances have to be taken into account in order to ascertain whether there is an association between exposure and effect. The number of persons included in an investigation is of great importance. The possibility of *confounding*, e.g., confusing causes, has to be considered. An example of this is the increased risk of lung cancer in individuals with a high alcohol consumption. The explanation is not the consumption of alcohol but that the frequency of smoking is higher among these individuals and that smoking increases the risk of lung cancer. Moreover, alcohol is often consumed in environments where passive smoking is common.

Epidemiological investigations can be designed in different ways. One way is to start with exposed and unexposed individuals, and to compare these individuals with regard to the occurrence of disease. An example: individuals who had been exposed to substances suspected to cause cancer were followed up in a register of death causes. The number of deaths caused by cancer was 32. The expected number of deaths due to cancer in a group with the same age distribution was 24.2. The quotient of 32 and 24.2 is 1.32. Therefore, there was an excessive mortality due to cancer in the studied group. This supports the suspicion that they were exposed to carcinogenic substances. A confidence interval is often stated, within which a true value lies with a specified probability. The width of such an interval depends, inter alia, on the number of individuals in the investigation.

7.2.3 Statistics of Injuries

In many countries, statistics of injuries and ill-health are collected. Often the statistics of occupational injuries are especially detailed and extensive. They are used to survey the development of risks and to give priority to efforts in the preventive work. The International Labour Organization, ILO, collects and publishes data on occupational injuries from a large number of countries (ILO, 2005). The objective of the European Agency for Safety and Health at Work in Bilbao is to collect, analyze and promote occupational safety and health information online within the European Union. Work-related statistics from individual EU members can be accessed from this agency (OSHA, 2008).

A restriction in the usefulness of statistics about injuries is that the specific national system of compensation has a decisive influence on what is registered and on the number of injuries reported. The correlation between the changes in the numbers and types of reported injuries and changes in the rules of compensation illustrates this fact. The trade conditions also have a significant influence on the types and numbers of injuries as well.

7.2.4 Data on Exposure and Uptake in the Body

Data on exposure to an environmental factor alone will say nothing about the risk of health effects. In order to assess the risk, it is necessary to know the association between exposure and effect or response. The relationship between the size of the dose and the kind or degree of effect upon an individual is called the *dose-effect* association. The relationship between the size of the dose and the fraction of individuals who have been subjected to a certain effect is called the *dose-response* association.

Data on exposure to chemicals in the work environment consist primarily of information on the concentration of substances in the inhaled air. The concentrations of substances or products of biotransformation in body liquids (blood, urine) can be used as a measure of the uptake of substances in the body. For example, this holds for substances processed in a workplace. Together with information on trade and type of work, such data are important in epidemiological investigations.

Historical information relating to occupational exposure in a large number of occupations in EU countries is available in a European Health and Safety Database (HASTE), which is administered and maintained by the Finnish Institute for Occupational Health (HASTE, 2008).

7.2.5 Examination of Materials and Equipment

One way of learning about harmful exposures at work or elsewhere is to examine materials and equipment used with regard to emissions of, for example, noise, air pollutants, and radiation. Emission data have to be combined with other data to provide risk information. However, in choosing the least hazardous materials and equipment, emission data can also be used directly. Emission data for a number of risk factors in work environments are published for example by the Health and Safety Commission (HSC) in Great Britain (HSC, 2008).

7.2.6 Assessment of Chemical Health Risks in the Environment

The objective of risk assessment of a substance is to describe probable health effects, including possible injuries. The assessment of risks of chemicals is carried out in three steps: hazard assessment, exposure assessment, and risk characterization.

The *hazard assessment* concerns the inherent potential of chemical substances to cause injuries to the health of human beings and to the environment. The assessment is founded on the biological and physicochemical properties of the substance. Among the physicochemical properties, the substance's persistence and tendency to be bio-accumulative in the environment need special attention. The hazard assessment also involves the establishment of the quantitative dose (concentration) – response (effect) relationship. The biological properties of a substance include toxicokinetics, biotransformation (metabolism) and distribution within the body, acute effects (acute toxicity, tissue irritation, and corrosivity), sensitization, repeated dose toxicity, carcinogenity, mutagenicity and toxicity for reproduction (teratogenicity).

The *exposure assessment* rests on the concentration of the substance in air, soil, and water. The exposure assessment should consider the use of the substance, the properties (e.g., volatility, mobility, biodegradability) and the nature of the environment in which the substance may spread, intentionally or unintentionally. The exposure assessment includes a description of the manufacture and use of the substance during its life cycle. Exposure assessment may also take into account engineering control measures to reduce the exposure of humans and the environment. In historical exposure estimations, it is of great value if records of operating conditions and earlier measurements are available.

The *risk characterization* should consider the human populations exposed and the types of effect, their duration, and their severity.

7.3 Chemical Health Risks in the Environment and in the Work Environment

Already in antiquity, it was found that work was associated with certain chemical health risks, e.g., lead poisoning in workers in lead mines, noted by Hippocrates and mercury poisoning among the slaves in Almaden described by Plinius. Bernadino Ramazzini, designated as the father of occupational medicine, described many occupational risks in a famous book, *De morbis artificum*, published in 1700. His followers were Linnaeus and some of his pupils. Linnaeus observed that there were only women sitting in the pews in Orsa, Sweden. They were widowed because their men were struck with silicosis when working with grindstones. He called the illness the "Orsa-disease."

7.3.1 Uptake of Xenobiotics in the Body

Of all the organs in the human body, the lungs have the largest contact surface with the environment, about 60 m^2. Through the lungs, we are exposed to about 10 m^3 of inhaled air every 24 hours. Therefore, large amounts of *xenobiotics* (substances foreign to the body) may enter the body through the lungs. The lungs are protected by several defense mechanisms. Coarse particles are stuck in the phlegm and transported upwards by the cilia in trachea and bronchi. Small particles reach the lungs and are engulfed and transported by phagocytes. However, some particles will remain in the lungs for a long time, e.g., quartz particles. Quartz particles poison the cells and cause silicosis, expressed as scar formation in the lung tissue, destroying the capacity for oxygen uptake.

When we swallow, substances foreign to the body and contained in the phlegm from the airways and the lungs enter the gastro-intestinal tract. Usually this way of absorption is, however, of less importance.

Lipid-soluble substances can be absorbed into the body via the skin, its total surface being about 1.7 m^2. The risk is largest in places where the epidermis is thin, e.g., on the upper side of the hands and on the arms. These parts of the body are also those most exposed to skin contact with pollutants. Substances that can be absorbed in significant amounts through the skin are, for example, hydrogen sulfide, benzene, and insecticides based on esters of phosphoric acid.

7.3.2 What Are the Chemical Effects on the Body?

The tissues are irritated when they come into contact with strong acids and bases in high concentration or with substances that emit oxygen, e.g., potassium permanganate, or reducing substances (inter alia substances that easily take up oxygen, e.g., sodium disulfide). Substances that can transform other substances are able to change biotic substances (produced or caused by the body) and thus influence the functions of the body. Such substances may harm the genes. A lesion is normally repaired, but it may also be lasting and cause mutations, resulting in tumors, fetal damage, and hereditary damage as well. The complex mechanism behind tumor genesis explains the long latency time. It may take many decades from the contact with a carcinogenic substance until the break-out of the disease. Many xenobiotics, e.g., heavy metals, may bind to receptors in the body important for normal biochemical reactions (e.g., cadmium to the renal cortex). In this way, they may prevent the normal function of the organ.

A defense mechanism of the body against xenobiotics is to *transform* them, i.e., biotransformation. Usually this will result in less harmful and more water-soluble substances which are excreted via the urine or feces. Sometimes the biotransformation will result in substances that are more toxic than the precursors. One example of this is the harmful effect of methanol on the eyesight. The toxic mechanism involves the biotransformation of methanol to formic acid.

The *additive effect* of two or more toxic substances means that they affect the same organ or function in the same way. For example, they may have a narcotic effect or bind to the same receptors. Substances may also interact, so that they increase or reduce the effect of each other, *synergy* and *antagonism*, respectively. Synergy means that the total effect of two or more substances is greater than the sum of their effects one by one. One example is the increase in the carcinogenicity of asbestos by smoking. The uses of antidotes against toxic substances are examples of antagonism.

The immunological system recognizes xenobiotics and attempts to make them less harmful. Sometimes this ability of the body will result in sensitization, i.e., a change in the response of the body to the presence of a substance. Sensitization causes allergic symptoms, allergic contact eczema, nasal catarrh, eye moist with tears, or allergic asthma. After a human has become sensitive to a substance, contact with a very small amount of the substance can produce an allergic reaction. A problem in investigation of allergies is the large number of possible sensitizing substances present in the human environment. Another problem is that the time and concentration necessary to cause sensitization may vary between individuals. For example, it may take 20 years for a baker to be sensitized to flour dust.

The time between contact with the foreign substance and the first appearance of a health effect, the *latency time*, may vary from minutes to decades. Naturally, the risks associated with fast-acting substances have long been known. An intermediate latency time is found in the case of carbon monoxide forming carbon monoxide hemoglobin in the red blood corpuscles and obstructing the normal uptake of oxygen in the body. Other examples are found in water-soluble organic solvents, e.g., ethanol, absorbed in relatively large amounts and thus requiring rather a long time for decomposition and excretion.

7.3.3 Chemical Health Risks of Current Interest

The elements may be combined into chemical substances in a very large number of ways. Millions of substances have been synthesized by man. A large number of these were not originally present in nature. The CAS registry (Chemical Abstracts Service, a division of the American Chemical Society) covers substances identified and reported in the scientific literature and contains more than 30 million organic and inorganic substances. Approximately 4 000 new substances are added each day (CAS, 2008).

The European chemical substances information system (ESIS) is an IT system that provides information on chemicals. The system contains an inventory (European Inventory of Existing Commercial Chemical Substances, EINECS) with not less than 100 000 substances. The current number of substances produced or imported in quantities of at least 1 000 tonnes per year in the EU is 2 767 (year 2007). The number of substances produced or imported in the EU with a tonnage between 1 000 and 10 tonnes per year is 7 802. In effect, these figures are relevant for the global development as well, since the global development is reflected in Europe. Most substances are used in small volumes and within narrow fields of application. Lists on limit values for air pollutants in workplaces contain a few hundred substances. Of these only a few dozen are of interest in most industries (ECB, 2008).

During the last few decades, the main emphasis on chemical health risks has shifted concurrently with changes in production and production methods. A few decades ago, research was directed towards chemical health risks in mines, heavy manufacturing industry and chemical industry. It was a matter of exposure to gases like carbon monoxide and nitrogen oxides, dust, especially quartz dust and asbestos, heavy metals, e.g., lead and mercury, and substances causing skin diseases. Problems from such exposures have by no means been solved in all parts of the world. The focus of research at present is, inter alia, on effects of complex exposures, e.g., to exhaust from internal combustion engines, passive smoking, and biomarkers in body liquids. Substances contributing to the emergence of so-called sick buildings have been identified as monomers from plastic materials and mucotoxins and endotoxins from microorganisms thriving under moist conditions. One notable group of substances is volatile organic substances, e.g., organic solvents. Some of them can injure a given organ, e.g., the liver. Chronic effects on the central nervous system of very different volatile substances have been discovered. Sensitizing substances, e.g., nickel, chemicals used in cleaning, and components of plastics like isocyanates, imply an important risk of injury.

7.3.4 Health Effects of Fine Particles

There is a growing interest in health effects of fine particles regardless of their composition and origin. Airborne particulate matter, PM, consists of many different substances suspended in air in the form of solid particles or liquid droplets which vary widely in size. The potential hazard depends on composition, particle size, and mass concentration. The health effects are associated with material deposited in the different regions of the respiratory tract. A simple definition of PM10 is particulate matter with an aerodynamic diameter of 10 μm or less and PM2,5 is particulate matter with an aerodynamic diameter 2.5 μm or less. More accurately speaking, PM10 is particulate matter which passes through a size-selective inlet with 50 % efficiency cut-off at 10 μm aerodynamic diameter and likewise PM2,5 is particulate matter which passes through a size-selective inlet with a 50 % efficiency cut-off at 2.5 μm aerodynamic diameter (EU, 1999).

The *aerodynamic diameter* of a specific particle is the diameter of a spherical particle with density 1 g/cm^3 and the same fall velocity in air as the particle in question, regardless of its real size, form, and density.

The reference method for PM10 is based on the collection on a filter of the PM10 fraction of ambient particulate matter and a gravimetric mass determination (CEN, 1999).

Major concerns for human health from exposure to PM10 include: effects on breathing and respiratory systems, damage to lung tissue, cancer, and premature death. The elderly, children, and people with chronic lung disease, influenza, or asthma, are especially sensitive to the effects of particulate matter. Acidic PM10 can also damage man-made materials and is a major cause of reduced visibility in many cities.

Table 7.1. The impact of PM10 on post-neonatal mortality per 100 000 children between ages 1 month and 1 year (WHO ECEH, 2005)

Post-neonatal mortality	PM10 reduction, annual mean level [μg/m^3]	Number of attributable cases per year	95 % confidence interval
Total	By 5	23.2	10.7–36.0
	To 20	56.6	24.9–88.9
	To 40	15.3	6.9–24.3
Respiratory	By 5	4.7	2.3–7.2
	To 20	13.1	5.3–24.8
	To 40	6.7	2.9–11.6
Sudden infant	By 5	6.7	3.9–9.4
death syndrome	To 20	9.3	5.4–13.3
	To 40	0.7	0.4–1.1

Table 7.2. Decrease in morbidity outcomes in children, for reduction scenarios of daily PM10. For example, a reduction in exposure from current levels to 20 μg/ m^3 would lead to 7 % (in the confidence interval between 3.6 % and 8.6 %) lower incidence of coughs and lower respiratory symptoms and to a 2 % (in the confidence interval between 1.0 % and 2.9 %) decrease in admissions to hospital of children aged under 15 years with respiratory condition (WHO EHEC, 2005)

Morbidity	PM10 reduction, daily levels [μg/m^3]	Attributable fraction [%]	95 % confidence interval [%]
Cough	By 5	2.0	1.0–2.5
5–17 years	To 20	7.0	3.6–8.6
	To 50	3.7	1.9–4.5
Lower respiratory	By 5	2.0	1.0–2.9
Symtoms	To 20	7.0	3.6–10.1
5–17 years	To 50	3.7	1.9–5.3
Hospital respiratory	By 5	0.5	0.0–1.0
Admissions	To 20	1.8	0.0–3.8

Exposure below current limit values of PM10, 50 μg/m^3, in the USA as well as in the EU (see below) has entailed a need for acute hospital treatment of elderly persons and difficulty in breathing of children with asthma. The health impact of exposure to PM10 has been summarized in terms of the number of *annually* attributable cases per 100 000 that could be reduced for different scenarios of PM10. In Table 7.1, the potential effect on post-neonatal mortality of different *annual* mean levels of PM10 reduction is shown, and in Table 7.2 the decrease in morbidity outcomes in children for different reduction scenarios of *daily* PM10 is presented. Thus, for instance, reducing the level of PM10 to 40 μg/m^3 decreases the morbidity by about 15 per year and 100 000 children (Table 7.1). The figures in the two tables are calculated under the condition that all other factors remain similar (WHO ECEH, 2005).

In work environments, it is common to define fractions of three particle sizes, based on particle sizes in selective sampling. To give an approximate idea of the aerodynamic diameters involved in the three fractions, the aerodynamic diameters at a collection efficiency of 50 % in the sampling is given within brackets; *inhalable particulate mass* (100 μm) for those materials that are hazardous when deposited anywhere in the respiratory tract, *thoracic particulate mass* (10 μm) for those materials that are hazardous when deposited within the lung airways and the gas-exchange region and *respirable particulate mass* (4 μm) for those materials that are hazardous when deposited in the gas-exchange region. For quantitative definitions of the corresponding particle sizes, the reader is referred to the literature, for example The American Conference of Governmental Industrial Hygienists (ACGIH, 2006).

A number of publications report on the association between exposure and increased mortality (Ostro, 1993; Brunekreef et al., 1995; Styer et al., 1995), lung function effects (Koenig et al., 1993), acute airway symptoms (Larson and Koenig, 1994), and increased need of medication and hospital care (Schwartz, 1995). Increased mortality and the need for hospital care for asthma have been reported at concentrations of PM10 below 100 μg/m^3. Impaired lung function and acute airway symptoms were observed at PM10 concentrations below 115 μg/m^3 (Brunekreef et al., 1995). A relative risk of 1.1 (95 % confidence interval 1.03–1.17) for acute hospital care of elderly people has been reported for an increase in the particle concentration of 50 μg/m^3 (Schwartz, 1995). Obviously, the risk is small but statistically significant. A study found that low levels of PM10 were significantly associated with decreased lung function in children with asthma; at levels of a maximum of 60 μg/m^3 of air the breathing of these children was affected (Timonen and Pekkanen, 1997).

7.3.5 Nanoparticles

A nanoparticle is a microscopic particle with a diameter of 10 nm (1 nm (nanometer) = 10^{-9} meter) or less. The fabrication, the properties including health effects, and the potential applications of nanomaterials are currently an area of intense scientific research. Fine particles are more irritating than large particles. The surface-to-volume ratio increases with decreasing size, making the particles more reactive. The forces keeping molecules adsorbed onto surfaces decrease when the surface radius decreases. Therefore, substances adsorbed on small particles will be more available to the tissues than substances adsorbed on larger particles. In animal experiments, effects have been shown for particles hitherto assumed to be inert. Ultra-fine titanium dioxide (TiO_2) particles (particle diameter in the nm range) not only induce a greater acute inflammatory reaction in the lower respiratory tract in rats than larger TiO_2 (250 nm diameter) particles, but can also lead to persistent chronic effects in the deep lung, as indicated by an adverse effect on the alveolar-macrophage-mediated clearance function of particles and beginning interstitial fibrotic lesions (Oberdörster et al., 1992; Ferin and Oberdörster, 1992). In vitro studies with lung cells (alveolar macrophages) also showed that ultra-fine TiO_2 particles have a greater potential than larger TiO_2 particles to induce mediators which can adversely affect other lung cells (Oberdörster et al., 1992).

The health risks of nanomaterials may be of special concern in the work environment, since occupational exposure usually occurs at higher levels than in the ambient air. On the whole, this field is unregulated, but given the limited amount of information about the health risks associated with occupational exposure to nanoparticles, The National Institute for Occupational Safety and Health (NIOSH) in the USA recommends measures to minimize worker exposure (NIOSH, 2005; NIOSH NRC, 2007).

7.3.6 Air Quality Standards of Fine Particles

A health-based national air quality standard for ambient air in the USA with regard to PM10 is 50 $\mu g/m^3$ (measured as an annual mean) and 150 $\mu g/m^3$ (measured as a daily concentration) (EPA, 2008). The current federal US standard for daily PM10 is 150 $\mu g/m^3$. The current and future limit values for PM10 in EU are presented in Table 7.3.

Table 7.3. Limit values for particulate matter (PM10) for the protection of human health. Stage 1: Limit value is to be met by January 1, 2005. Stage 2: Limit value is to be met by January 1, 2010. Indicative limit values to be reviewed in the light of further information on health and environmental effects, technical feasibility, and experience in the application of Stage 1 limit values in the Member States (EU, 1999)

	Averaging period	Limit value
Stage 1	24 hours	50 $\mu g/m^3$ not to be exceeded more than 35 times a calendar year
	Calendar year	40 $\mu g/m^3$
Stage 2	24 hours	50 $\mu g/m^3$ not to be exceeded more than 7 times a calendar year
	Calendar year	20 $\mu g/m^3$

The limit values of PM in work environments depend on the composition. They are usually much higher than limit values for ambient air. The difference is justified by the fact that the ambient air limit values refer to particles with unspecified composition and exposure in a work environment is limited in time compared to ambient air exposure.

The EU issues so-called indicative limit values related to the work environment for a few substances. Some of these apply to PM. As an example, the limit value for 8 hours′ exposure to chromium in metallic form or inorganic chromium (III) is 2 mg/m^3 (EU, 2006). An example of recommended limit values for PM in the form of "inert" dusts in work environment is 3 mg/m^3 for respirable particles and 10 mg/m^3 for inhalable particles (ACGIH, 2006).

7.3.7 Diesel Exhaust

In spite of the fact that diesel engines have many advantages, e.g., a high energy output per unit of consumed fuel and the possibility of using fuels of many kinds, there are also disadvantages. In standardized testing programs, only so-called regulated emissions are measured: carbon monoxide, hydrocarbons, nitrogen oxides, and particles. The number of components in diesel exhaust is, however, very large, almost 15 000 substances. Many particles are also formed in the combustion process, because the combustion process in a piston engine is such that the mixing of fuel and air is incomplete, causing an occasional local shortage of air.

The particles make the surroundings dirty, but the major problem is that they contain a large number of adsorbed hazardous substances. Exposure to diesel exhaust has been shown to cause acute inflammatory changes in the lungs. To begin with, a temporary impairment of the lung function may arise, but repeated, temporary effects gradually lead to chronic effects. This has been shown in workers in garages for diesel buses and in children living in densely populated areas. Signs of asthma have also been found in association with exposure to diesel exhaust particles.

Diesel exhaust exposure increases the risk of cardiovascular diseases. Investigations of the effects of diesel exhaust exposure on the respiratory system, the central nervous system, and the genitals indicate serious problems of a more general kind.

The combustion process in a diesel engine leads to a large number of poly-aromatic hydrocarbons, adsorbed on the diesel particles. These substances react with ozone and nitrogen oxides in complex reactions, some of which are photochemical. Heavier poly-aromatic substances and nitrated poly-aromatic substances, nitroarenes, are carcinogenic.

The International Agency for Research on Cancer (IARC) has established that diesel exhaust is classified in group 2A, i.e., probably carcinogenic to humans (IARC, 2007). The development of lung cancer is a slow process. Therefore, it is not possible to attribute human carcinogenicity to any individual exhaust component in epidemiological investigations alone. In order to develop such knowledge, other sources of information have to be consulted, such as carcinogenicity testing or genetic toxicity testing of diesel exhaust components in animal experiments.

The capacity of a substance to induce mutations, *mutagenicity*, is used to predict carcinogenicity in animals and humans. Mutagenicity is examined by genetic toxicity testing. In the *Ames test*, bacteria are used. Bacteria cells are mixed with the chemical to be tested in a nutrient solution composed so that only mutated bacteria can grow. The number of colonies counted after a certain incubation time is a measure of the mutagenicity of the test chemical. Sometimes a non-mutagenic substance may be transformed by the enzymes in the body into a mutagenic product. In order to simulate this mechanism, homogenized rat liver is added to the nutrient solution. The demonstration that an agent can induce gene and chromosomal mutations in mammals in vivo indicates that it may have carcinogenic activity. Negative results in tests for mutagenicity in selected tissues from animals treated in vivo have less weight, partly because they do not exclude the possibility of an effect in tissues other than those examined (Vainio et al., 1992).

To set immission limit values (see Section 7.4) for occupational exposure to diesel exhausts based on their effects on health has been a challenge to occupational hygiene for a long time. One reason for the difficulties is the complex composition of diesel exhaust. At present, there are no criteria available for the accumulated effect of all diesel exhaust components. To establish such criteria is difficult, considering the fact that the composition of diesel exhaust varies with type of engine, fuel, and operation parameters. One way to approach the problem is to use an indicator of exposure. Such an indicator should be a component in the exhausts, which is easy to determine. The indicator could be hazardous in itself, but at least its concentrations should be correlated with those of the critical components. Nitrous oxides have long been used as an indicator, but their suitability for the purpose is doubtful. Another indicator could be soot.

Soot particles are always present in diesel exhaust, but their concentration varies. Many health effects of diesel exhaust are associated with soot particles, since hazardous substances are adsorbed onto their surfaces, but there are many disadvantages in using soot as an indicator of diesel exhaust exposure. It is difficult to determine the concentration of soot particles and to discriminate them from other particles. Soot particles from diesel engines vary in size, and particles of different sizes have different health effects. So far, the most effective way of limiting exposure to diesel exhaust is to use emission limit values, the goal being to decrease the emissions as far as possible by improving the engine construction and the fuel, and by exhaust treatment, e.g., filtering the exhaust.

7.3.8 Indoor Environment Causing Ill-health

People's exposure to indoor air pollution is determined by the concentrations of pollutants in the indoor environment and by the time that individuals spend in the polluted environment. In developing countries the pollutants are often determined by the type of fuel and stove used in the kitchen and its location.

The indoor pollution may be much less in industrially developed countries, but it can still cause health problems. The environment indoors may be polluted by volatile substances from building materials, substances from human metabolism, substances from microbiological activity, textile dust (containing for instance mites) and other biological dust (for instance pollen and hair from domestic animals). Tobacco smoking results in a number of air pollutants, and not only the smoker but also the non-smoker, is exposed to health risks. The so-called side-smoke contains higher concentrations of hazardous substances than the smoke inhaled by the smoker. The most important effect of smoking is an increase in the risk of lung cancer. Diseases of the airways, cardiovascular diseases, and lung cancer have been associated with passive smoking in a large number of investigations.

Ill-health caused by unhealthy indoor environments has been called *sick-building syndrome*, including symptoms such as irritation of the eyes, nose, and throat, dry mucous membranes and skin, rash, fatigue, headache, nausea, and allergic reactions. Approximately 800 substances have been identified in the air in non-industrial indoor environments. These include a large number of volatile organic substances described as VOCs (volatile organic compounds), for example formaldehyde, phenyl cyclohexene (from glue), diethyl amino ethanol (from air-moistening units), benzyl chloride and benzal chloride (from vinyl mats) and a large number of organic solvents, e.g., from paints, insulation materials, mat glues, wallpaper, adhesives, cleaning detergents, insecticide sprays, marking pencils. Ozone may develop when electrical appliances are used. Many types of dust are said to cause ill-health. Examples are pollen (from potted plants), mold spores, paper dust, tobacco smoke and mineral fibres (from lagging material).

In most cases, the combined concentration of all these substances is too small to cause irritation if their effects are additive. Therefore, the possibility of synergistic effects between the components has been discussed (see Section 7.3.2).

Part of the explanation of the sick-building syndrome may be exposure to mycotoxins. Fungi excrete substances belonging to a large and very heterogeneous group of biotransformation products called mycotoxins, i.e., fungal poisons. Bad ventilation is a common feature of sick buildings. Problems with moisture and mildew seem to be more common in countries with low outdoor temperatures in winter than in many other countries. This may be due to poor indoor ventilation during the cold seasons. High humidity, e.g., after water leaks, in combination with heat, results in favorable conditions for the growth of microorganisms.

7.4 Limit Values

In this chapter the term *limit value* is generally used to describe the standards set to regulate the human exposure in ambient air or in a work environment. The main reason for this choice is that the EU uses this term in its regulations. Other terms are used by other organizations: The Occupational Safety and Health Administration in the USA (OSHA) use Permissible Exposure Limits. The Health and Safety Executive in Great Britain (HSE) use Occupational Exposure Standards and Maximum Exposure Limits. The ILO's Encyclopaedia of Occupational Health and Safety employs Occupational Exposure Limits and alternatively Exposure Standards.

A large number of organizations and authorities in various countries publish hygienic limit values for air contaminants. More about their use and the setting of limit values (or exposure limits) is given in Chapter 8.

It is necessary to distinguish between *immission*, i.e., the occurrence of pollutants in the environment regardless of the source and the manner of distribution, and *emission*, which is the discharge of pollutants from various processes and sources in general. The purpose of immission limit values primarily is to protect humans against exposure to substances in inhaled air, drinking water, and food. Immission limit values are designed in different ways, depending on how they may be supervised (by sampling and measuring) and how the pollutants or their effects may be limited or controlled. With the help of immission limit values for air pollutants in densely populated areas, it is possible to warn sensitive persons against staying in polluted air. If the source of the pollutants is motor vehicles, occasional traffic controls may be used to decrease the concentrations of the substances in the streets.

The ILO publishes compilations of limit values for working sites issued in different countries. In the USA, limit values for working sites are issued by the federal authority OSHA and by the private association ACGIH. The latter collection of limit values is published annually and takes into consideration both chemical and physical factors. It is widely distributed and used internationally, although it lacks official status. In most European countries, the responsible authorities publish limit values for air contaminants in the work environment.

Limit values concerning substances in the environment are issued in many countries by authorities responsible for, e.g., the natural environment, housing, building and planning, and food legislation. Immission limit values can be established, but in contrast to what applies for the work environment, it is difficult to design control measures directed towards pollution sources for substances in the atmosphere, in lakes and the sea, or in the soil. The association between emission and immission is complex and influenced by, for example, geographical, topographical, and meteorological factors. The natural concentrations of elements in the earth's crust vary geographically within wide limits. Pollutants may be transported over long distances, accumulate, and biomagnify (increase in concentration) in food chains in a way that makes it difficult to tell when a certain concentration found in one place can be considered safe in the whole ecosystem.

On the other hand, emission limit values are much used. They can be supervised. They may be based on medical criteria, i.e., on what can be considered safe in a given situation, for instance limit values for the concentration of cadmium in discharged effluents. Emission limit values are usually used strategically and are based on what is practically feasible. The short-term aim is to limit emissions to what can be achieved with the best available technology. The responsible authority may then reduce the emission limit values gradually so that improved technology will be promoted and at best eventually eliminate the emissions.

7.5 Physical Risk Factors

7.5.1 Radiation

We are exposed to radiation from natural sources (e.g., the sun, cosmic radiation, the ground, radon in buildings) and from artificial sources (e.g., in medical treatment and in industry). See Figure 7.2.

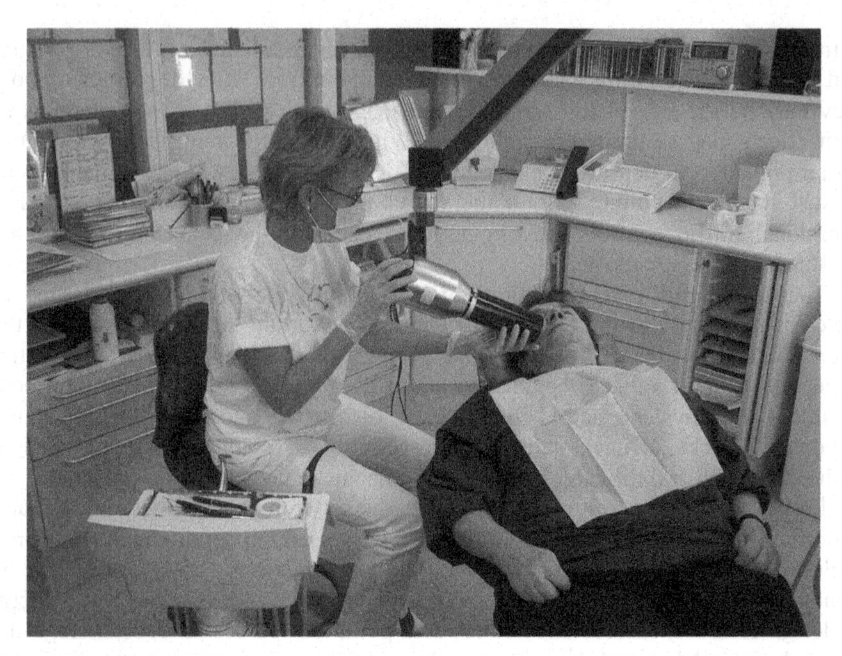

Figure 7.2. A dentist taking an X-ray picture of a patient's teeth. (Photo: Ulf Ulfvarson)

The radiation may consist of a wave motion of electric and magnetic fields (electromagnetic radiation) or a flow of particles (particle radiation). Some kinds of radiation may harm the body due to its ability to tear off electrons from atoms and molecules in its way. This will result in positively charged ions and therefore this kind of radiation is called *ionizing radiation*, see Table 7.4.

Table 7.4. Ionizing radiation

Sources	Type of radiation	Consists of	Examples of exposures
Radioactive substances and nuclear reactors	Alpha (α)	Helium nuclei	Air in mines, radon in buildings
	Beta (β)	Electrons	Tritium- and carbon-14 labeling
	Neutron radiation	Neutrons	Nuclear reactions
	Proton radiation	Protons	Nuclear reactor
	Gamma (γ)	Photons	Industrial level indicators
Accelerators	X-ray	Photons	X-ray diagnostic
	Electron	Electrons	Radiation treatment

As a rule, *non-ionizing radiation* does not have the same ability to damage the body as ionizing radiation. This is because its energy is too small to extract electrons from atoms and molecules. Information about wavelength and frequencies is presented in Table 7.5. The basic unit hertz (Hz) is the number of oscillations per second.

Table 7.5. Non-ionizing radiation

Type of radiation	Wavelength	Frequency
Ultraviolet radiation (UV)	10–400 nm	3 000–750 THz
Visible light	400–800 nm	750–375 THz
Infrared radiation (IR)	1 μm–1 mm	300 THz–300 GHz
Microwaves	1 cm–1 dm	30–3 GHz
Radio waves	1 cm–1 km	30 GHz–30 kHz
Power frequency fields		≈ 50–60 Hz
Static field		0 Hz

1 μm (micrometer) = 10^{-6} meter; nm (nanometer) = 10^{-9} meter; k (kilo) = 10^{3}; M (mega) = 10^{6}; G (giga) = 10^{9}; T (tera) = 10^{12}

7.5.2 The Influence of Radiation on the Body

The ability of ionizing radiation to penetrate objects varies with its type. The effects on the body therefore depend on the type of radiation. Alpha particles are stopped, for instance, by wallpaper. Nevertheless, an alpha-particle-emitting substance entering the body with, for instance, radon in water or air is capable of harming the body. This is because the radiation may last for a long time close to vulnerable parts of the body, such as the bone marrow.

Ionizing radiation with a high penetration capacity, e.g., gamma radiation and X-rays, has the largest health effects. Protection against gamma radiation and X-rays is offered by a lead layer several centimeters thick, by a concrete layer one decimeter thick or by several meters of water. Of course, high doses of ionizing radiation are very dangerous to the health as is illustrated in Table 7.6.

The part of the radiation energy that is absorbed is called the *radiation dose.* Together with the radiation dose, the type of radiation is also important for the effect on the body. In order to estimate the effect of radiation, the concept of *effective radiation dose* has been introduced. This measure considers the biological effect of various types of radiation on the human organs. The unit is sievert (Sv), or more commonly millisievert (1 mSv = 0.001 Sv); 1 Sv has the same biological effect regardless of the source of radiation.

The annual global per capita effective dose due to natural radiation sources is 2.4 mSv (UNSCEAR, 2000). In Table 7.7, per capita effective doses from natural and artificial sources are presented. As can be seen in this table, the greatest contribution to exposure comes from natural background radiation. The exposure to radon in indoor air varies from location to location and is very significant in some places (UNSCEAR, 2000).

Apart from the immediate effects there is also the risk of cancer, fetal damage, and hereditary injuries. There are, for instance, indications that children of male workers in the nuclear power plant in Sellafield have an increased risk of leukemia and lymphatic gland cancer if their fathers were exposed to radiation before conception (Gardner et al., 1990; Sever et al., 1997).

Table 7.6. Health effects of high doses of ionizing radiation. Adapted from Fentiman et al. (unknown year)

Dose [mSv]	Effects
50–200	Possible latent effects (cancer), possible chromosomal aberrations
250–1 000	Blood changes
> 500	Temporary sterility in males
1 000	Double the normal incidents of genetic defects
1 000–2 000	Vomiting, diarrhea, reduction in infection resistance, possible bone growth retardation in children
2 000–3 000	Serious radiation sickness, nausea
> 3 000	Permanent sterility in females
3 000–4 000	Bone marrow and intestine destruction
4 000–10 000	Acute illness and early death (usually within days)

Most types of cancer develop 10 years or more after a carcinogenic exposure. Many different factors interplay when a tumor builds up. The probability of getting cancer is proportional to the radiation dose. A greater risk of cancer has been observed when people have been exposed to radiation doses higher than 100 mSv. In most cases, it is not possible to distinguish between the effects of radiation, the effects of other environmental factors and the effects of smoking.

Fetuses are particularly sensitive to radiation and X-ray investigations of pregnant women must therefore be limited to a minimum. Hereditary injuries may develop if a gamete is damaged by radiation, but this kind of damage has not been found in man, not even after the atomic bombs in Japan.

7.5.3 The Chernobyl Accident

The accident in the Chernobyl nuclear power station in 1986 was a tragic example of an event that has provided knowledge of the acute and chronic health effects of radiation. The accident caused high doses of radiation with acute effects in emergency and recovery operation workers. Altogether 600 000 workers took part in the cleaning-up operations. The doses received ranged up to more than 500 mSv, with an average of about 100 mSv according to the State Registries of Belarus, Russia, and Ukraine. Only a small fraction of these workers were exposed to very high levels of radiation.

A total of 116 000 people, who were evacuated in 1986 from the highly contaminated zone, received an average dose of 33 mSv. Two hundred and seventy thousand people who were residents in the areas of strict radiological control, 1986–2005, were exposed to not less than 50 mSv and up to several hundreds mSv, while a total of 5 000 000 residents of other contaminated areas in 1986–2005 received 10 to 20 mSv (Chernobyl Forum, 2005; IAEA, 2006).

Radiation injuries in those who worked with fire fighting immediately after the accident caused the death of 28 people within a few months. Two persons died from other causes. Nineteen more died in 1987-2004 of various causes, not directly attributable to radiation exposure.

Table 7.7. Annual per capita effective doses in 2000 from natural and man-made sources (UNSCEAR, 2000)

Source	Worldwide annual per capita effective dose [mSv]	Range of trend in exposure
Natural background	2.4	Typically –10 mSv, depending on circumstances at particular locations; with sizeable population also 10-20 mSv
Diagnostic medical examinations	0.4	Ranges from 0.04 to1.0 mSv, at lowest and highest levels of health care
Atmospheric nuclear testing	0.005	Has decreased from a maximum of 0.15 mSv in 1963. Higher in the northern hemisphere and lower in the southern hemisphere
Chernobyl accident	0.002	Has decreased from a maximum of 0.04 mSv in 1986 (average in northern hemisphere). Higher at locations nearer the accident site
Nuclear power production	0.0002	Has increased with expansion of the programme, but has decreased with improved practice

It was estimated that, among those who lived within a radius of 30 km from the accident site, the thyroid gland took up 70 mSv in adults and about 1 000 mSv (1 Sv) in children. This uptake was due to direct radiation, inhalation of radioactive iodine isotopes in the passing radioactive cloud, and the consumption of contaminated milk. About 4 000 people, who were children in 1986 living in areas close to the site of accident, contracted cancer of the thyroid gland. Nine of them had died of the disease up to 2005.

Among the general population exposed to the Chernobyl radioactive fallout, the radiation doses were relatively low and acute radiation syndrome and associated fatalities did not occur. An international expert group predicts that among the 600 000 persons exposed to more significant amounts (>500 mSv), the possible increase in cancer mortality due to radiation exposure might be up to a few per cent, representing up to 4 000 fatal cancers. Since approximately 100 000 fatal cancers are expected due to all other causes of cancer in this population, the increase would be very difficult to detect. This is even truer for the expected increase in cancer mortality of less than 1 % among the 5 million persons residing in other contaminated areas that received an average exposure to 10–20 mSv (Chernobyl Forum, 2005).

A radioactive substance is a substance which decays emitting ionizing radiation. To determine the strength of the radioactivity, the number of decays per second is measured. The unit becquerel (Bq) is defined as one decay per second. An area of more than 200 000 km^2 in Europe was contaminated with radioactive cesium (above 0.04 MBq of $^{137}Cs/m^2$), of which 71 % was in the three most affected countries (Belarus, the Russian Federation, and Ukraine). The deposition varied greatly since it was influenced by rain when the contaminated air masses passed. Countries in the vicinity of Chernobyl, in Scandinavia, for instance, also suffered significant pollution. Today, more than 20 years after the accident, there is still strontium and plutonium to be found, but mainly within 100 km from the reactor. From the aspect of radiation protection, only the deposited cesium-137 is of any importance.

Agriculture suffered heavily during the first few years after the accident. Today cesium-137 in milk and meat contributes most to the radiation dose via foodstuffs from agriculture. There are still highly polluted areas where the concentration of cesium-137 is above the limit value of 100 Bq per litre. The concentrations of cesium-137 in such food are, however, generally below the limit values in all three countries.

In products from the forest, there may still be relatively high levels of radioactive cesium in some localities and during some periods. The concentration may be above the limit value for products for sale (1 500 Bq/kg).

7.5.4 Non-ionizing Radiation

The kind and extent of the effects of non-ionizing radiation depend on the amount of energy transmitted and on how the radiation is absorbed by the body. IARC classifies extremely low-frequency magnetic fields as possibly carcinogenic to humans (Group 2B), whereas static electric and magnetic fields and extremely low-frequency electric fields as not classifiable regarding their carcinogenicity to humans (Group 3) (IARC, 2002).

Some examples of the health effects of non-ionizing radiation and the sources of such radiation are listed in Table 7.8. For a long time, there have been apprehensions about the carcinogenic effects of electric and magnetic fields from power transmission lines, but this has mainly not been supported by epidemiologic studies performed on exposed groups.

Electric and magnetic fields from cathode ray tubes in monitors have been under suspicion for causing pregnancy disturbances and electromagnetic hypersensitivity, i.e., symptoms and subjective inconvenience in the skin, the nervous system and the eyes. So far, no association between computer terminal work and pregnancy disturbances has been found, nor do research findings indicate any association between skin troubles and the electric and magnetic fields around monitors. The recent change of technology to flat-panel monitors means that it will be difficult to make new epidemiologic studies of groups exposed to monitors with cathode ray tubes.

There are many indications that electromagnetic hypersensitivity can be attributed to several factors other than exposure to electric and magnetic fields. Other factors of possible importance are allergies to other agents, sensitivity to light, flickering fluorescent lamps, chemicals, individual factors and bad work organization.

Considering the uncertainty about the causes of effects of low-frequency electric and magnetic fields, especially the risk of cancer, authorities in some countries recommend that the precautionary principle (in a broad sense) be applied. There are various formulations available. The most well known is the "Rio declaration" (UNCED, 1992). "Where there are threats of serious or irreversible environmental damage, lack of full scientific certainty shall not be used as a reason for postponing cost-effective measures to prevent environmental degradation."

Table 7.8. Examples of effects of non-ionizing radiation injurious to health

Type of radiation	Examples of health injurious effects	Examples of sources
Ultraviolet radiation (UV)	Skin cancer, cataract, burn injury	Sun, solarium
Visible light	Burn of retina	Sun, laser, welding arc
Infrared radiation (IR)	Injury of the cornea, cataract	Melt, e.g., steel, glass
Microwaves	Effects of heat	Plastic welding, microwave oven
Radio waves		Mobile phone
Electric field		Electric apparatus
Magnetic fields		Electric apparatus, high-voltage power line

For some time, the possibility that radiation from mobile phones could cause cancer has been discussed. The radiation from the antennae of the base stations is of the same kind as that from television transmitter antennae. Although the latter radiation is about one thousand times stronger than that from the base stations of mobile phones, it is not considered to involve a risk. To date, a number of well-designed, internationally accepted epidemiologic investigations have, with a reasonable degree of certainty, ruled out any causal association between mobile phones and cancer (Boice and McLaughlin, 2002).

7.5.5 Limit Values for Radiation

The International Commission on Radiological Protection (ICRP) recommends that all use of radiation should be justified. Moreover, the exposure should be as low as can reasonably be achieved considering economic and social factors. According to the limit values published by ICRP, the effective radiation exposure in a work environment should not exceed 50 mSv during a single year and should not exceed a total of 100 mSv during five consecutive years.

We may swallow radioactive substances via our food. After the Chernobyl accident, temporary limit values were adopted for radioactivity in food valid within EU. According to these, the accumulated maximum radioactive level in terms of cesium-134 and cesium-137 is 370 Bq/kg for milk and milk products and foodstuffs intended for the special feeding of infants during the first four to six months of life, and 600 Bq/kg for all other products. The regulation has since been extended several times (EEC, 1990).

Limit values for different frequency ranges of non-ionizing radiation are published by the International Commission on Non-ionizing Radiation Protection (ICNIRP). For electric and magnetic fields with high frequency, 3 MHz to 300 GHz, the limit values are based on the capacity of the field to induce heat. This is of particular interest in special industrial applications and for microwave ovens. In a modern technological society, almost all persons are exposed to the lower ranges of frequencies, including the power frequency (50–60 Hz). The International Radiation Protection Association (IRPA) and authorities in several countries publish limit values for these frequencies.

7.6 Noise

A human being with normal hearing can perceive sound with frequencies between 20 Hz and 20 000 Hz. Of course, we want an environment in which we decide the sound level ourselves and where we are not disturbed by undesired sound, i.e., what is called noise. Noise is one of the most frequent and most serious work environment problems, because it is injurious to hearing and causes stress.

The threshold of hearing, the lowest level of sound required to be perceived by a normal ear, varies with the frequency. Therefore, when analyzing noise it is necessary to know the frequency distribution. This is achieved by measuring the noise level in standardized octave bands. An octave is an interval whose higher tone has a sound-wave frequency of vibration twice that of its lower tone. Thus, the not A vibrates at 220 Hz (cycles per second), while its octave, A′, vibrates at 440 Hz.

The intensity of sound or noise is measured by its sound level, the unit of which is decibel, dB. The decibel scale is logarithmic, which means that if two similar sources of sound at the same distance from the measuring point each have a sound level of 80 dB, the total sound level will be about 83 dB (10 times the logarithm of $2 \approx 3$).

In order to judge the risk of impaired hearing, a noise filter is used in the measurement to imitate the sensitivity of the human ear. The most widely used sound level filter is the A scale. Using this filter, the noise is reduced for low frequencies and to some extent for high frequencies. The noise is somewhat intensified around 1 kHz. Measurements made on this scale are expressed as dBA. The C scale is practically linear over several octaves, and is thus suitable for subjective measurements only for very high sound levels and impulse sounds, for instance firing range noise. The unit is dBC.

Since noise often varies over time, it is necessary to describe the noise situation in a simplified manner. This is done with the *equivalent sound level*, an average value for the total acoustic energy during a certain period, for instance a workday. There are various ways of forming the average. The International Standards Organization (ISO) has established a standard for the calculation so that, for instance, the noise level for a given time of exposure is equivalent as regards impaired hearing to half the time of exposure at a noise level 3 dB higher. The risk of impaired hearing from exposure at 100 dB during one hour is thus considered to be the same as that from 103 dB during half an hour.

Noise exposure and the hearing impairment from this exposure have long been well documented. In the outdoor environment, noise comes first from traffic. Other important noise sources are household appliances, hobby tools, garden tools, and pleasure boats. Noise harmful to hearing may also come from music in various forms, live music and music through loudspeakers and earphones (e.g., portable CD or MP3 players). The World Health Organization (WHO) recommends that the sound exposure of the audience at music events, measured as the equivalent sound level during a four-hour period, should be below 100 dBA. The maximum sound level should never exceed 110 dBA. Unfortunately there are many examples of situations were these levels are exceeded.

7.6.1 Influence of Noise on Health and Well-being

An elevated noise level of short duration may cause a temporary impairment of hearing. A single short, but sufficiently strong sound impulse may lead to permanent hearing impairment. This also holds for noise at a high level (see Section 7.4) over a longer time. The stronger the noise and the longer the time of exposure, the more pronounced is the risk of hearing impairment. Children run a higher risk of suffering hearing impairment than adults, since they have a shorter and narrower auditory meatus. In children, the maximum amplification by the resonance in the auditory meatus occurs higher in the treble area than in adults. This is the area where the risk of injury from noise is most pronounced. The sound level should be kept particularly low to protect the hearing of small children.

Apart from the risk of hearing impairment, a noisy environment also has a psychological influence. We experience discomfort and are unable to perform as well as usual, especially if the tasks demand great concentration. This is more so for people with hearing impairment. In addition, there is the risk that people in a noisy environment will not notice approaching vehicles or alarm signals. Many people feel discomfort from traffic noise. At a given sound intensity level, the source of the noise is of great importance for how the noise is perceived.

According to a calculation by the WHO, there are around 120 million people with hearing disablement. The disablement may depend on a hearing injury or to a change due to old age. Hearing impairment conditioned by age results in a gradual loss of hearing, especially in the upper frequency area, treble (1–8 kHz) and this leads to difficulties in perceiving voiceless consonants, so that conversation is difficult to catch, particularly when there is disturbing noise in the background. Many people suffer hearing injuries in their work. The WHO states that 15 % of the employees in Germany are exposed at work to noise levels injurious to hearing. Hearing injuries may involve hypersensitiveness to sound, tinnitus, or an impaired ability to perceive certain sound frequencies. According to statistics from USA, almost 12 % of men who are 65 to 74 years of age are affected by tinnitus. Approximately 28 million Americans have a hearing impairment (NIDCD, 2008).

7.6.2 Limit Values for Noise

According to recommendations from the WHO, continuous sound indoors should not exceed 30 dBA. Most people may avoid hearing injuries as long as the average continuous noise exposure is below 70 dB. An adult may cope with occasional noise levels up to 140 dB, but children should never be exposed to levels above 120 dB.

The WHO recommends 85 dB as a limit for occupational exposure during an eight-hour day. If the limit is not exceeded most people should be able to withstand this exposure without permanent hearing injuries after 40 years at work. Fluctuating noise is judged according to the equivalent sound level.

National limit values in various countries usually follow these international recommendations. Compliance to the limit values varies considerably between different countries.

7.7 Environment and Health in a Holistic Perspective

Already in the nineteenth century, researchers asked the question: how does the environment influence people? Charles Darwin's research into the effects of the environment on the development of species inspired later scientists and philosophers to investigate the influence of the environment on human beings. Is the changed environment created by man a benefit or a disadvantage from a health point of view? During his development in his natural environment, man adapted to low levels of noise, radiation, and substances in air, water, and food of natural occurrence. Millennia of man's striving to satisfy his needs have radically changed the environment for most people. Without doubt, people in the developed part of the world live considerably longer than early man did. From the Paleolithic to the late medieval period, the mean life expectancy of humans increased from between 20 and 30 years to 30 to 40 years (MacLennan and Sellers, 1999). Nevertheless, in the eighteenth century, the average length of life was no longer than 30–35 years.

Cold and humidity were part of our ancestors' lives and so were periods of famine, illness, especially infections, and accidents. All this shortened their lives. Over the last 150 years, there has been a dramatic improvement in hygienic conditions, state of nutrition and medical treatment. In the whole world, the mean life expectancy of humans at birth is 66 years, divided into 65 years in the developing countries and 77 years in the developed countries (US Census Bureau, 2007). The difference depends to a high degree on the fact that diseases have been pressed back in the developed countries. In the developing countries, the decrease in infectious diseases has led to a population explosion, and this has increased the pressure on natural resources and has resulted in erosion and lack of water, and sometimes also a shortage of food. Nevertheless, the average length of life is also increasing in many developing countries. In Sub-Saharan Africa, the mean life expectancy at birth is 50 years. Nowadays HIV/AIDS influences the average length of life in many countries.

On the other hand, man's improved control of his environment has caused environmental problems, which may more or less directly influence health negatively. In certain well-defined groups, effects on health of exposure to substances and radiation can be discerned. This is obvious in, for example, occupational groups in developing countries. In developed countries, there are small differences in mortality and morbidity between different socio-economic groups and different regions. Cardio-vascular diseases take the lives of 12 million people each year (WHO, 2005). Life-style factors are responsible for most of the risk factors involved. The majority of deaths in tumor diseases are due to diet and way of life. Only a small part of the tumor diseases is related to environment pollution and radiation (Doll and Peto, 1981).

The conclusion is that, from a health perspective, man's control of the physical environment, outdoors and indoors, has so far been successful. Side-effects of this development are, inter alia, an increased exposure to substances, radiation, and noise. There is sufficient knowledge available to prevent ill-health due to this exposure in workplaces and in other environments.

References

ACGIH (2006) Threshold limit values for chemical substances and physical agents & biological exposure indices 2006. Adopted appendices C. Particle size-selective sampling criteria for air borne particulate matter. The American Conference of Governmental Industrial Hygienists (ACGIH), Cincinnati, OH

Boice Jr. JD, Mclaughlin JK (2002) Epidemiologic studies of cellular telephones and cancer risk. SSI report 2002:16. Swedish Radiation Protection Authority (SSI). www.ssi.se/ssi_rapporter/pdf/ssi_rapp_2002_16.pdf

Brunekreef B, Dockery D, Krzyzanowski M (1995) Epidemiologic studies on short-term effects of low levels of major ambient air pollution components, Environmental Health Perspectives 103 (Suppl 2):3–13

CAS (2008) Chemical Abstract Service (CAS). www.cas.org/

CEN (1999) Air quality – field test procedure to demonstrate reference equivalence of sampling methods for the PM10 fraction of particulate matter – Reference method and

field test procedure to demonstrate reference equivalence of measurement methods. EN 12341:1999, European Committee for Standardization (CEN)
Chernobyl Forum (2005) Chernobyl's Legacy: Health, environmental and socio-economic impacts and recommendations to the governments of Belarus, the Russian Federation and Ukraine. Second revised version, The Chernobyl Forum: 2003–2005. www.iaea.org/Publications/Booklets/Chernobyl/chernobyl.pdf
Doll R, Peto R (1981) The causes of cancer: quantitative estimates of avoidable risks of cancer in the United States today. Journal of National Cancer Institute 66:1193–1308
ECB (2008) European chemical substances information system (ESIS). European Chemicals Bureau (ECB). www.ecb.jrc.it/ESIS/
EEC (1990) Council Regulation (EEC) No 737/90 of 22 March 1990 on the conditions governing imports of agricultural products originating in third countries following the accident at the Chernobyl nuclear power station. www.ec.europa.eu/energy/nuclear/radioprotection/doc/legislation/90737_en.pdf
EPA (2008) US Environmental Protection Agency (EPA) – Particulate Matter Standards. www.epa.gov/PM/standards.html
EU (1999) Council Directive 1999/30/EC of 22 April 1999 relating to limit values for sulphur dioxide, nitrogen dioxide and oxides of nitrogen, particulate matter and lead in ambient air. Official Journal of the European Communities L 163/41, June 1999. www.eur-lex.europa.eu/LexUriServ/LexUriServ.do?uri=OJ:L:1999:163:0041:0060:EN:PDF
EU (2006) Commission directive 2006/15/EC of 7 February 2006 establishing a second list of indicative occupational exposure limit values in implementation of Council Directive 98/24/EC and amending Directives 91/322/EEC and 2000/39/EC. Official Journal of the European Union L 38/39, February 2006. www.eur-lex.europa.eu/LexUriServ/site/en/oj/2006/l_038/l_03820060209en00360039.pdf
Fentiman AW, Smith M, Meredith JE (unknown year) What are the health effects of ionizing radiation? RER-24, The Ohio State University Information. www.ag.ohio-state.edu/~rer/rerhtml/rer_24.html
Ferin J, Oberdörster G (1992) Potential health effects of fume particles on the crew of spacecrafts. SAE PAPER 921387, 22nd International Conference on Environmental Systems, Seattle, WA, July 13–16, Society of Automotive Engineers (SAE)
Gardner MJ, Snee MP, Hall AJ, Powell CA, Downes S, Terrell JD (1990) Results of case-control study of leukaemia and lymphoma among young people near Sellafield nuclear plant in West Cumbria. British Medical Journal 300:423–429
HASTE (2008) The European Health and Safety Database (HASTE). www.ttl.fi/internet/partner/haste
HSC (2008) Health & Safety Commission (HSC), UK. www.hse.gov.uk/index.htm
IAEA (2006) Environmental consequences of the Chernobyl accident and their remediation: twenty years of experience. Report of the Chernobyl Forum Expert Group 'Environment'. Radiological assessment reports series, International Atomic Energy Agency (IAEA). www-pub.iaea.org/MTCD/publications/PDF/Pub1239_web.pdf
IARC (2002) Volume 80 non-ionizing radiation, part 1: static and extremely low-frequency (ELF) electric and magnetic fields. Summary of data reported and evaluation. IARC Monographs on the Evaluation of Carcinogenic Risks to Humans. International Agency for Research on Cancer (IARC). www.monographs.iarc.fr/ENG/Monographs/vol80/volume80.pdf
IARC (2007) Agents reviewed by the IARC monographs Volumes 1–98. International Agency for Research on Cancer (IARC). www.monographs.iarc.fr/ENG/Classification/Listagentsalphorder.pdf

ILO (2005) Global estimates of fatal work related diseases and occupational accidents, World Bank Regions 2005 (most data collected in 2001). International Labour Organization (ILO).
www.ilo.org/public/english/protection/safework/accidis/globest_2005/index.htm

Koenig JQ, Larson TV, Hanley QS, Rebolledo V, Dumler K, Checkoway H, Wang SZ, Lin D, Pierson WE (1993) Pulmonary function changes in children associated with fine particulate matter. Environmental Research 63:26–38

Larson TV, Koenig JQ (1994) Wood smoke: emissions and non-cancer respiratory effects. Annual Review of Public Health 15:133–156

MacLennan WJ, Sellers WI (1999) Ageing through the ages. Proc Royal College of Physicians Edinburgh 29:71–75

NIDCD (2008) Statistics about hearing disorders, ear infections, and deafness. National Institute on Deafness and other Communication Disorders (NIDCD).
www.nidcd.nih.gov/health/statistics/hearing.asp

NIOSH (2005) Strategic plan for NIOSH nanotechnology research: filling the knowledge gaps. National Institute for Occupational Safety and Health (NIOSH).
http://www.cdc.gov/niosh/topics/nanotech/strat_plan.html

NIOSH NRC (2007) Progress toward safe nanotechnology in the workplace. NIOSH Nanotechnology Research Center (NRC), Publication No. 2007–123, National Institute for Occupational Safety and Health (NIOSH).
www.cdc.gov/niosh/docs/2007-123/pdfs/2007-123.pdf

Oberdörster G, Ferin J, Finkelstein J, Baggs R, Stavert DM, Lehnert BE (1992) Potential health hazards from thermal degradation events: particulate versus gas phase effects. SAE PAPER 921388, 22nd International Conference on Environmental Systems, Seattle, WA, July 13–16, Society of Automotive Engineers (SAE)

OSHA (2008) European Union occupational safety and health statistics. European Agency for Safety and Health at Work (OSHA). www.osha.europa.eu/statistics

Ostro B (1993) The association of air pollution and mortality: examining the case for inference. Archives of Environmental Health 48:336–342

Schwartz J (1995) Short term fluctuations in air pollution and hospital admissions of the elderly for respiratory disease. Thorax 50:531–538

Sever LE, Gilbert ES, Tucker K, Greaves J, Greaves C, Buchanan J (1997) Epidemiologic evaluation of childhood leukemia and paternal exposure to ionizing radiation. National Institute for Occupational Safety and Health, Health Related Energy Research Branch (HERB), Cincinnati, OH

Styer P, McMillan N, Gao F, Davis J, Sacks J (1995) Effect of outdoor airborne particulate matter on daily death counts. Environmental Health Perspectives 103:490-497

Timonen KL, Pekkanen J (1997) Air pollution and respiratory health among children with asthmatic or cough symptoms. American Journal of Respiratory and Critical Care Medicine 156:546–552

UNCED (1992) Rio declaration on environment and development. The United Nations Environment Programme (UNEP).
www.unep.org/Documents/Default.asp?DocumentID=78&ArticleID=1163

UNSCEAR (2000) UNSCEAR 2000 report volume 1: sources and effects of ionizing radiation. UNSCEAR 2000 Report to the General Assembly, with scientific annexes. United Nations Scientific Committee on the Effects of Atomic Radiation (UNSCEAR).
www.unscear.org/unscear/en/publications/2000_1.html

US Census Bureau (2007) World population profile 2007. US Census Bureau International Data Base (IDB). www.census.gov/ipc/www/idb/

Vainio H, Magee P, McGregor D, McMichael A (eds) (1992) Mechanisms of carcinogenesis in risk identification. IARC Scientific Publications, Lyon

WHO (2005) Avoiding heart attacks and strokes. Don't be a victim – Protect yourself. ISBN 92 4 154672 7. World Health Organization (WHO).
www.who.int/cardiovascular_diseases/resources/cvd_report.pdf
WHO ECEH (2005) Implementing environment and health information system in Europe (ENHIS). Final technical report. WHO European Centre for Environment and Health (ECEH), Bonn.
www.ec.europa.eu/health/ph_projects/2003/action1/docs/2003_1_28_frep_en.pdf

8

Safety Factors and Exposure Limits

Sven Ove Hansson

8.1 Numerical Decision Tools

Numerical decision tools are abundantly employed in safety engineering. Two of the most commonly used tools are safety factors and exposure limits. A safety factor is the ratio of the maximal burden on a system not believed to cause damage to the highest allowed burden. An exposure limit is the highest allowed level of some potentially damaging exposure.

8.2 Safety Factors

Humans have made use of safety reserves since prehistoric times. Builders and tool-makers have added extra strength to their constructions to be on the safe side. Nevertheless, the explicit use of safety factors in calculations seems to be of much later origin, probably the latter half of the nineteenth century. In the 1860s, the German railroad engineer A. Wohler recommended a factor of 2 for tension. In the early 1880s, the term "factor of safety" was in use, hence Rankine's *A Manual of Civil Engineering* defined it as the ratio of the breaking load to the working load, and recommended different factors of safety for different materials (Randall, 1976).

In structural engineering, the use of safety factors is now well established, and design criteria employing safety factors can be found in many engineering norms and standards. Most commonly, a safety factor is defined as the ratio of a measure of the maximum load not inducing failure to a corresponding measure of the load that is actually applied. In order to cover all the major integrity-threatening mechanisms that can occur, several safety factors may be needed. For instance, one safety factor may be required for resistance to plastic deformation and another for fatigue resistance.

The other major application area for safety factors is toxicology. Here, the use of explicit safety factors is more recent. Apart from some precursors, it dates from the middle of the twentieth century (Dourson and Stara, 1983). The first proposal for a safety factor for toxicity was Lehman's and Fitzhugh's proposal in 1954 that an ADI (Acceptable Daily Intake) be calculated for a food additive by dividing the chronic animal NEL (maximum No Effect Level) in mg/kg of diet by 100. They thus defined a safety factor as the ratio of an experimentally determined dose to a dose to be accepted in humans in a particular regulatory context. If the NEL is 0.5 mg/kg body weight, the application of a safety factor of 100 will then result in a maximum allowed dose of 0.005 mg/kg body weight. This definition is still in use. Their value of 100 is also still widely used, but higher factors such as 1 000, 2 000, and even 5 000 are employed in the regulation of substances believed to induce severe toxic effects in humans. Compare with Figure 8.1.

Toxicological safety factors are often based on products of subfactors, each of which relates to a particular "extrapolation." The factor 100, for example, is described as composed of two factors of 10, one for the extrapolation from animals to humans and the other for the extrapolation from the average human to the most sensitive members of the human population (Weil, 1972). For ecotoxicity, factors below 100, such as 10, 20, and 50, are widely in use. Lower factors are, of course, associated with a higher degree of risk-taking.

Figure 8.1. Japanese factory workers inspect packages of processed foods containing roasted peanuts imported from China at a confectionary factory in Niigata city, Japan, March 28, 2008. Following news reports of lethal chemicals found in imported Chinese food products, Japanese companies are facing tighter scrutiny from consumers and health ministry inspectors to insure the quality of their food products. (Photo: Everett Kennedy Brown/EPA/Scanpix)

In addition to safety factors, the closely related concept of a safety margin is used in many applications. The essential difference is that, whereas safety factors are multiplicative, safety margins are additive. Airplanes are kept apart in the air; a safety margin in the form of a minimum distance is required. Surgeons removing a tumor also remove the tissue closest to the tumor. Their safety margin (surgical margin) is defined as the distance between the tumor and the lesion. Typical values are 1–2 cm (Kawaguchi, 1995). The notion of a safety margin is also sometimes used in structural engineering, and it is then defined as capacity minus load.

Independently of the area of application, safety factors and safety margins can be divided into three categories (Clausen et al., 2006):

1. Explicitly chosen safety factors and margins. Safety factors in this category are used, e.g., by the engineer who multiplies the foreseen load on a structure by a standard value of, say, 3 and uses this larger value in his or her construction work. Similarly, the regulatory toxicologist applies an explicitly chosen safety factor when she divides the dose believed to be harmless in animals by a previously decided factor such as 100, and uses the obtained value as a regulatory limit. Explicitly chosen safety factors are also used in, e.g., geotechnical engineering, ecotoxicology, and fusion research (for plasma containment). As already mentioned, explicitly chosen safety margins are used in air traffic and in surgery. They are also used in radiotherapy to cope with set-up errors and internal organ motion.

2. Implicit safety reserves. These are safety factors or margins that have not been specifically chosen, but can, after the fact, be described as such. They have their origin in human choice, but in choices that are not made in terms of safety factors or margins. As one example of this, occupational toxicology differs from food toxicology in that allowable doses are usually determined in a case-by-case negotiation-like process that does not involve the use of generalized (fixed) safety factors. However, it is possible to infer implicit safety factors; in other words, a regulatory decision can be shown to be the same *as if* certain safety factors had been used (Hansson, 1998). Another example can be found in traffic safety research. The behavior of drivers can be described *as if* they applied a certain safety margin to the distance between their car and the car nearest ahead. This margin is often measured as the *time headway*, i.e., the distance divided by the speed (Hulst et al., 1999).

3. Naturally occurring safety reserves. These are the safety reserves with regard to natural phenomena that can be calculated by comparing a structural or physiological capacity to the actually occurring load. These safety reserves have not been chosen by human beings, but are our way of describing properties that have developed through evolution. As in the case of implicit safety reserves, naturally occurring safety reserves can often be described in terms of safety factors or margins. Structural safety factors have been calculated for mammalian bones, crab claws, shells of limpets, and tree stems. Physiological safety factors have been calculated, e.g., for intestinal capacities such as glucose transport and lactose uptake, for hypoxia tolerance in insects, and for human speech recognition under conditions of speech distortion (Clausen et al., 2006).

The reason why safety factors can be applied in descriptions of natural phenomena is that when we calculate loads – whether of natural or artificial origin – we do not consider unusual loads. Resistance to unusual, unforeseen loads is as important for the survival of an organism as it is for the continued structural integrity of a man-made artifact. For example, the extra strength of tree stems enables them to withstand storms even if they have been damaged by insects. On the other hand, there is a limit to the evolutionary advantage of excessive safety reserves. Trees with large safety reserves are better able to resist storms but, in the competition for light reception, they may lose out to tender and high trees with smaller safety reserves. In general, the costs associated with excessive capacities result in their elimination by natural selection. There are at least two important lessons to learn from nature here. First, resistance to unusual loads that are sometimes difficult to foresee is essential for survival. Secondly, a balance must nevertheless always be struck between the danger of having too little reserve capacity and the cost of having a reserve capacity that is never or rarely used.

8.3 What Do Safety Factors Protect Against?

In characterizing the sources of failure against which safety factors provide protection we need to consider the decision-theoretical distinction between risk and uncertainty. A decision is said to be made under risk if the probabilities of the relevant outcomes are known or are assumed to be known. Otherwise, it is made under uncertainty. Uncertainty comes in different forms. Sometimes it is due to a lack of reasonable probability estimates for identified outcomes. On other occasions, there may also be a considerable uncertainty about what outcomes are in fact possible (Hansson, 1996).

In structural engineering, safety factors are intended to compensate for five major sources of failure: (1) higher loads than those foreseen, (2) worse properties of the material than foreseen, (3) imperfect theory of the failure mechanism in question, (4) possible unknown failure mechanisms, and (5) human error (e.g., in design) (Moses, 1997). The first two of these can possibly be described in terms of probabilities, whereas the last three concern uncertainty rather than risk (Hansson, 2007a).

In toxicology, safety factors are typically presented as compensations for (1) various extrapolations such as that from animals to humans, (2) intraspecies variability, (3) lack of data, and (4) imperfection in the models used for interpreting the data (Gaylor and Kodell, 2002). At least the last two of these refer primarily to uncertainty rather than to risk in the probabilistic sense.

In structural engineering, in particular, it has often been proposed that safety factors should be replaced by specifications expressed in terms of probabilities. This means that instead of building a bridge with a specified safety factor it should be constructed in a way that conforms with a specified maximum probability of failure. However, it has turned out to be difficult to replace safety factors by probabilities. The major obstacle is that safety factors are intended to protect not only against risk (in the probabilistic sense) but also against uncertainties for which no meaningful probabilities are available (Clausen et al., 2006).

8.4 Exposure Limits

An exposure limit is a restriction on the allowed exposure to some agent having undesired effects. Exposure limits are usually expressed in terms of some physical or chemical measurement, such as dB for noise, Sievert (Sv) for ionising radiation, and mg/m^3 for the inhalation of chemical pollutants. The most well-developed systems of exposure limits are those for occupational chemical exposure. (Exposure limits or limit values are also dealt with in Chapter 7.)

The first occupational exposure limits were proposed by individual researchers in the 1880s. In the 1920s and 1930s, several lists were published in both Europe and the USA, and in 1930 the USSR Ministry of Labor issued what was probably the first official list. However, by far the most influential occupational exposure limits are the *threshold limit values* (TLVs) issued by the American Conference of Governmental Industrial Hygienists (ACGIH). In spite of its name, the ACGIH is a voluntary organization with no formal ties to government or state authorities in the USA or elsewhere. In 1946, the ACGIH adopted a list of exposure limits, covering approximately 140 chemical substances. The list has since then been gradually extended and revised, a new edition being published annually (Cook, 1985).

In the 1940s and 1950s, the ACGIH and the American Standards Association (ASA) competed for the position of leading setter of occupational health standards. The values of the ASA and those of the ACGIH did not differ greatly in numerical terms, but the ASA values were ceiling values below which all workplace concentrations should fluctuate, whereas the ACGIH values were (and still are, with few exceptions) upper limits for the average during a whole working-day. The ASA standards thus provided greater protection to exposed workers. The ACGIH won the struggle and became in the early 1960s virtually the only source of exposure limits that practitioners looked to for guidance.

In 1969, the federal US government adopted the 1968 TLVs as an official standard. Subsequently, the Occupational Safety and Health Administration (OSHA) and state authorities have developed exposure limits of their own. In most other countries, the TLVs have similarly been the starting-point for national standards on occupational chemical exposure. This applies for instance in Argentina, Australia, Austria, Belgium, Brazil, Canada, Chile, Denmark, Germany, Holland, India, Indonesia, Ireland, Israel, Japan, Malaysia, Mexico, the Philippines, Portugal, South Africa, Spain, Sweden, Switzerland, Thailand, the United Kingdom, Venezuela, the former Yugoslavia, and probably many other countries as well. In most of these countries, however, independent national exposure limits have gradually been developed after the initial adoption of the TLVs (Hansson, 1998).

8.5 Dose–response Relationships

Exposure limits are based on the estimated dose–response relationships for toxic effects, i.e., the relationship showing how the frequency of the toxic effect in a population is related to the dose or exposure. Dose–response relationships are graphically represented by *dose–response curves.*

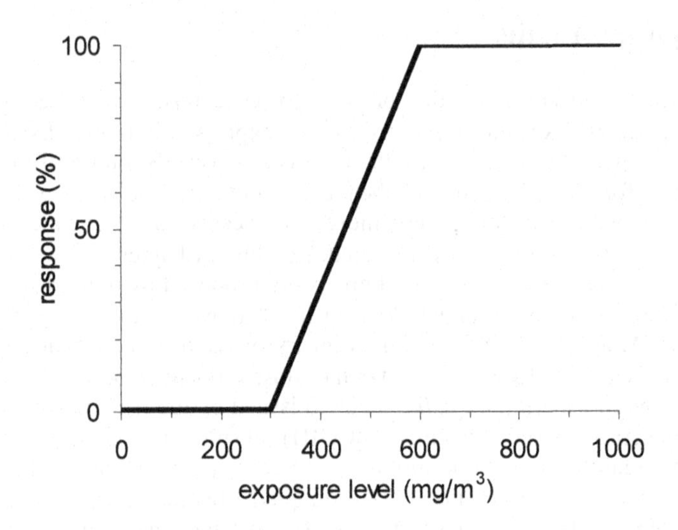

Figure 8.2. The dose–response curve of a hypothetical substance

Figures 8.2 and 8.3 are the dose–response curves for two (hypothetical) substances that both induce an all-or-nothing effect in humans. The comparatively steep curve in Figure 8.2 admits of a fairly accurate prediction of the effect in most individual cases. A person who inhales 300 mg/m^3 is sure not to be injured, whereas an exposure in excess of 600 mg/m^3 will almost certainly lead to injury. Only when the dose is between these two values is the outcome uncertain. Figure 8.3 has much less predictive power for individual cases. Over a wide range of doses, all that can be predicted for an individual exposure is a probability. Unfortunately, for many toxic effects of chemicals, the shape of the dose–response curve is closer to that of Figure 8.3 than to that of Figure 8.2.

The non-deterministic nature of a dose response relationship is well-known from the most important environmental source of cancer: tobacco. Smokers run a drastically increased risk of lung cancer (and many other diseases). This risk is higher the more a person smokes, but there is no way to know in advance whether or not a particular smoker will have cancer. Some very heavy smokers will not contract the disease, whereas others who smoke much less become victims.

If the highest dose level with zero frequency of a particular effect is above zero, then it is called a threshold. Figures 8.2 and 8.3 represent dose–response relationships with thresholds, whereas Figure 8.4 represents one without a threshold. The existence of a threshold has an obvious regulatory relevance. If there are thresholds for all toxic effects of a substance, and if these thresholds can all be determined, then a regulation can eliminate adverse effects without altogether prohibiting exposure to the substance. On the other hand, when there is no threshold for a toxic effect, then no exposure limit above zero offers complete protection against that effect. In other words, any such limit represents a compromise between health protection and other interests, such as economic and technological demands.

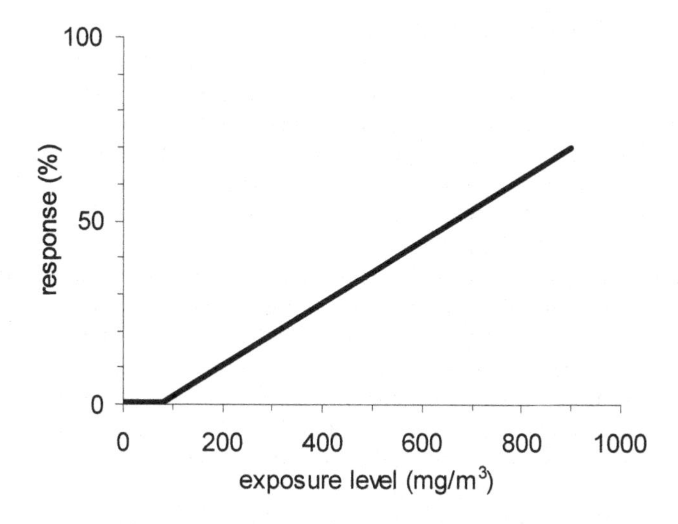

Figure 8.3. The dose–response curve of a hypothetical substance

There are biological reasons for believing that many toxic effects do have thresholds. Many biochemical processes are so structured that small perturbations give rise to compensating mechanisms that repair damage and restore normal physiological conditions. However, in other cases a relationship such as that shown in Figure 8.4 may exist. This is generally believed to be true of mutagenic substances and of those carcinogens that act through damage to the genetic code.

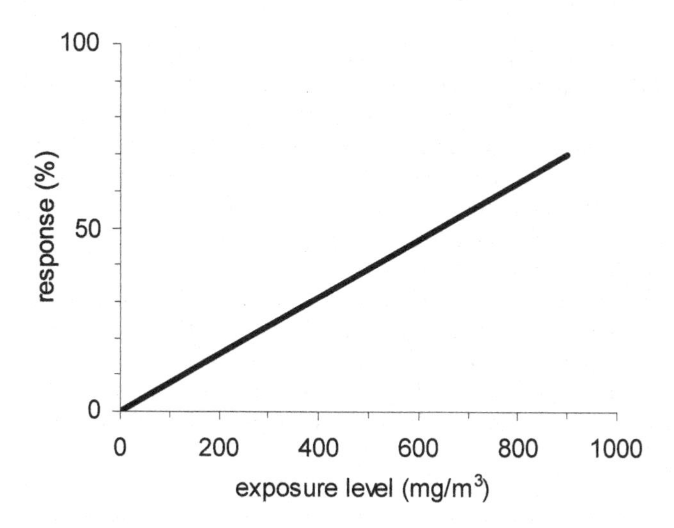

Figure 8.4. The dose–response curve of a hypothetical substance

8.6 Collective Dose Limits

Exposure limits for ionizing radiation differ in important ways from those for chemical exposure. Although radiation protection is regulated in national legislation, just like protection against chemical exposure, radiological exposure limits are in practice subject to international harmonization through the recommendations of the International Commission on Radiological Protection (ICRP). Compare with Figure 8.5.

Furthermore, whereas chemical exposure limits refer only to the exposure of an individual person, radiological protection employs a combination of individual and collective dose limits (as exposure limits are called in this context). The reason for this is that the risk of cancer due to radiological exposure is assumed to be proportional to the dose (as in Figure 8.4). This means, for instance, that exposing five persons to 1 mSv each will, presumably, lead to the same statistically expected number of cancer cases as exposing one person to 5 mSv.

In radiological work, although individual doses can often be reduced by distributing the work task among a larger number of persons, such a “dilution” of doses is associated with an increase in the expected number of radiation-induced cancer cases, because a larger number of workers are involved. In order to prevent this from happening, radiation protectors keep track of both individual and collective doses (Hansson, 2007b). The same argument is applicable to some chemical carcinogens (primarily mutagenic carcinogens), but in practice collective doses are seldom used in the regulation of chemical risks.

8.7 Remaining Uncertainties

Unfortunately, we do not in practice have the knowledge required to set non-zero exposure limits that offer complete protection against negative health effects. Most substances have not been subjected to extensive toxicological investigations, and even for those that have, uncertainties remain as to the nature of their health effects and about the dose–response relationships. Even after fairly thorough investigations, the possibility remains that a substance may have negative effects that have not yet been detected. In addition, surprisingly large effects can be undetectable for statistical reasons.

For a simple example of statistical undetectability, suppose that a certain exposure increases the frequency of a disease from 0 % to 3 %. Furthermore, suppose that a study is made of ten exposed individuals. Obviously, we cannot then be at all sure of seeing a case of the disease. (To be more precise, the probability that we will see a case is only 0.26.) Next, suppose instead that the frequency of the disease is 20 %, and that exposure leads to an increase of that frequency to 23 %. It is of course impossible to distinguish, in a sample consisting of ten persons, between a frequency of 20 % and one of 23 %. In other words, the increase caused by the exposure is statistically undetectable.

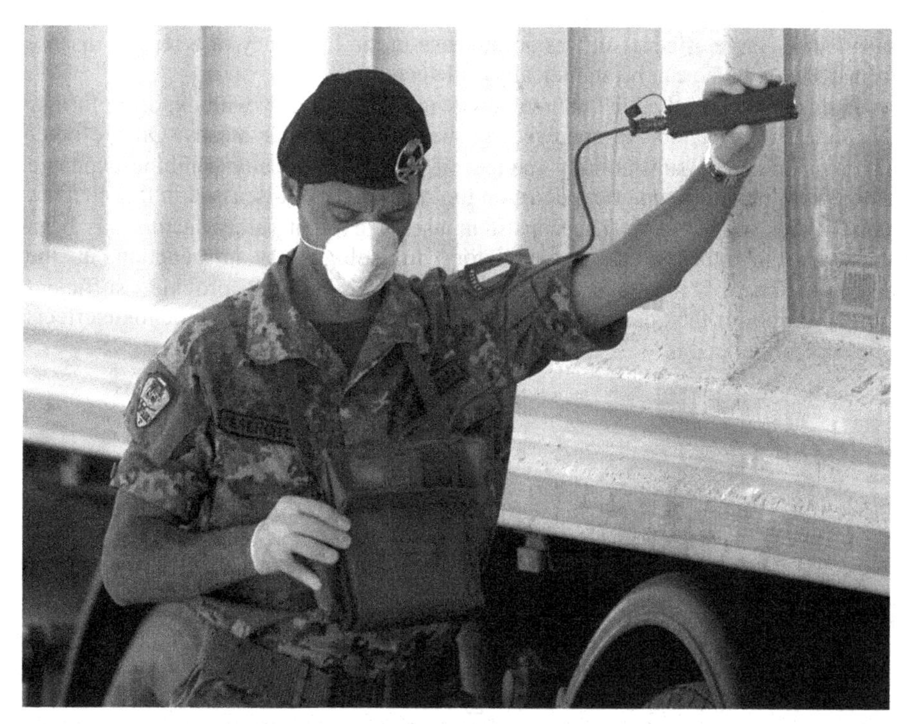

Figure 8.5. An Italian soldier checking radiation levels on a truck carrying refuse to a garbage dump in Savignano, near Naples June 16, 2008. Italian troops, who are controlling the entry to the dump, earlier blocked a truck containing traces of radioactive material thought to be medical waste. (Photo: Stefano Renna/Reuters/Scanpix)

The chance of detecting harmful effects increases when larger groups are used, but unfortunately the chance does not increase as much as one might have wished. Quite large disease frequencies may go undetected in all studies of feasible size. As a rough rule of thumb, epidemiological studies can only reliably detect excess relative risks that are about 10 % or greater. The more common diseases in a population tend to have frequencies up to about 10 %. Therefore, even in the most sensitive studies, lifetime risks smaller than 1 % cannot be observed. In other words, if an exposure increases the frequency of a disease such as myocardial infarction from 10 % to 11 %, we will not be able to discover this increase in any (epidemiological) study of the exposed population (Hansson, 1999).

In many cases, health effects that cannot be detected in studies on humans can be discovered in animal experiments. In this way, uncertainty about health risks can be reduced, but it cannot be eliminated. The problem of statistical undetectability is also present in animal experiments. In addition, the extrapolation from animal models to human health is associated with considerable uncertainty.

In summary, the scientific basis for regulatory toxicology is fraught with uncertainties: Although it can often be proved beyond reasonable doubt that a substance has a particular adverse effect, it can seldom be proved beyond reasonable doubt that it does *not* have a particular adverse effect, and in practice it is never possible to prove that it has no adverse effect at all. Furthermore, surprisingly large effects, such as an increase in the frequency of a serious disease from 10 % to 11 %, can be statistically undetectable.

Is this problem solvable? Is there a way to achieve safety with exposure limits? Absolute safety cannot be achieved by these or any other means. On the other hand, risks can be substantially reduced if we systematically combine exposure limits with the other numerical decision tool that we have discussed in this chapter, namely the safety factor. The consistent use of explicit safety factors has been successful in other areas of regulatory toxicology. Its introduction in the determination of occupational exposure limits could provide sufficient compensation for the unavoidable uncertainty in our knowledge of the toxic effects of chemical pollutants.

References

Clausen J, Hansson SO, Nilsson F (2006) Generalizing the safety factor approach. Reliability Engineering and System Safety 91:964–973

Cook WA (1985) History of ACGIH TLVs. Annals of the American Conference of Governmental Industrial Hygienists 12:3–9

Dourson ML, Stara JF (1983) Regulatory history and experimental support of uncertainty (safety) factors. Regulatory Toxicology and Pharmacology 3:24–238

Gaylor DW, Kodell RL (2002) A procedure for developing risk-based reference doses. Regulatory Toxicology and Pharmacology 35:137–141

Hansson SO (1996) Decision-making under great uncertainty. Philosophy of the Social Sciences 26:369–386

Hansson SO (1998) Setting the limit – occupational health standards and the limits of science. Oxford University Press, Oxford

Hansson SO (1999) The moral significance of indetectable effects. Risk 10:101–108

Hansson SO (2007a) Safe design. Techne 10:43–49

Hansson SO (2007b) Ethics and radiation protection. Journal of Radiological Protection 27:147–156

Hulst MVD, Meijman T, Rothengatter T (1999) Anticipation and the adaptive control of safety margins in driving. Ergonomics 42:336–345

Kawaguchi N (1995) New method of evaluating the surgical margin and safety margin for musculoskeletal sarcoma, analysed on the basis of 457 surgical cases. Journal of Cancer Research and Clinical Oncology 121:555–563

Moses F (1997) Problems and prospects of reliability-based optimisation. Engineering Structures 19:293–301

Randall FA (1976) The safety factor of structures in history. Professional Safety (January):12–28

Weil CS (1972) Statistics vs safety factors and scientific judgment in the evaluation of safety for man. Toxicology and Applied Pharmacology 21:454–463

9

How Dangerous Is It to Travel?

Torbjörn Thedéen (introduction and road traffic), Evert Andersson (rail traffic), Lena Mårtensson (air traffic), and Olle Rutgersson (sea traffic)

9.1 Introduction

In this chapter, we shall mainly consider risks to persons and only briefly consider risks to the environment (compare with Chapter 6). Risks connected with traffic by road, by rail, in the air, and at sea are considered.

The economic progress of society is accompanied by an increase in transportation, which has not been replaced by telecommunication. On the contrary, telecommunication has given rise to more travel. The travel mode distribution has changed from a dominant role of travel by sea, horse, and on foot to include car, rail, and air traffic. The transportation of individuals has increased dramatically in the Western world. In the European Union (EU), for example, the daily travel distance per person has increased by 20 % during the last decade and was, at the beginning of the twenty-first century, 36 kilometers per day (EU, 2007). The modal split of the traffic load in EU is given in Table 9.1.

The risk in the EU, measured as the number of fatalities per 100 000 persons per annum, is for air traffic 0.010, rail traffic 0.013, and road traffic 8.6. The number of fatalities on German roads per year is 5 300 and on US roads 42 600 (IRTAD, 2008). Figure 9.1 shows additional data for road fatalities. The regular international air traffic as well as railway and bus traffic is very safe compared with other transportation modes. Nevertheless, the relatively few air crashes cause many people to refrain from flying but not from the much more dangerous use of cars.

The environment is affected both locally and regionally by direct emissions of, e.g., nitrogen oxides, sulfur dioxide, and carbon monoxide. The greenhouse gas carbon dioxide may change the climate of the earth. The performance of different modes of transport with regard to their greenhouse gas emissions varies greatly both worldwide and within the EU (EEA, 2008).

These emissions come directly from the vehicles or indirectly from energy production for the transportation sector. Part of the traffic risks and the environmental effects will hit other persons than the car drivers and passengers. This is one of the reasons why society must control the risks, which can be done by decreasing the traffic load and by changing the distribution of the transportation modes. One can also, by rules and fees, control how the means of transportation are used. The locations of our homes and of our places of work and the size of domestic and international trade determine the transportation volume. Regional planning can be used to manage the transport sector. The choice of transportation mode depends on the alternatives available and on their costs. To some extent we take the risks into account in making that choice, but normally individuals do not regard the hazards as very important. Even though the risk for the individual is small, the risk load for society (the collective risk) is substantial. That risk, as well as the environmental one, particularly the climate threat from the use of fossil fuels, has to be handled by society as a whole. It can be managed by taxes and a better public transportation, which will in turn improve total traffic safety. Other means are information campaigns, and stricter rules and controls of both vehicles and drivers.

Table 9.1. Modes of passenger transportation in the European Union (EU, 2007)

Transportation mode	Share [%]
Car	73
Air	9
Bus and coach	8
Rail	6
Two-wheel vehicles	2
Tram and metro	1
Sea	1

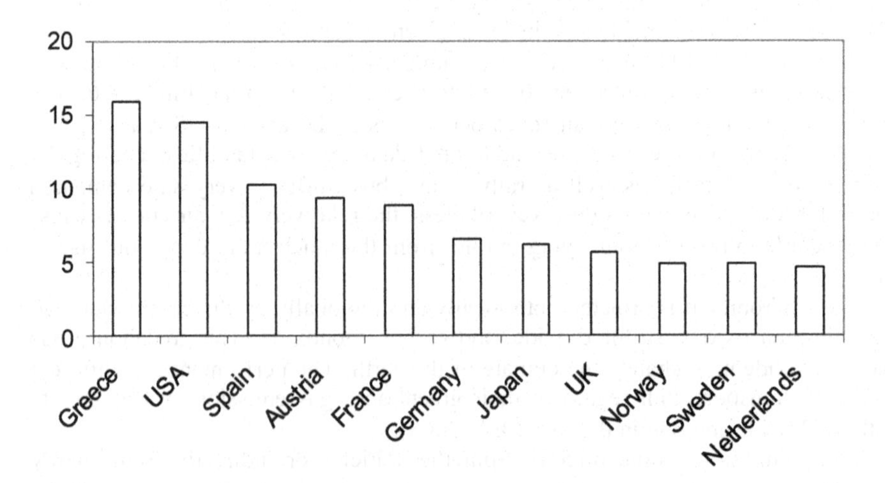

Figure 9.1. Number of road fatalities per 100 000 inhabitants in some countries in 2006 (IRTAD, 2008)

9.2 Risks Associated with the Road Traffic System

Car transportation accounts for the main traffic volume and, at the same time, the greatest traffic risks (Figure 9.2). Some studies indicate that it is the time from the start of the "car age" which explains the risk variation between different countries. It is, however, too simple just to consider the risk per kilometer. If one plans to travel between two places it is usually necessary to combine several transportation modes. Both rail and air transport demand traveling to and from the stations and airports, and this adds to the risks.

How are the risks distributed? Data from the US shows that it is 18 times more risky per billion person-kilometer for a motorcyclist than for a car driver (NCSA, 2006). Comparing the risk for different age groups per billion person-kilometer the risk is 3 times larger for the 16–24-year group than for the 40-year group (those with the smallest risk). People more than 84 years old are subject to a risk 9 times greater than that to which the 25–69-olds are exposed.

It is typical for road transportation and the accompanying risks that there is a lot of data available for the analysis of safety-related measures. The situation is different from that of other traffic modes as rail, sea, and air transportation. The road accidents also hit a large number of individuals. Many people have personal experience of incidents and minor accidents, which explains the large interest in road traffic and traffic safety.

Measures to improve traffic safety are directed to the road, to the driver, and to the vehicle. The effect of some measures to improve safety can be assessed by empirical data and sometimes by regional experiments. For ethical and economical reasons, it is difficult to evaluate the effects of possible measures using available data, but theoretical traffic models for traffic on real or virtual roads can be used. Previously this has been done for simple situations, see, e.g., Haight (1960), and Thedéen (1976). In more realistic situations, such models are analyzed by simulation models (see Chapter 11).

The roads can be safer – more lanes, separation of traffic of different kinds, better curve planning, better maintenance and snow clearance. Many of these measures are expensive, and it is then necessary to weigh the effects against the cost. The number of accidents is large and the relation between the road parameters is well known, and this motivates the use of cost benefit analysis. When considering a certain road improvement, one has to take into account all the benefits and all the costs including saved lives. A saved human life is (in Sweden) currently valued at about USD 3 million (compare with Chapter 16).

The vehicles have become much safer with time. The brakes are better and more reliable, and (in some countries) headlights are used also in daylight. The consequences of an accident are less severe due to car belts and air bags. The compulsory control of the vehicles and drivers also improves the safety.

The relation between alcohol and road accidents has been proven by many studies. The control of inebriated driving using blood or breath tests has been shown to decrease the number of accidents. Speed limits and general police control also improve road safety. Other measures try to homogenize the traffic with respect to kinetic energy and speed in order to decrease the number of dangerous situations in traffic. Using separate lanes for bicycles and cars, imposing both upper and lower speed limits, and limiting the number of small dangerous cars all improve road safety.

Figure 9.2. Aerial view of road accident near Junction 19A of the M1 motorway on July 15, 2008. The accident, which resulted in one dead and several injured people, happened close to where the M1 meets the M6 at Catthorpe interchange near Rugby, United Kingdom. (Photo: Steve Woods/EPA/Scanpix)

In some countries, many drivers use studded tires that decrease the stop time on icy roads, but a mixture of cars with and without studs causes the traffic to be more heterogeneous and this in turn will increase the collision risks.

Lower speeds decrease both the number of accidents and the resulting damage. In many countries lower speed limits are being discussed, but one has to weigh in the effect of longer travel times.

The electronic revolution will soon influence both road accessibility and road safety. The driver will in real time receive information about the traffic situation on main roads, e.g., about collisions, slippery roads, and congestion. The driver can even be advised about the best approach to his destination. The speed limits can be adjusted to the actual traffic situation and the weather. It is also possible to check whether the driver is sober using breath tests, which, if they indicate too high a value, will prevent the vehicle from starting. The actual condition of the vehicle, e.g., the status of the brakes, can be controlled by a computer. In the future, cars will be equipped with radar that will indicate when they are too close to fixed or moving hindrances. It should be most efficient if all cars had the same standard. During a transition period of maybe 20 years most cars will still be of a conventional type. The mixture can increase the risks. Traffic at sea and by rail have similar problem with a lack of homogeneity.

Besides these future measures, a better design of the inner car milieu will give less severe injuries. The same is true of the environment adjacent to the roads, such as hedges, trees, and barriers. Traffic, like most technical systems, is highly dependent on the behavior of the individual. The automation of road traffic lies far in the future. In the mean time, traffic safety can be improved by training, control, and information.

9.3 Risks Associated with Rail Travel

9.3.1 How Safe Is Rail Travel?

The risks related to traveling by rail should be judged in relation to the risks associated with traveling in general, i.e., compared with other modes of transport (car, bus and air). Risks should also be judged in relation to other accepted risks in our society.

Firstly it should be stated that the risks associated with public transport are in general very low compared with risks related to traveling in a private car. Secondly, it can be stated that the risks related to rail travel are not generally lower than those associated with other common modes of public transport. Table 9.2 shows the risk of being killed when traveling by different modes of transport.

Table 9.2. Number of travelers (passengers or drivers) killed per billion passenger-km. Risks for train crew not included

Transportation mode	Fatalities [per 10^9 passenger-km]
Car (EU-15, 2004)	4.1
Bus (EU-15, 2004)	0.35
Air (EU-25, 2000-04)	0.16
Rail (EU-15, 2000-04)	0.31

Source: EU (2007). EU-15 refers to the 15 first EU countries and EU-25 to 25 EU countries

If the risk figures for rail travel are reported as the risk of a fatal accident per hour, i.e., the risk of being killed per hour of activity, rail travel is still very safe. It is one of the safest activities in which a human being can participate; on average it is at least as safe as staying at home.

However, there are categories of people other than passengers who are affected with respect to safety by railways and trains. One such category is the railway staff, including the train crew. Another category is motorists who collide with trains at level crossings. This class of accident is not, however, related solely to rail passenger services but also involves rail freight service and road traffic.

9.3.2 What Are the Possible Causes of Accidents?

The railway system is complex with a large number of failure modes related to technical systems (track, signal, and rolling stock) as well as to human factors.

Collisions at level crossings between road and railway are one of the major causes of accidents and fatalities. In most cases only the road traveler suffers, but it has happened that a train has been derailed with serious consequences. In many railway networks there is an ongoing process to reduce the number of level crossings. In high-speed rail operations (say above 200 km/h) level crossings are completely eliminated.

A train passing a stop signal has been another cause of accidents, particularly on railway systems where automatic train protection (ATP) is not in place. (ATP automatically brakes the train if the driver does not.) In the old days, manual train dispatch also sometimes caused dangerous movements of trains in relation to each other.

Faulty tracks, such as broken rails, switch damage, thermal buckling, undermined embankments due to flooding, etc, are another source of unsafe train operation. Strong cross winds have in fact also caused a train to blow off the track.

Faulty train carriages, particularly broken axles, wheels, etc, are possible causes of derailment with subsequent overturning of the train. Faulty brakes are another cause of train accidents, if the train is not able to stop before it hits another train beyond a stop signal.

Fire on a train is a potentially very dangerous situation. The risk is particularly high on diesel-motored trains with fuel tanks, in the event of a collision with a road vehicle or another train. The risk is also high if fire breaks out in a tunnel.

The train itself, as well as the operational conditions, will usually influence the outcome of an unsafe situation. The risk of serious consequences of collisions, derailments and train turnover is more serious if the speed is high and the if carriage body structure (surrounding passengers and crew) is weak. Some train vehicle designs have the ability to hold the train almost centrally aligned and upright above the track after a derailment; a property called derailment worthiness.

9.3.3 Safety Has improved

The safety records for most transport modes have improved for a long time. This is particularly true of rail travel. There are several reasons behind this improvement.

- The introduction of a modern safety and traffic control system, relying on machinery and thus avoiding human error in manual train dispatching.

- The introduction of ATP.
- The taking out of service passenger trains with weak car bodies.
- The establishment of independent railway inspectorates requiring more strict safety rules and procedures from infrastructure managers, railway operators, and the suppliers of equipment.

An important issue is whether high-speed rail transport (from 200 km/h and above) can be considered safe. According to statistics so far (2007), the answer is that high-speed rail is very safe. From the mid-1960's onwards, speeds from 200 km/h and above have been introduced in a large number of countries, starting with Japan, followed by France, Germany, Spain, Korea, Taiwan, Italy, USA, Great Britain, Belgium, Sweden, Norway, Finland etc. The total number of passenger-km carried in high-speed rail operations is estimated to at least 3 000 billion, i.e., 3×10^{12} (author's estimate).

Figure 9.3. The train crash in Eschede, in northern Germany, on June 3, 1998, resulted in 101 dead and circa 90 injured. Twelve of the 14 carriages left the tracks at high speed and crashed into the support pillar of the overpass which then collapsed. (Photo: Ingo Wagner/Pressens bild/Scanpix)

During the history of high-speed rail travel, few major accidents have occurred. In Eschede, Germany in 1998, 101 persons were killed (almost all were passengers). See Figure 9.3. However, in relation to the number of passengers carried by high-speed rail, the statistical safety level is very high. In this context it should be pointed out that a substantial part of the high-speed rail operations has been on dedicated high-speed railway tracks, where there are no freight trains or level crossings.

Based on what has actually happened, the number of persons killed in high-speed rail operations onboard a train is of the order of 0.03–0.04 per billion passenger-km.

9.4 Risks Associated with Aviation

9.4.1 Aircraft Accidents

Within international aviation, safety has increased from 60 accidents with fatalities per million departures during the 1960s to 1 accident per million departures during the 1980s, according to statistics from Boeing. The decrease from the 1960s is associated with the introduction of jet aircraft. It also reflects the improvements due to equipment of a higher quality, advanced simulation equipment and increased experience of civil aviation. However, the level of one fatal accident per million departures seems to represent some kind of barrier. An increase in traffic volume may lead to more accidents.

The aim of accident investigation is to analyze aircraft accidents in order to avoid them in the future. It is well known that 70–80 % of aircraft accidents occur due to human error. The remainder is due to technical malfunctions of the aircraft (Wiegemann et al., 2000). Technical malfunctions may be due to problems with the engines, brakes, landing gear, fire, etc. Accidents may also be due to collisions, e.g., with other vehicles on the ground, or to a diversion from the runway, which may, however, also be due to human error.

"Controlled flight into terrain," CFIT, is said to be caused by the pilot if there is no technical malfunction. Deviation from procedures, lack of supervision, and misjudgment of distances, speed, and altitude belong to this category. Great efforts are made within aviation to reduce the number of CFIT accidents which have unfortunately increased over the past few years. Statistics for the period 1993–2002, where the first six months of each year have been analyzed, say that out of 18 accidents in 2002, nine were CFIT accidents (Flight International, 2002). During this 10-year period, there were 20 accidents during the first six months of each year. In 1994 there were 741 fatalities and in 2002 there were 716 fatalities. During the intermediate years there were fewer fatalities, from 114 to 609. Since 2002, around 800 fatalities have occurred per year worldwide. This should be compared to the much higher figures for car accidents. However, the development of the non-industrialized part of the world is worrying when it comes to aviation safety.

9.4.2 Surprised by Automation at Gottröra

Research has shown the importance of automation in the work of the pilot. The concept of "automation-induced surprises" has been coined by Wiener (1989) and further elaborated by Sarter and Woods (1992) and Wickens (1994).

"Automation may sometimes be perceived to fail, when the complexity of its algorithms causes it to do things that a pilot does not expect … because the pilot did not possess an integrated awareness of all the preconditions that might have triggered the particular form of aircraft behavior. These forms of automation-induced surprises, experienced by a substantial proportion of pilots can, when they occur, create a potential for the pilot to intervene with a non-optimal and unsafe response" (Wickens, 1994).

New technology that creates bottlenecks during busy, high-tempo, high-criticality event-driven operations – but gives benefits during routine, low workload situations – has been called clumsy automation by Wiener (1989), referring to a study of pilots flying B-757s. After one year, 55 % were still surprised by the automatic system, and 20 % did not understand the automation system modes. Their four most common questions were: "What is it doing now?", "Why did it do that?", "How am I going to stop it from doing that?", and "What will it do next?" The discrepancy between the intentions of the designer and the possibility for the pilot to understand the system is a great issue for research, not only in aviation. The opacity of the computer system is considerable, and it is of course a critical issue to be dealt with in safety-critical systems like aircraft and nuclear power plants. The pilots at the Gottröra accident were surprised by automation.

The weather conditions on December 27, 1991 for flight SK 751 bound for Copenhagen were those of a snowy, windy winter morning with a temperature of approximately 0°C. The captain found snow and patches of ice on the wings and also a 3 mm layer of frost under the wings. The captain therefore asked the mechanic to order an extra de-icing procedure.

The takeoff was normal and the instruments indicated normal takeoff power. However, after 25 seconds, bangs, vibrations, and jerks were felt in the aircraft and the first surge of the right engine was recorded by the flight computer. The altitude was then 1 000 ft. After 64 seconds, the first surge of the left engine was recorded and, after 77 seconds, both engines had permanently lost their power.

An off-duty captain, who was a passenger on board, had been alerted by the sounds from the engines. He realized the crew was having problems and entered the cockpit to assist. The co-pilot made contact with Stockholm Control: "We have problems with our engines, please, we need to go back to Arlanda." The aircraft came out of the clouds after three minutes and 45 seconds and the pilots found a field where they could land, as they could not make it back to the airport. The co-pilot then announced "and Stockholm, SK 751, we are crashing to the ground now!"

The aircraft hit some trees, causing the right wing to be torn off and the aircraft to roll over along its longitudinal axis. The aircraft was broken into three pieces in the landing (Figure 9.4). All 129 people on board survived.

Figure 9.4. The Gottröra accident 1991. View of the crash site near Gottröra, north-east of Stockholm/Arlanda Airport. The fuselage of the Scandinavian Airlines System plane Dana Viking of the type DC-9 -81 was broken into three pieces after an emergency landing, due to failure to remove clear ice on the wings prior to take-off. All 129 on board survived. (Photo: Lennart Isaksson/DN/Scanpix)

According to the Board of Accident Investigation (1993), the obvious reason for the crash was clear ice, broken off from the wings, causing damage to the engine fan stages, which led to engine surges which destroyed the engines. Their report said:

"The accident occurred because the instructions and routines of the airline company were inadequate to ensure that clear ice was removed from the wings of the aircraft prior to takeoff ... Contributory causes were that the pilots were not trained to identify and eliminate engine surging. ATR (Automatic Thrust Restoration), which was unknown within the company, was activated and increased the engine power without the pilot's knowledge."

The function of ATR was not known within the company, and thus came to the pilots as an "automation-induced surprise" (Wickens, 1994). Neither the manufacturer of the engines, nor the designer of the software, nor the US Federal Aviation Administration, FAA, had foreseen the chain of events that led to the total destruction of the engines. The situation was aggravated by the fact that none of the instruments, all set in alarm mode, gave the pilots the support they needed in the emergency situation. Warnings were given, not only by the flashing lights and different warning sounds, but also by the synthesized voice of the autopilot which the pilots experienced as cacophony. So, while pilots can cope with poor design under normal operating conditions, the situation may deteriorate in an emergency, when single bits of information can come to play a crucial role (Mårtensson, 1995; 1999).

9.4.3 Überlingen Midair Collision

A midair collision over Überlingen in Germany could also be referred to as an automation issue. At 21.35, July 1, 2002 an air traffic controller at Zürich ACC had two aircraft on collision course on his radar screen, a TU154M with Russian school children on board and a B757-200 freight aircraft.

Since the traffic volume at the time was small, the air traffic controller was operating two work stations simultaneously and thus had to constantly shift position between them. At the other work station, there was a communication problem with an approaching A320 that demanded attention. Coming back, the controller identified the conflict at flight level (FL) 360 (approximately 36 000 feet) and issued a clearance for FL 350 to the TU154M. The crew started a descent but did not read back the clearance. When the crew did not answer, the controller issued the clearance again more emphatically. At that point, the airborne Traffic Collision Avoidance System, TCAS, issued a Resolution Advisory that the TU154M should climb and that the B757-200 should descend. The crew in the B757-200 had no conflicting information, so they started the descent. The TU154M on the other hand had already started a descent and was now instructed to initiate a climb. As TCAS was not mandatory equipment in the Russian federation and the company had scarce resources for simulator training, the crew of the TU154M had little experience of TCAS. The radio communication occupied the frequency and, when the B757-200 reported that they were following a TCAS-descent, the controller had again turned his attention to the approaching A320 on the other work station. Both aircraft continued descending and eventually collided, and 71 people died (BFU, 2004).

This last example shows the importance of coherence and clarity of information where several operators with overlapping responsibilities are involved. Controllers are primarily responsible for the separation of aircraft, but their responsibility may be revoked by automation (TCAS). As the pilots of one aircraft had two contradicting instructions, decision-making was hampered. New technology may increase information exchange in the aviation system, and this puts even more focus on responsibility issues (Barchéus and Mårtensson, 2006).

9.4.4 What Is Being Done to Improve Flight Safety?

There are several international organizations for the supervision of safety in the air. The airline companies cooperate within IATA, the International Air Transport Association. The international trade union of pilots is the International Federation of Airline Pilots' Associations, IFALPA. The International Civil Aviation Organisation, ICAO, is an organization within the United Nations for civil aviation, with more than 180 member countries, and the Joint Aviation Authority, JAA, in Europe is equivalent to the American Federal Aviation Administration, FAA. These organizations are all working on regulations in aviation including the issue of safety, a demanding task considering the challenge to increase the safety in the third world as well as to cope with the increase in flying in the industrialized world.

One of the main issues in aviation safety is the training of the pilots. After the accident in Tenerife, described in Chapter 15 on the human aspects of risks, many airlines realized the importance of using all the resources of the crew. "Crew resource management" is a mandatory course for pilots and other staff in many airline companies.

Technical research to improve aviation safety is carried out on a continuous basis at universities and research institutes. After the Gottröra accident, research to improve warning systems has been carried out (Mårtensson and Singer, 1998; Ulfvengren, 2003). Automation issues are a big research area and today there is technical equipment to prevent the accident at Überlingen from being repeated. The air traffic management, ATM, will see big changes in the future. Technology changes for communication, navigation and surveillance, CNS/ATM, are being developed (Galotti, 1998). Cockpit Displays of Traffic Information, CDTI, for example, makes it possible for the pilot to see the surrounding aircraft and increases the possibilities for a closer cooperation with the air traffic controller (Barchéus and Mårtensson, 2006).

In 1996, the American Federation of Aviation Administration, FAA, established a Human Factors Team to study the human effects of automation, (FAA, 1996). Having made analyses of several accidents, the team concluded that the automation issue is not an individual one in terms of human or technical errors, but should be dealt with on a systems level.

The team has given the authorities 51 recommendations to improve aviation safety. These concern design, training, organization, human – machine systems and cultural differences, where it is not only language barriers that constitute problems in communication between cockpit and air traffic control. Communication between other stakeholders such as manufacturers, authorities, researchers, airline companies and pilot corporations also needs to be improved. Some of these recommendations have given rise to new regulations considering human issues as well as technical matters.

In the implementation of new rules, regulations and technology, the end-user should play an important role. The experiences of airline pilots transporting passengers all over the world, in many kinds of weather, communicating with air traffic controllers speaking all different kinds of English are invaluable. The experiences have been gained not only under normal operation but also during malfunctions and emergencies, experiences which must be communicated to the aircraft designers. Mistakes made at the drawing board or at the computer are difficult to amend considering the long life cycle of an aircraft, more than thirty years. "Keep the pilot in the loop" is a well-known expression when considering the user in the technology development.

Furthermore, technical and organizational issues must be integrated together for greater aviation safety. A research project within the 6th Framework Programme of the European Union deserves to be mentioned. Human Integration into the Life Cycle of Aviation System, HILAS, has set out to use human factor knowledge in all stages of the aviation system. Forty partners are involved, among them eight airline companies, manufacturers, electronics companies, universities and research institutes, all working together for four years to design and implement the best practice of flight operation, flight deck technology and maintenance. In the case of flight operations, the objective is to develop and implement a new methodology for monitoring and evaluating overall system performance to support flight safety, operational risk management, and process improvement (HILAS, 2005).

There are worldwide activities going on to improve flight safety. The terror attack of September 11, 2001, has, however, put flying into a completely new dimension. The event in July 2007 with a car bomb at the Glasgow airport shows that terrorist attacks can occur anywhere at any time. Security control of the passengers before they enter the aircraft has increased enormously, the cockpit door is locked, which makes communication between pilots and cabin attendants more difficult, and many other security issues have become part of every-day life in aviation. The security is thereby just as important as safety on board.

There are many ways of considering the risk of flying. One of the more unusual is to compare flying with living. An English researcher has come to the conclusion that "to fly as a passenger is not more risky than to be 55 years of age" (O'Hare and Roscoe, 1990).

9.5 Risks Associated with the Sea

9.5.1 From Daredevil Enterprise to Safe Cruising Traffic

Traveling by boat across large open lakes, bays, or seas has been regarded as a risky business since time immemorial. The trading journeys of the Vikings in the East and in the West were dangerous, and this was not only on account of the warlike atmosphere which sometimes prevailed. The "long ships" developed in Scandinavia were probably the most seaworthy ships of their time, and they were handled by acknowledged skillful sailors, but the risks associated with travel in open boats on open waters were great, and unexpected storms and tempests claimed many victims. Those sailing on board the East Indiamen to the Orient and China in the seventeenth century to trade in textiles, spices, and porcelain knew that there was considerable risk that the enterprise would fail because these sailing vessels were sensitive to rough weather.

For both the Vikings and those sailing to the orient, it was the possibility of carrying on successful business and becoming rich which drove these people out onto obviously risky enterprises. For the emigrants, in the early nineteenth century, the driving force was the possibility of a new and more tolerable life on the other side of the Atlantic, and they traveled in wooden sailing ships which were not really intended for passenger traffic. The ships were small in relation to the waves which they met, their rigging was easily damaged if struck by a heavy sea, and division of the hulls into water-tight compartments was not even contemplated. Nevertheless, these ships laid the foundation for today's large passenger vessels.

With the arrival of the steam engine and the rapid developments in the nineteenth century of inventions such as the blade wheel, the screw propeller, the steam turbine and ships of steel, the possibility of building larger and larger vessels increased. In the twentieth century, the diesel engine and all-welded steel hulls then laid the foundation for large and strong ships which could ride out storms and face higher and higher waves, and which were to make comfortable and relatively safe journeys possible across the oceans of the world. Today, transport and travel on the open sea are no longer risky enterprises but are a part of the daily movements of people and goods which are the foundation of our modern welfare.

9.5.2 Accident Statistics

The statistics of shipping accidents are best discussed from an international viewpoint, since shipping usually has an international character. Lloyd's Register of Shipping (LR) in London, which is the oldest and most traditional of the classification societies, has taken upon itself the responsibility of keeping and publishing statistics of shipping accidents all over the world. Table 9.3 shows the number of fatalities worldwide during 1998–2007 in connection with completely wrecked ships over 100 gross register tons.

Table 9.3. Number of fatalities in the world for vessels totally lost (vessels of 100 GT and above)

Vessel type	Number of fatalities										Total
	98	99	00	01	02	03	04	05	06	07	
Crude				15					2		17
Products	7	6	14	30	2	7	9		3		78
Chemical/Product		2					40	6			48
Pure Chemical		3	5		1			1		14	24
LPG	1	5			2		5				13
Other tanker					3	2				7	12
Bulker	111	3	65	68	5		28	7	37	41	365
General cargo	107	257	106	117	82	130	185	170	148	194	1496
Other dry		5	18				1	6	36		66
Container		9									9
RoRo (Roll-on/Roll-off	2			43	2	7			4	11	69
Ferry	241	74	90		1088	33	289	196	1461	24	3496
Cruise	31	8		19	1					2	61
Offshore	3	3	3				5			17	31
Service	11	14	3		9	19		3	6	9	74
Fishing	44	65	71	40	47	10	44	35	70	8	434
Miscellaneous									4		4
Total	*558*	*454*	*375*	*332*	*1242*	*208*	*606*	*424*	*1771*	*327*	*6297*

Source: Lloyd's Register – Fairplay

The mean total over this 10-year period was thus 630 victims per year in the world as a whole. Combined passenger and RoRo cargo-ships (which are often called car ferries or RoRo-ferries) are, as can be seen, over-represented in these statistics and the deaths on these ships alone represent almost half of all the victims within this shipping sector. This type of ship is especially common in Northern Europe. Pure passenger ships accounted for only six deaths per year during the same 10-year period. The combining of passenger traffic and RoRo-transport has evidently considerably increased the risk to the passengers. Another problem is the increasing accident frequency involving elderly ships. Statistics from the late 1900s show that the total losses of ships world-wide increases markedly from ships which are 10–14 years old to ships which are 20–24 years old. The ships in the 20–24-year age group are relatively few, amounting to circa 10 % of the whole shipping fleet, but they nevertheless account for more than 25 % of all the ships lost.

The explanation of this development is to be found in factors such as ship maintenance, the commercial and technical aging of the ships and changes in ownership. Since the requirements of the classification societies regarding inspections increase when a ship is 20 years old, many serious ship-owners sell their ships before they reach this age. Unfortunately, these old ships often end up in traffic in countries with less serious ship-owners, where there are lower standards with regard to operation and maintenance.

9.5.3 Authorities and International Organizations

Shipping is an international activity, and this fact greatly influences the organizations and the systems of regulations which have been developed. Today's organizations originate from the eighteenth century, when international trade required a well-developed insurance system in order to spread the risks associated with trade by sea. To support the insurance companies, classification societies were formed to stand as independent guarantors of the strength, seaworthiness, and maintenance of the ships which were granted a classification certificate. These classification societies have been extended into international organizations which also today represent the ship-owner's knowledge base with respect to the strength and maintenance of the hull. In addition, they use their knowledge base as the starting-point for investigations on a consultant basis. The largest classification societies also have the confidence of the authorities and carry out inspections on the authorities' behalf. Thus, they have acquired the unique role of representing at the same time both the authorities and the competitively exposed consultant companies. These double roles demand a great measure of integrity on the part of the classification societies, and they have had to tolerate the fact that their role in shipping development has sometimes been questioned.

With their systems of rules and regulations, the national maritime authorities act as a flag state for those ships that are registered under the national flag. International shipping has an international competitive situation that has meant that many ship-owners register their ships in those flag states which offer the most favorable financial terms.

The International Maritime Organization (IMO) is a permanent organization under the UN which is concerned with maritime questions, and which has the task of developing international agreements in the form of conventions and resolutions regarding safety at sea and environmental protection. The IMO lacks sanctioning powers, but those member countries which ratify its conventions undertake to incorporate them into their national regulations. Since an imbalance between different national regulations leads to a distortion of the international competitive situation, most national authorities strive to follow and to influence the development of the regulations by the IMO. This often leads to an unacceptable inertia in the adaptation of the marine safety regulations to meet the needs associated with the development of new types of ships and other problems which require change. For example, the changes in the regulations relating to leak stability in RoRo ships after the Estonia accident have had to be implemented as a local agreement. The so-called Stockholm Rule applies for the time-being only in the North European countries where this problem was felt to be especially acute. In other countries which are members of IMO, the acceptance of the regulations has been delayed since it is considered that they make ferry traffic unnecessarily expensive. Databases for the analysis of accidents and incidents within shipping are created by shipping companies, maritime administrations, and the IMO.

9.5.4 Regulations for the Safety of Ships

SOLAS (Safety Of Life At Sea) is an international set of regulations which is the result of an effort to create uniform rules for the design particularly of passenger ships. The work, which began in the early part of the twentieth century, was speeded up by the sinking of the RMS Titanic in 1912, and was available in its first version in 1914. The currently valid SOLAS 1974 can be regarded as the foundation for the establishment of safety on board and is the result of the IMO's work. SOLAS 1974 includes regulations for everything from stability and the division of the hull into compartments by water-tight bulkheads to the design of life-boats and life-saving equipment. To safeguard that there is sufficient freeboard, all ships must also follow the International Convention on Load Lines which is under the responsibility of the IMO, the so-called Plimsoll line, amidships on the hull of the ship, which shows how deeply the ship can be loaded depending on the area to be trafficked and on the season of the year.

A completely new approach to the question of safety within the shipping sector has been achieved with the establishment of the International Safety Management Code (the ISM Code). This code means that one now not only adopts a holistic approach to the ship and its owner but also considers the total interaction between human beings, technology, and organization. The code had applied to European passenger traffic since 1996, and it emphasizes the commitment and responsibility of the shipping management and the country organization for the establishment of safety on board the ships. No ship-owner can assert any longer that he was not informed and is not therefore responsible for the safety on board. The ISM Code has led to an intensified commitment on the part of the shipping companies to develop routines and plans of action which distribute responsibility in the safety organization on land and on board the ships.

9.5.5 The Estonia Accident and Its Consequences for the Future Safety at Sea

On 28 September 1994, 856 persons died when MS Estonia sank in heavy seas on route from Tallinn to Stockholm (Figure 9.5). The ship capsized within a few minutes and sank in less than half an hour. After three years' work, the international commission of inquiry established that the immediate cause of the accident was that large quantities of water penetrated onto the open RoRo deck through the damaged bow door in the stem. On this deck, which on this type of ship completely lacked both longitudinal and transverse bulkheads, the penetrating water gathered along one side and caused the instability which resulted in a rapid capsizing. Most of the passengers who were in their cabins (the accident took place after midnight) were caught in the interior of the ship when it capsized and had no possibility of getting out.

Figure 9.5. MS Estonia with open bow visor in the harbour of Tallin, Estonia 1994. (Photo: Li Samuelson/Scanpix)

The Estonia accident will probably be as important for the future development of safety measures within the shipping sector as was the sinking of the Titanic. In 1912, there was as yet no joint set of regulations for how a safe ship should be designed. It was therefore natural to invest everything in the development of regulations for how ships should be designed and built in order to be safe, and modern ships are to a high degree based on the result of this work in the form of SOLAS 1974. The fact that the Estonia sank in spite of this set of regulations underlines the earlier realization that regulations as to how ships shall be built from a purely physical standpoint do not solve all the safety problems. The Estonia accident has thus functioned as a strong reminder to the shipping industry, which now feels the need to show its ability to act and to develop a more modern line of approach to the problem of ensuring safety. After the accidents with Herald of Free Enterprise (1987), Scandinavian Star (1990), and Estonia (1994), the work of achieving greater safety has the following main directions:

- The regulations relating to the leakage stability of a ship are being reviewed. The effect of water which can enter onto the RoRo deck, as the result of a collision or in some other way, can be limited by dividing the deck according to the "Stockholm Rule" into water-tight departments with the help of bulkheads and doors.
- The insight into the importance of the organization for safety on board has resulted in the so-called ISM Code, which forces the ship-owners to establish a safety organization and to actively participate in the task of safety management. For the shipping company to receive permission to carry on its activity, the organization must also submit to safety audits at regular intervals.
- Risk-based methods for the analysis of ship systems and designs are being introduced as a supplement to the traditional sets of regulations.
- The human factor has been identified as a risk element and projects for training and research within this sphere have been initiated.
- IT-based systems to support the navigation and operation of shipping vessels are being introduced.

Henceforth, the starting-point for the maritime safety work will probably be to consider how the organization and engineering can be adapted to match human abilities in contrast to the previous attitude that the human being should be required to adapt to the technology. See also the report from the Joint Accident Investigation Commission of Estonia, Finland and Sweden (Joint Commission, 1997).

9.5.6 A Few Reflections on Survival Equipment and Attitudes

Survival equipment such as lifejackets, lifeboats, and the like must be regarded as the last resort of the safety system on board. The management of ship safety deals in the first place with how accidents can be avoided, in the second place with how the effect of an accident which has occurred shall be limited, and only in the third place with how a damaged or burning ship shall be evacuated or abandoned.

The survival equipment shall be used only when all other measures have failed, yet discussions regarding ship safety are often interpreted as being concerned with the function of the survival equipment such as life jackets, survival dresses, and lifeboats.

In the case of an accident in rough weather, in cold waters and with such a rapid development as that with Estonia, it is clear that the currently available survival equipment functions very poorly. In calm weather, in warm waters and in a ship which has not suffered a list, modern systems of lifeboats or rubber rafts function acceptably well. Under other conditions, which are not unusual in northern waters, there is currently no well functioning survival equipment available for passenger vessels.

On merchant vessels, where there are only a small number of persons to be saved, so-called free-fall lifeboats are being used to an increasing extent. In this type of lifeboat, usually fixed in a rack at the stern of the ship, the whole crew embarks and they strap themselves into special chairs similar to those in a spacecraft. Thereafter, the lifeboat is shot loose and falls freely, stem first, 20–30 m down into the water. This type of launching has been shown to function in practically all types of weather and even when the ship has developed a relatively severe list to one side. To adapt this type of lifeboat for use with 2 000 persons on a passenger ship has, however, so far been considered impossible.

Other solutions which are being discussed for passenger ships are water-tight compartments high up in the superstructure which can prevent capsizing and which mean that the ship is able to stay afloat until evacuation can be carried out directly to a rescuing ship, but the problem of evacuating a passenger ship remains for the time being as one of the great maritime engineering challenges.

Traveling on a passenger ferry has developed into a recreational activity which we associate with restaurant visits and shopping and where the actual traveling is often of minor importance. In this pleasure-oriented type of shipping, it has not been considered opportune to inform about evacuation routes and survival equipment. Neither the passengers nor the ship operators seem to feel a need for this, in spite of the Estonia accident. In an aircraft, however, we accept the demonstration of life jackets at each take-off, even though their value in a rescue from a crashing aircraft is probably more symbolic than in a rescue from a wrecked ship. From the viewpoint of the shipping operators, the attitude of the passengers can, however, be interpreted as positive. To travel by ship is a pleasure to which one devotes oneself without any great worry, which is after all indicative of confidence.

References

Barchéus F, Mårtensson L (2006) Air traffic management and future technology—the views of the controllers. Human Factors and Aerospace Safety 6(1):1–16

Board of Accident Investigation (1993). Air traffic accident on 27th of December 1991 at Gottröra, AB county, Case L-124/91

Bundestelle für Flugunfalluntersuchung (BFU) (2004) Report of the accident (near) Ueberlingen/Lake of Constance/Germany 1 July 2002, Investigation Report no AX001-1-2/02, Braunschweig

EU (2007) Energy and transport in figures 2007. Part 3: transport. European Union (EU), Directorate-General for Energy and Transport in co-operation with Eurostat. http://ec.europa.eu/dgs/energy_transport/figures/pocketbook/doc/2007/pb_3_transport_2007.pdf

European Environment Agency (EEA) (2008) No 1/2008. Climate for a transport change. TERM 2007: indicators tracking transport and environment in the European Union. http://reports.eea.europa.eu/eea_report_2008_1/en

Federal Aviation Administration (FAA) (1996) The interfaces between flightcrews and modern flight deck systems. Human Factors Team, New York

Galotti VP (1997) The future air navigation system FANS: communication, navigation, surveillance, air traffic management. Aldershot, Ashgate
Flight International July 30 –august 5 (2002) Safety review: CFIT is back
Haight F (1960) Mathematical theories of traffic flow. Academic Press, New York
Human Integration into the Life-cycle of Aviation Systems (HILAS) (2005) Annex 1: Description of work programme. Project number 516181, European commission—6th Framework programme
IRTAD (2008) Selected risk values for the year 2006. International Traffic Safety and Analysis Group (IRTAD), OECD and International Transport Forum. http://cemt.org/IRTAD/IRTADPublic/we2.html
Joint Commission (1997) Final report on the capsizing on 28 September 1994 in the Baltic Sea of the ro-ro passenger vessel MV ESTONIA. Edita Ltd., Helsinki
Mårtensson, L (1995) The aircraft crash at Gottröra: Experiences of the cockpit crew. The International Journal of Aviation Psychology, 5(3):305-326
Mårtensson L (1999) Are operators and pilots in control of complex systems? Control Engineering Practice, 7:173-182
Mårtensson L, Singer G (1998) Warning systems in commercial aircraft – an analysis of existing systems, KTH Report, ISRN KTH/IEO/R-98/1-SE, Royal Institute of Technology, Stockholm
NCSA (2006) Traffic safety facts 2006. The National Center for Statistics and Analysis. www.nhtsa.gov
O'Hare D, Roscoe, S (1990) Flightdeck performance – the human factor. Iowa State University Press, Ames
Sarter NB, Woods DD (1992) Pilot interaction with cockpit automation I: Operational experiences with the flight management system. Int. Journal of Aviation Psychology 2: 303–321
Thedéen T (1976) Freely flowing traffic with journeys of finite lengths. Transportation Research 10:13–16
Ulfvengren P (2003) Design of natural warning sounds in human-machine systems. Doctoral thesis, ISBN 91-7283-656-3, Royal Institute of Technology, Stockholm
Wickens C (1994) Designing for situation awareness and trust in automation. I G Johannsen (ed) Preprints of the IFAC conference on integrated systems engineering, pp 77–82. Pergamon, Oxford
Wiegemann D, Rich A, Shapell S (2000) Human error and accident causation theories frameworks and analytical techniques: An annotated bibliography (Tech. Rep. ARL-oo-12/FAA-oo-7). Savoy, IL: University of Illinois, Aviation Research Lab.
Wiener E (1989) Human factors of advanced technology (glasscockpit) transport aircraft. NASA CR 177528, NASA Ames Research Center, Moffett Field

10

IT – Risks and Security

Viiveke Fåk

10.1 Computers and IT Systems

In its most basic form, a computer is just a programmable processor, with some means for input and output of the treated data. More complex computers have a far more complicated architecture, with special units for contact with the environment, such as keyboards, displays, communication line interfaces, readers for special media like USB and DVD, etc. They often also contain large data storage areas, such as an internal hard disk. Today, computers are often connected via networks, such as specialized local networks, as in a car, or completely general networks like the Internet. The distinction between a single computer and a cluster of computers becomes more and more blurred. The shared property of all these diverse systems is that they are all technical systems for the treatment of information. Thus, today we often talk of a more general concept of information technology (IT) systems.

The basic control of an IT system is carried out in the processor, which is a piece of fixed hardware. However, that piece of *hardware* takes in data and sends out resulting signals as directed by other data, the *software* program. At the bottom of all software handling is the *operating system* (OS), which in turn governs how other pieces of software can use and share a processor and all other parts of the system. These other software parts can perform general services like database management, communications coordination, etc. Finally there are the applications, the services actually used by the IT system owners, such as control programs for a steel mill, or e-mail for a home user. And at all these levels, one part of the software can change its own behavior or rewrite another piece of software governed by the data it treats.

From a security point of view, this complexity introduces difficulties. Attempts to prevent some situations arising in the application program may fail because of a failure in the OS or in the service program. OS limitations introduced for security reasons may be circumvented because of an effect in another part of the OS, which was in turn deliberately or accidentally triggered by an application program. Fifty years ago, computer programmers knew exactly what was supposed to go on during the execution of their programs, and they could check it. Their programs may have contained design flaws, but when discovered, the flaws could easily be understood, and, given enough time, one could test almost every possible situation which could arise in the computing. This is impossible with today's complex IT systems.

Today, powerful computers are found not only in offices and factories, but everywhere, in your car, telephone, paper copier, etc. So what are the risks ensuing from our increasing dependence on these machines?

10.2 Computer-related Risks

First, it should be noted that even though the number of IT systems that control life-critical equipment is increasing at a somewhat alarming rate, the majority of all IT systems are still being used for convenience and efficiency, in more or less administrative activities. Failure in these systems will naturally often have economic consequences, some of which can be very severe. Nevertheless, when discussing security and risks for IT systems, we shall distinguish between risks to human life and health, risks regarding economy, and risks that lead to no direct financial costs but cause trouble in efforts to make corrections, loss of personal reputation, etc.

IT systems treat data and they output data. These data are supposed to correspond to something in the physical world: an abstraction of the actual properties of some physical entity, but they have no effect and thus pose no risk until they are used, either in controlling machinery or by humans making decisions. It is the result of this decision or the action of the machinery that can pose a real risk.

When IT system security is publicly discussed, there is seldom any distinction between errors which probably only cause inconveniences and errors which may be life-threatening. Due to the diversity of situations in which computers are used and the complexity of the systems, discussions tend to pick out the worst possible case or the case closest to the author's own experience, and to present that as the risk associated with a security failure. The complexity and diversity make it impossible, however, to make any firm statements about the severity of a generic security failure in an IT system. Risk analysis presupposes that the actual use is known, but this is not true of IT systems as a general category. A computer virus or an OS failure can delay your attempt to sort last summer's photos or incapacitate some life-saving system.

One major point to consider is whether or not the risk associated with an IT system has any immediate effect in the physical world. In administrative uses, i.e. uses where no physical process is directly controlled by the computer, the first-hand result is just that some facts are misrepresented. This can be rectified once the error is detected. In the meantime, we may have lost time, we may have spent energy on unnecessary tasks, and we may have suffered a blow to the economy. We may even have made a seriously wrong decision based on the erroneous data, but the result is seldom irrevocable in the same sense as when an IT system directly controls health and life, the environment, etc. without any human intervention.

When estimating the severity of a risk, it is thus necessary to estimate the way in which the data affected by the IT system failure are being used. Systematic treatment of risks must always take into account that the total risk depends on many disparate factors in the environment, but in the case of IT systems this is taken to the extreme.

There is in fact no way to predict exactly when an IT system will be used and for what purpose. Even embedded control systems may be developed, extended with further functions, ported to new versions of the equipment, etc., until the initial conditions, which were taken for granted by the original programmer, are no longer valid. In the case of more general systems, there is of course no way of predicting where they will be used. The following three examples of IT system failures serve as an illustration of this:

- In 1985–87, three people were killed because of erroneous behavior in the control program of the radio-therapy system Therac-25 (Levenson and Turner, 1993).
- In 1996, the 400 million dollar space ship Ariane 5 was blown up immediately after being launched because of design and specification errors in a computer program, namely the software of the inertial reference system (Nuseibeh, 1997). See Figure 10.1.
- In 1997, a major US warship became absolutely immobile when its computer control system was fed a zero into a division calculation by mistake (Yorktown, 2008).

Figure 10.1. The Ariane-5 rocket takes off from its launch pad in Kourou in French Guyana, June 4, 1996. The 400-million-dollar space ship was blown up 37 seconds after being launched because of design and specification errors in the software of the inertial reference system. (Photo: Patrick Hertzog/EPA/Scanpix)

Another issue to consider with regard to IT-related risks is that the consequences are difficult to evaluate in economic terms, even in what at first glance may appear to be a simple situation. One sometimes sees estimates of the cost of the latest computer virus outbreak, the latest spam flooding, the latest bank frauds, etc. The reliability of these figures is, however, often very low. Even if you know how much time your computer administrators spend on virus removal, their working duties are such that you cannot say that the salary spent is an extra cost, unless you can point exactly to things that were not done or were seriously delayed due to the virus cleaning. And then you must estimate what you would have gained, had these, probably lower priority, duties actually been carried out. If an e-commerce site is shut down for a specific time by a denial of service attack, can you really say that the loss is the gain you on the average get from the same period of ordinary business? Or will the customers come back in an hour or so, and then buy whatever they intended to buy before?

The time scale and customer behavior are totally different for IT systems and for more classical shopping. Even if you know down to the last cent how much money was illegally stolen from accounts in a fraud against e-banking customers, you do not know how much effort customers had to put into discovering that they were victims, and contacting the bank, and you do not know what they could not pay as required in time, etc. These are just three examples of this difficulty in estimating damage. For every new situation where an IT system is used, we need new knowledge and experience before we are able to estimate what a failure means in economic terms.

When can an IT system pose a really serious risk? The answer is obvious from the above; either when their output data directly influence the automatic behavior of some critical system, or when people are making critical decisions based on the IT system output data. There is actually a third possibility, which is slightly more philosophical in nature: Humans can use IT system data as an excuse for a catastrophic decisions, even when there were alternatives with total reliance on the IT system data. But this is not really a risk caused by the IT system as such. Even when no technology is involved, people can choose to rely on biased or outdated information because of laziness, ignorance, overly strict rules, or pure lack of resources to investigate all the alternatives.

When do we in fact have to take IT-system-related risks into account? Obviously the answer is totally dependent on the use of the IT system. The basis of risk related to IT is that computers are just "raw material" for all kinds of complicated systems. Just as steel is the basis for needles, axes, bridges, and skyscrapers, computers are the basis of tools for authors, autopilots, control systems for steel mills, and payment systems. A weak base involves serious risks for any system built on that base. The specific risk with IT systems is that those building on them as a basic part have few opportunities to discover their basic flaws.

10.3 Where Do We Find Computers and Related Risks?

The answer to this question is very simple: everywhere. To be more specific, computers are a critical part in every complex system like the power grid, hospital equipment, taxation systems, payment systems, theater lighting, music players, cars, etc. Many of the examples of risks discussed in other chapters in this book may have a computer failure as their primary cause.

We have built up a reliance on computers to a degree of which we are seldom aware. In industrialized countries, farming is no longer a matter of manual labor combined with personal knowledge and experience. Such methods do not pay enough. Modern farms function with computers to do the planning, to assist all the paper work, to automatically feed the right quantities to the right cow, etc. Theaters depend on the correct functioning of their computerized lighting systems. If these fail, there is no performance until the computer system is operational again. Payment in developed countries is more and more a matter of computerized notices of transactions, altering account information without any physical transfer of money being involved. If the IT system does not work as expected, we cannot pay our bills as required. On the other hand, without an IT system the receiver of the payment cannot check whether we have paid!

One specific application of IT systems which has received more and more attention during recent years is the SCADA (supervisory, control and data acquisition) systems controlling critical infrastructure. The generation and distribution of power, drinking water, logical and physical communications, etc. is nowadays normally completely dependent on computerized control systems. It is obvious that in these systems, an IT failure can cause severe disruptions in society.

Much effort has been devoted to predicting possible failures, finding weaknesses, and building resilience into these systems as a whole. However, the IT security aspects of SCADA systems have only lately started to receive attention. Examples from this area were not included among the three IT system failures previously mentioned, but there are a number on record (see Figure 10.2 for example). Temporary problems in phone systems have delayed ambulance assistance. Train traffic has been stopped for hours due to infections by malicious code. Water cleaning has been sabotaged via the control system (NIST, 2008). But these failures, for various reasons, have not received the extensive media coverage they probably deserve.

Sometimes there is no strict difference between IT-enabled systems and other systems. We have programmable computers, more fixed electronics and just electrically powered gadgets blurring together in a way that defies strict definition. It can be hard to decide what is really a system based on IT. For the user this is often an academic discussion. As a user, I do not care whether my car suddenly accelerates by itself because of a mechanical failure, an electrical failure, or a computer failure. I just want to know that spontaneous acceleration does not occur. For the designer of the car, however, the difference is extremely important. The designer must know when and how computers can fail, so that fail-safe precautions can be built into the total system.

Figure 10.2. Employees work in the control room at Browns Ferry Nuclear Plant in Athens, Alabama, June 21, 2007. In April 2007 the US Nuclear Regulatory Commission (NRC) reported that a recirculation pump controller failure on August 19, 2006, had caused operators at Browns Ferry, Unit 3, to manually scram the unit following a loss of both the 3A and 3B reactor recirculation pumps. The root cause of the event was a malfunction of the pump controller device (non-responsiveness) due to excessive data traffic on the plant industrial control system network. More information about the Browns Ferry incident can be found in NRC (2007). (Photo: Saul Loeb/AFP PHOTO/Scanpix)

10.4 In What Ways Can a Computer Fail?

IT systems treat data and only data. These data represent information, but that is important only when the data are to be used, either in the system or as output to a human or other system. Correct system behavior is described in some kind of instructions that we have tried to feed into the computer, based on our intentions with the system. These instructions can refer to what computations should be carried out, what aggregations and selection of data should be made, which users should have access to data, etc. It is obviously not possible to analyze all the possible errors in all these steps, but we must nevertheless make some analysis of the risks if there is a malfunction. Fortunately, there are actually only three basic ways in which a computer can fail:

1. The computer does not deliver the data on which we are dependent.
2. The computer delivers data with an erroneous value.
3. The computer delivers data to someone who is not authorized to receive those data.

These three types of errors are the bases for the three basic computer security criteria:

1. Availability of data when we need them.
2. Correctness of data.
3. Confidentiality for restricted data.

In general, there is, no way for a computer system to check the correctness of a value, unless it is something derived from other data, which must then in their turn be correct. Correctness must in general be ensured through thorough checks when the data are entered. What can be ensured within the computer is that no unauthorized changes in these carefully entered values take place. This property is called data integrity. For historical reasons, and to facilitate memorization, the three requirements are usually listed as (1) confidentiality, (2) integrity, (3) availability, or in short CIA.

Often we see additions to this triplet, like auditing, user authentication, etc., but these are merely examples of the fact that some of the internal data have specific, security critical requirements regarding their adherence to the CIA. User authentication, for example, means that there must be a procedure to reliably capture the user identity (correctness of these specific data). Audit means that we have to keep the authentication information available and related to all subsequent actions of that user, and *also* that relationships between actions can be followed.

From these three basic kinds of IT failure, it is simple to obtain a rough estimate of what risks there are in a given situation. Consider, for example, the engine control in a car. The last type of failure, availability, i.e., no control signal, is the easiest both to recognize as a possible failure and to guard against. The engine should work in such a way that, if there is no signal, the engine gets just a necessary minimal amount of fuel, and so that the control signal can only increase this amount. Thus, the worst that can happen if the IT control signal is unavailable is that we lose power just as we are overtaking another car. This is of course not a desirable situation but, compared to other risks with other strategies, it is definitely to be preferred.

The risk associated with erroneous data is more worrying in this case, since this can cause an unexpected, involuntary acceleration. Actually, this error could occur in electronic fuel control system which was used in many types of automobiles in the 1980s. The combination of a broken wire and an external disturbance resulted in full acceleration, and this did in fact cause accidents, one of which was fatal (Risks, 1988; Gunnerhed, 1988). Non-intuitive control system responses can also be dangerous. If the driver thinks that the system is not responding correctly, her attempts to counter this can lead to further instability in a spiral that leads to an accident. This is what happened when a Swedish fighter aircraft crashed during an exhibition flight in Stockholm in 1993 (JAS39, 1993).

The risk associated with the revelation of data to unauthorized recipients is not relevant in the engine control case. At most, one could think of the radiation surrounding all electrical activity, but then the problem is that signals disturb other IT systems, causing erroneous data in them, not that the signals reveal secrets.

If, however, we apply the same simple criteria to another system which analyses samples in a hospital, we get a very different result. The unavailability of a test result is of course undesirable but, under normal circumstances, the only effect is that a new sample has to be taken. We may think of situations where the delay results in too late a discovery of a life-threatening condition, but the alternative – not to use IT-controlled equipment – may delay the discovery of the condition further due to a lower efficiency.

Errors in the provided data are, however, also very serious in this case. An undetected disease is of course serious, but so are errors incorrectly indicating a serious condition when there is none. Medication, anxiety, etc. may cause serious damage to the health of the victim of such an error.

Finally, if the result of a test is received by an unauthorized person, this can also be serious. For most persons it is an unwarranted inconvenience, but if the person holds some public position, the consequences can be very serious. The simple information that a test was made may in itself grow into a major problem in, for example, a precarious political situation.

These two examples illustrate that it is not difficult to assess the consequences of an IT failure, once the exact use of the IT system is known and the simple CIA failure structure is kept in mind. What we need is knowledge of the system where the IT system is being used, not knowledge of computers and telecommunications as such. It is not, however, sufficient to know what happens once a risk has materialized. We need to know what caused the IT failure so that we can estimate the probability and take the necessary precautions against its happening, etc. This does require knowledge of the IT system itself.

10.5 Why Do IT Systems Typically Fail?

One obvious cause of an IT system failure is a severe disturbance to the physical IT system. Computers are unbelievably resilient, but there are limits. The most vulnerable parts are actually the communications to humans and the peripheral parts. One example is the hard disk in a typical home computer. This is not a necessary component of a computer, and it is by far the most common cause of breakdown in computers equipped with one, and thus the reason for the unavailability of that computer's data. The alternative, however, when it comes to these types of computers, is a work station having all its programs and data stored on a distant server. This configuration is extremely sensitive to any disturbance in the communication with that server. Thus we must know about these two possible sources of unavailability of data, and we must weigh the risks of the one against the risks of the other.

The most common cause of an availability failure is, however, a "system crash," some kind of error in the program logic which halts all further execution of programs and demands a total restart of the IT system before anything will work. This is still just an accident, an unintended mistake in the programming, but its effects can be far more severe and widespread than a similar mistake in the construction of physical machinery. Such program bugs (called thus because, when one of the very first computers started to exhibit unexpected program behavior some 60 years ago, the cause turned out to be an insect, which had died in a sensitive position in the machine) are extremely common in programming logic. They can lie undetected for years, until suddenly an unusual combination of input values steers the programming into a hitherto unused state, which triggers the bug. The sheer mass and complexity of all the program code in any "simple" computer makes it impossible to discover all the bugs. Thus one must always be prepared for unexpected outcomes of computer system use.

Programming mistakes are also the most common source of erroneous data. It is not a question of single values that go wrong once. It is more a question of seemingly inexplicable errors that turn up every now and then. One example is a program for carrying out and registering medical analyses. Under certain conditions, it reset a pointer and registered the current value under the identity of the patient who was first analyzed on that day.

Finally, the release of confidential data to unauthorized recipients is often also caused by simple mistakes. Here, however, the ultimate cause is usually a human typing in the wrong recipient, hitting the wrong button with the cursor on a screen, etc. Computers can be blamed here only if their user interfaces are cluttered, non-standard, and demanding the user to use and remember codes instead of everyday designations.

The complexity of computer programs makes it almost certain that every application contains some programming error, no matter how conscientiously the programmer has tried to do everything correctly. Nevertheless, these common errors are not what are normally referred to when the news media discuss risks concerning IT systems; antagonistic attacks on IT systems are far more often described and warned against.

10.6 Deliberate Attacks on IT Systems

Even though deliberate attacks are not the most common cause of IT system failures, they must still be considered. One reason is that computers attract immature enthusiasts, who want to show off by carrying out attacks on systems just for the fun of it. Thus it is not necessary for a system (or system owner) to have enemies in order to suffer an attack. Even more important is the fact that the same complexity which causes system mistakes also makes it virtually impossible for most users to know what is going on in their computers, and they are thus vulnerable to both random and directed attacks.

Normally the attacker uses some kind of attack program, "malware," like Trojan horses, viruses, or scripts using known program errors in general system code, to crash or take over a computer, etc. (See Table 10.1 for explanations of these.)

Table 10.1. Examples of attack tools

Trojan horse	A program that seems to do something useful, but which also contains a code attacking the system in which it is run, like searching for what look like credit card numbers and sending them to the program author
Spyware	A program which analyses the system set up, surfing behavior or something else, and sends this information to the spyware source without the user's permission
Viruses	Parts of software, which copy themselves and attach these copies to something else, like another piece of software, an attachment to an e-mail, etc
Botnets and zombies	A computer infected by software which makes it attack other computers is called a zombie, and a set of such computers is called a botnet

The malware program or script can be directed towards a specific user, but it is often just sent out at random over the Internet, hitting whoever happens to come in its way. Since everyone is almost equally close to someone else in the IT world, anyone can be a victim. If you are worried about random violence against your property, or even against yourself, your first precaution is normally to choose a better neighborhood, to avoid unsavory places when travelling, etc., but in the IT world there is no such thing as an "unsavory neighbourhood." The closest you come to this concept in the IT world is that you should be careful what sites you visit in the network. You normally do not become infected with spyware, viruses, and such unwanted phenomena through IT contacts with authorities, major companies, etc., but there are no signs in the IT world that indicate what may be a haunt for criminals and not a nice neighborhood e-shopping site. Random attacks are just as likely to find your network address as any other.

An exception to this is the specific attacks against availability which are called "denial of service" attacks, DoS. These are normally directed at specific targets, often for some political or economical end. In some cases, they are also just pure revenge, often for petty reasons such as a "wrong" opinion statement from the site owner. Such attackers often take advantage of the fact that far too many owners do not protect their computers against intrusion, thinking that they are not likely targets or have little to lose from an attack. The attackers simply plant their attack tools in these unprotected computers, and this gives them far more attacking power while at the same time hiding who the real attacker is.

One typical example is when Estonia in 2007 decided to move a monument dedicated to soldiers from the Red Army who had fallen in WWII. The move was from the city center to a more remote location. The monument was a reminder of what to the Estonians was an occupation force, although to the Russians it was a symbol of the heroism in their just fight against evil, i.e., the Nazi forces. This conflict caused some persons to direct a massive DoS attack against Estonian authorities and public services as a protest. The attack did not completely cripple the attacked targets, but it came rather close to doing so. Estonia has very quickly embraced IT as a means of providing efficient citizen services, but this also makes the country very vulnerable to successful IT attacks. The event was, however, described in the media as something comparable to direct war, attacks with physical bombs, a major disaster, etc. Even though it is definitely serious if the infrastructure becomes unavailable for long periods of time, a comparison with war seems too strong, and repair is far easier than if the damage had been due to a physical attack. More balanced accounts of the event are found in SEMA (2008) and in Schneier (2007).

The attacks are made possible, and often even easy, through the unfortunate history of computer development. The first computers were highly specialised machines used by a small workforce, where people knew each other. Systems had no protection at all against misuse, since misuse was inconceivable. When the usage spread, attacks were still very rare, and efficiency was the highest priority. When we emerged into today's interconnected IT world, protection against intrusion, etc., was still introduced and used mainly when it did not disturb efficiency and convenience. Risks could be paid for, mainly by customers, or simply ignored.

One example is the way in which computer user identity is usually authenticated. When a payment is involved, one might think that it is paramount that the person withdrawing funds from an account is indeed authorized to do so. In the old paper-based systems, a payment required lots of paperwork, signatures, proof of identity through photo-cards and badges, etc., whereas, over the network, e-shopping often requires nothing more than the knowledge of someone's credit card number; the same number that must be given to any clerk, waiter, etc., whom you pay. This naturally opens up endless opportunities for false transactions, where the onus is on the credit card owner to prove he did not make a specific purchase. E-banking, which may present an opportunity to empty your account directly and completely, is also sometimes the victim of too little an understanding of IT-related risks. Insecure cards, with magnetic strips that are easy to read and copy, are still the only type used by most banks in ATM systems. This has led to so-called "skimming", installing a small extra reader in front of the slot of a legitimate ATM or reprogramming shop card readers in order to catch and store card numbers. And even worse, some systems rely on a simple password to allow a user to start transferring large amounts from one account to another. This has resulted in widespread so-called phishing, websites and e-mail, to fool users into entering their ID and authentication password at the attacker's site instead of their bank's. One-time passwords from a list do not help here, since the attacker simply catches the next valid password and uses this before the victim himself contacts his bank.

One result of all these attacks has been a tendency to directly or indirectly blame the user of an insecure system. "You should not click on attachments from unknown users, since they may contain viruses." The truth is that you are just as likely to get a virus from a friend, who had a friend who was not sufficiently careful. "You should not visit suspicious sites." But how do I differentiate between a suspicious site and an innocuous one selling just what I need? "You must understand that the bank would never send an e-mail requiring you to log in at a new address." But how am I to understand that, if I have no knowledge of bank routines in these new IT days? Of course the computer users must be reasonably sensible in their actions, but what is really needed is a realization among IT system creators that IT risks exists, and it is the constructors who must implement the necessary precautions. And at the same time, the IT system customers must learn that there are precautions and must demand that they are implemented – just as the market has driven the car industry towards modern safety systems as standard.

10.7 Counter-measures Against IT-related Risks and Failures

The methods and tools listed below are the most well-known ones and are not a complete list. One concept which is highly relevant, but is not a direct method or tool is *defense in depth*, a strategy in which multiple layers of defense are placed throughout the IT system. The basic idea goes back to what was said at the start of this chapter about the complexity of IT systems and the impossibility of predicting their ultimate use during construction. The obvious conclusion is that it is not enough to raise a perimeter defense, because the source of a risk may lie inside that perimeter, or can have penetrated a weak point. At the same time, we cannot wait to place the defense at a point very close to the ultimate attack target, because we cannot predict exactly where the strike is going to be, and this may expose many weak points allowing the attack to circumvent this last line of defense. The "defense in depth" concept is also planned to delay and create time to forestall an attack.

When implementing the tools and methods below, the concept of defense in depth should therefore be kept in mind. We must also take into account that many modern systems are only further developments of much older systems that were then designed with far fewer security precautions. As a result, many necessary security functions are late additions and not parts of the original design, and this has in turn made them far less efficient than if they had been implemented as part of the core of the original system. In some cases resources are lacking to place them where they are needed, and this makes it necessary to use further security tools elsewhere in the total system.

Back-up

Back-up should be the first security precaution. This addresses availability risks. If we have copies of the data, it is possible to restore crashed registers, mutilated web pages, etc., and the risk is reduced to a slight delay and some extra costs instead of a possible catastrophe. In the same way, redundant equipment, communication paths, etc., ensure that operations can continue even with a serious hardware failure or attack.

Proper Program Development and Testing

The most important risk, program bugs, can hardly be avoided completely. We have slowly learned the importance of thorough development procedures, testing, etc., but there is so much program code already in existence and so much continuously being produced that there is no way that we can hope to get bug-free programs. This subject has been high on the agenda for responsible program developers for decades, but the situation is still very bad.

Automatic Program Updates

One solution that is very common today is that the major program and system providers offer automatic updates to their customers, making it possible to fix a security hole which is susceptible to an attack as soon as the company knows that there is a problem and has found a solution. But attacks hit some before the solution is there, and many do not realize the importance of using the update service. Every monthly security report from a major network such as a university, a major company, etc., is likely to contain details about some computer that was connected to the network without proper security updates being installed. A variant of this occurred in a major Greek wire-tapping scandal reported in, e.g., *IEEE Spectrum* (Prevelakis and Diomidis, 2007). One reason for the success of the wire-tappers was that the phone company had not yet installed the security analysis programs offered by the equipment providers. (Phone systems nowadays are of course normally digital and just a special version of an IT system.)

Physical Protection of Critical Computers

Computers are very likely to be stolen, if a thief can get at them. Computers are sensitive to bad electrical conditions, heat, excessive humidity, and direct fluids. Also, an unprotected computer can have malware installed simply by using the physical access and restarting the computer from some inserted false program medium. A stolen computer hard disk, with or without the computer it once resided in, can reveal critical, confidential data, and computers emanate electro-magnetic signals, which can be used for eavesdropping by serious attackers. The recommendation is that critical computers should be placed out of reach of unauthorized persons. They should also have a protected environment with regard to climate, electricity, etc., unless they are specifically designed for a harsher environment.

Cryptographic Techniques

Communicated data cannot be physically protected from unauthorized persons. Eavesdropping, whether passive or active (inserting new messages and altering passing ones), is always a risk. For example, a radio network, which is a very common technology today, can be accessed by anyone raising an aerial in the area. Portable computers are routinely equipped with an antenna and software to connect to any radio network in the vicinity. In the Internet, we must also either fully trust any organization providing a switching node, or we must protect our data. The only protection that works in this context is cryptography. It can transform confidential data into forms which can be reconstructed into meaningful information only by an authorized unit. Cryptography can add information, checksums, and digital signatures, which make it impossible to change or insert data under a false identity without detection, thus preserving data integrity. It can be used to abolish passwords and instead use cryptographic transformations of constantly changing challenges, thus making card skimming and phishing obsolete, but it requires that encryption keys are properly generated, stored, and properly attributed to the correct owner. This is currently solved with certificates. These techniques can never give 100 % security, but they can lower the risk level tremendously at a comparably low cost.

User Authentication
Weak user authentication is an Achilles heel in many systems today. Passwords dominate, and systems often lack precautions against weak choices of passwords, while users are supposed to remember at least a dozen if they e-shop, etc. Better methods involve cryptographic techniques combined with binding authentication cards, tokens, etc., to biometrics, the technique for simply measuring some unique property of the user, such as a fingerprint. Biometrics can also be fooled, for example by gelatine fingertips emulating the fingerprint of someone else, but they are still more secure in most applications than simple passwords, and much more convenient for the computer user.

Access Control
Logical access control involves routines to check which users can read or write which data and execute which program and computer services. It also includes firewalls, equipment which checks which network addresses are allowed to contact the protected computer or sub-network, and then what services can be called. Access control is, however, only as strong as the rules that the IT-system owner sets. If we run a computer as a public web server, we cannot deny the public access to the computer. If we do not block those services on our home computers that we personally do not need, nobody else is going to do it.

Detection Systems
One important tool today is *intrusion detection systems* (IDS). They are sophisticated analysis programs for communication logs, in real time or for stored logs, and they are supposed to discover traffic patterns that indicate that an attack is ongoing or is coming, that a user has some suspicious activity going, etc. This category of systems also includes such things as virus filters that are supposed to recognize all known viruses passing through the filter. The same principles can also be used to detect possible fraud, such as when someone has taken over a victim's phone account, credit card, etc. These types of tools are very useful for

professionals, but they often require too much work and skill in order to sort out real attacks from false alarms and still not miss new attack variants. Virus filters which only compare data contents to known parts of known viruses are so far the only type that does not cause much further work for the user. But this is at the expense of not detecting entirely new virus types.

10.8 Where Are We Heading?

Computer-related risks often receive considerable coverage in the media today. It is, however, difficult to obtain scientifically sound and verified information on the severity of the problem. Surveys presented in journals and magazines, are too often not very reliable. They typically indicate the answers of say 56 of the 100 companies in a certain sector to whom a questionnaire was sent, or they are the result of simply counting the answers to a question published on some professional paper's website. There is no way to say what selection of the possible victims actually answered the questionnaire, no way to know whether those who never had any trouble were those who did not answer or whether those with serious trouble did not want to admit it. It is also well known that victims of fraud due to their own negligence concerning security, etc., are very careful not to reveal their misfortunes if they can avoid it.

Another problem in assessing the situation concerning IT risks is that reports often do not look at the actual damage caused. The general low level of awareness concerning what real protection possibilities are available often leads to strange presentations of actual incidents. For example, one sees bold headlines for events, which are more or less due to very weak protection systems, while the headlines and articles below them paint a picture of unavoidable IT risks or bad human behavior. Sometimes pure cultural evaluations also cloud the issue.

A typical example is the presentation in *eWeek* of the "Worst data breaches ever" (eWeek, 2007). The cases presented all concerned US databases and they were all about the revelation of customer data or data about individuals in general. The designation "worst ever" depended in this case on the number of persons whose personal data was revealed. There was no mention of the real damage caused by this release of private data.

One example is the TJX system, where 45.6 million credit and debit card numbers were stolen around 2006. This was reported as costing the company considerable expected losses for the next 3 years. The reported cost was for the company, not the customers, and concerned the cost of giving out new numbers etc. Losses due to misused numbers are not mentioned in the report. The reason why this is being so severe is a system deficit, which makes it possible to purchase expensive goods and divert the payment to any person whose card number you happen to know. Would banks have paid out money over the counter 50 years ago, just because you happened to know someone else's account number? No. They demanded identification, written signatures, etc. So what is the difference between a bank account number and a credit card number? The answer is that a bank account number was seldom known to anyone except the bank, the customer, and sometimes an employer, whereas the credit card number is known to everyone you have paid with the card. If IT payment systems had been designed with a maximum of security from the start, as they very well could have been considering the technical possibilities, then this "worst event" would have been worth only a shrug. So is this an IT risk or a "bad banking routines" risk?

Another example from the same list concerns stolen laptops containing some 130 000 social security numbers. Knowledge of a social security number enables what in the US is now called "identity theft." This means that you get your actions, whether they are economic, health-related, or whatever, registered under another person's identity. You can take out a credit card, hire a car, get treatment for some illness, etc., and get everything attributed to another person. But is this an IT risk? It was definitely an IT event that the crucial information leaked but, again, the real weakness is that you can get an attribution to a person based on knowledge of a number that is available to thousands of persons as a legitimate part of what they must know about you in their daily work.

In the case of the third of these "worst events," it is even more questionable whether this was an IT risk. This concerns more general personal information, like address, family, etc., being revealed to thieves, but in many instances this same type of information is directly available as web services for the whole world. These services include phone books, public national registers, and information from companies about how to contact employees, and in particular the general search engines for the World Wide Web. Here cultural differences really come into play. What in the US is regarded as highly personal information can in other countries be available via government services and be regarded as absolutely harmless.

Although these much publicized "worst events" may not have caused any real damage to individuals, IT use has augmented real risks in sneaky ways. One typical example is persons who have an officially protected identity, such as persecuted women who are given a new identity to escape threats from ex-husbands. Too many people have direct access to these protected data via IT systems and may reveal them by mistake or from ignorance. Automatic couplings between IT systems may transport such data to unprotected registers, without anyone even knowing that this path existed before an assault on the protected person is suddenly reported.

Last but not least, we often have too little understanding of the risks associated with the power to sort, compare, and assemble information about individuals. Phone books for example may seem innocent, as long as you keep protected numbers out of them, but suppose that someone assembles the names and addresses of everyone within a region who has a military title. This is obviously a sensitive assembly of data and, the other way around, suppose that someone studies all the numbers called in your mobile phone. This can be highly personal information and critical in some cases. One such case is a person who was convicted for instigation of murder based on the contents of the mobile phone owned by the young woman he had under his spell.

The truth of the matter is that our IT awareness is still very immature. We implement IT systems without checking whether they incorporate the precautions that existed in their non-IT precursors. We blame persons when we should demand better systems. We let ourselves be fooled into thinking that some drawbacks are inherent in IT, and thus have to be accepted together with advantages, when the truth may be that system designers and owners were ignorant, lazy, or simply greedy when it came to adding security precautions. As long as customers accept the cost and do not complain (because they do not understand that they have reason to), why take costly precautions against IT risks?

The examples of severe physical events given at the start of this chapter, and the examples of the clumsy handling and evaluation of personal information in this latter part, clearly indicate that we still have a long way to go to get a good grip on the risks that IT has introduced into our lives.

10.9 Literature and Other Sources of Information

There are very few good books on IT-related risks, but there is more on risk assessment in an IT environment and on IT security. One authority on IT risks, and the general assessment of modern risk confinement, is Bruce Schneier; in for example *Secrets and Lies* (Schneier, 2000), and *Beyond Fear* (Schneier, 2003). He also has a blog and monthly newsletter about IT and related risks (Schneier, 2009). Ross Anderson is another much respected security researcher who among other things has published the book *Security Engineering* (Anderson, 2008). He is also the co-author of an interesting paper written for the EU Network and Security Agency, ENISA (Anderson, 2008). His homepage at Cambridge University contains links to much interesting material in the field.

Books about IT security are usually very technical. A few recent examples of recommended books are: Bishop (2005), Gollmann (2006), and Pfleeger and Pfleeger (2007). Books concentrating on risks and critical IT systems are much rarer. A classical reference is Levenson (1996), but Storey (1996) and Perrow (1999) are also worth reading.

Most sources on IT-related accidents and attacks are news articles or sources from web publications. This makes it difficult to find consistent and durable sources. Some stories are also transformed into modern folklore with lots of versions. One can for example read about the US warship mentioned above in many sources. They all agree that the cause was an attempt to divide by zero, and that the computer system was based on Windows NT, which was set to pass on its current task to a neighbor, if it failed and broke down. All sources agree that this caused all computers to cease working, thus immobilizing the ship. Nevertheless, the name of the ship and its position at the accident differ! The source here, Yorktown (2008), is the official US Navy site at the time of writing of this chapter.

Good information on current discoveries concerning security holes in software are the computer emergency response teams (CERT) that exist in many countries. Major companies like Microsoft also have security news bulletins on their websites. There are also websites which continuously add information about IT security events. Some of these are:

- ENISA (The European Network and Information Security Agency): http://www.enisa.europa.eu/
- US-CERT (The United States Computer Emergency Readiness Team): http://www.us-cert.gov/

- SecurityFocus (A vendor-neutral site with lots of information and useful links): http://www.securityfocus.com/
- Infosecurity (An on-line magazine): http://www.infosecurity-magazine.com/

References

Anderson R (2008) Security engineering: a guide to building dependable distributed systems. 2nd ed, Wiley, Indianapolis

Anderson R, Böhme R, Clayton R, Moore T (2008) Security, economics, and the internal market. European Network and Information Security Agency (ENISA). http://www.enisa.europa.eu/doc/pdf/report_sec_econ_&_int_mark_20080131.pdf

Bishop M (2005) Introduction to computer security. Addison-Wesley, Boston

eWeek (2007) Worst data breaches ever. August 2007 http://www.eweek.com/c/a/Security/Worst-Data-Breaches-Ever/

Gollmann D (2006) Computer security. Wiley, Chichester

Gunnerhed M (1988) Risk assessment of cruise control. FOA Report E 30010-3.3, Swedish Defence Research Establishment, Stockholm

JAS39 (1993) JAS 39 Gripen crash in Stockholm 1993 Aug 08, Report summary http://www.canit.se/~griffon/aviation/text/gripcras.htm

Levenson NC (1996) Safeware: system safety and computers. Addison-Wesley, Reading

Levenson N, Turner C (1993) An investigation of the Therac-25 accident, IEEE Computer 26(7):18-41

NIST (2008) NIST SP 800-82 – Guide to industrial control systems (ICS) security. Final public draft, National Institute for Standards and Technology (NIST). http://csrc.nist.gov/publications/drafts/800-82/draft_sp800-82-fpd.pdf

NRC (2007) Effects of Ethernet-based, non-safety related controls on the safe and continued operation of nuclerar power stations. NRC Information Notice 2007-15, US Nuclear Regulatory Commission (NRC). http://www.nrc.gov/reading-rm/doc-collections/gen-comm/info-notices/2007/in200715.pdf

Nuseibeh B (1997) Ariane 5: Who dunnit? IEEE Software 14(3):15–16

Perrow C (1999) Normal accidents: living with high-risk technologies. Princeton University Press, Princeton

Pfleeger CP, Pfleeger SL (2007) Securty in computing. Prentice Hall, Upper Saddle River

Prevelakis V, Diomidis S (2007) The Athens affair. IEEE Spectrum, July:18–25

Risks (1988) Risks digest: Audi involuntary acceleration. http://catless.ncl.ac.uk/Risks/7.25.html

Schneier B (2000) Secrets and lies: digital security in a networked world. Wiley, New York

Schneier B (2003) Beyond fear: thinking sensibly about security in an uncertain world. Copernicus Books, New York

Schneier B (2007) "Cyberwar" in Estonia. http://www.schneier.com/blog archives/2007/08/cyberwar_in_est.html

Schneier B (2009) Schneier on security – a blog covering security and security technology. http://www.schneier.com/blog/

SEMA (2008) Large scale Internet attacks: the Internet attacks on Estonia – Sweden's emergency preparedness for Internet attacks. Swedish Emergency Management Agency (SEMA), Stockholm

Storey N (1996) Safety-critical computer systems. Addison-Wesley, Harlow

Yorktown (2008) USS Yorktown (CG 48) Official Navy Homepage. http://www.navysite.de/cg/cg48.html

11

Vulnerability of Infrastructures

Torbjörn Thedéen

11.1 Introduction

Society depends on the functioning of working connections for the transportation of individuals and goods, mail and telephony, etc. The fast technical development has changed these systems into larger and automated ones. The number of individual journeys as well as their lengths has increased and this holds also for telecommunications. The average journey length in the European Union in 2004 was 12 000 km per capita and it has increased by 2 % annually during the last 10 years. Internal and external trade has increased dramatically, and this in turn has led to more transportation of persons and goods.

Contrary to the opinion of many experts the demand for personal meetings has not been replaced by electronic communication. Nevertheless there is also a negative side. Consumers and enterprises depend on a functioning power supply. Many enterprises need goods to be delivered from their subcontractors just in time. Infrastructures of particular importance for the function of the society are called critical. Examples of critical systems are transportation, power, and telecommunication. The critical networks are vital for society but they are also vulnerable to threats and hazards.

11.2 Critical Infrastructures

Let us consider some examples of critical infrastructures.

Electric Power Distribution

Modern society is dependent on a reliable electric power supply. This is not, however, just a national interest since we increasingly export and import large quantities of electric power. Parts of the grid can be out of order for a shorter or longer time hitting regions or the whole country. The threats can originate from extreme weather and technical failures but also from vandalism and antagonistic attacks. The power system is also very dependent on various forms of IT-based control systems.

Infrastructures for data and telecommunication
An increasing part of our societal activities depends on the communication of data. Besides being vulnerable to nature and technology-related threats the networks are also vulnerable to planned attacks. These can be realized at small costs and at large distances in time and space (Chapter 10).

Structures for the transport of goods and persons by road, rail, sea and air
We travel more and more and global trade is expanding rapidly. The transportation structures, not least in the big cities, are vulnerable to small disturbances with possible grave consequences for society.

Structures for water and sewage
Most households today rely on local water and sewage systems. In particular, the provision of clean water is necessary for life. Technical faults as well as toxic outlets can jeopardize that provision.

Figure 11.1. Traders work on the floor of the New York Stock Exchange (NYSE) in New York City. Almost all NYSE stocks can be traded via its so-called hybrid market system. Customers can send orders for immediate electronic execution or direct orders to the floor for traditional trade in the live auction market. The key advantage to electronic trades is speed, and more and more trading is being automated. The Internet has caused a major shift in the behavior of investors, and the trading of stocks is becoming an instantaneous electronic exchange of information between buyers and sellers. (Photo: Spencer Platt/AFP/Scanpix)

Structures for provision of heating and cooling
The provision of heating is very dependent on electric power. During the cold season it is in many regions necessary to guarantee a minimum amount of heating. In the hot season in some tropical regions there is a corresponding cooling problem.

Structures for financial transactions
We use cash to a decreasing extent. Trade and financial transactions rely instead on electronic transfer (Figure 11.1) and Internet banking.

All the structures mentioned are critical for the function of vital parts of society. Further, they are interdependent – all require the supply of electric power. Computer systems are also a necessary part of almost all infrastructures.

These infrastructures are critical not only for the civil society but are also for its defense, since antagonistic attacks on the infrastructures can threaten national security. National defense must not only maintain the integrity of national territory but also defend the critical infrastructures.

11.3 Examples of Infrastructure Collapse

In the preceding section, we have considered critical infrastructures in general terms, and it was pointed out that the provision of electric power is fundamental for the function of most critical infrastructures. This is further illustrated in the following examples.

The Ice Storm in Canada
In January 1998 Eastern Canada, in particular Quebec and Ontario, was hit by the most severe ice storm in modern times. Extreme ice formation on the electricity lines caused power cuts up to 4 weeks for about 1.6 million consumers (Figure 11.2). It became impossible to carry out common social and economic activities, and the situation became dangerous for humans and animals. The telephone services were out of order and so were the mobile networks, since many masts broke.

One can draw many conclusions from this example. Extreme weather can hit a whole region and have devastating consequences. It is in practice impossible to safeguard against power breaks. One can instead install reserve systems for heat and transportation, i.e., create a redundancy. It is also important to plan for fast recovery to a normal situation. That was the case in Canada, where the society was well prepared to handle the crisis (NPCC, 1998).

The Collapse of the Electricity Network in Auckland
At the end of January 1998, the electric power distribution in central Auckland broke down due to failure in a pair of insulated earth cables. In this case, the collapse was due purely to a technical fault, partly caused by the warm summer weather. The crisis was not over until the end of March. During that period, attempts were made to find temporary solutions. Since the collapse occurred in the

summer there were almost no consequences for human health and lives (in contrast to the ice storm situation in Canada). On the other hand, there were severe consequences for the economic and social life in Auckland. No electric power could be provided for cooking, hot water, etc. Cash dispensers and electronic trade did not function, nor did the air conditioning. Restaurants and shops as well as public communication experienced great difficulties. Using reserve generators, a limited distribution of electric power could start after some time. At the end of February, it became possible to get electricity from diesel generators in ships in the port.

Figure 11.2. High-voltage towers near St-Bruno collapsed on January 10, 1998, after a severe ice storm hit southwest Quebec. (Photo: Jacques Boissinot/NTB Pluss/AP Photo/Scanpix)

The Auckland example demonstrates how dependent a modern large city is on the provision of electric power and how most infrastructures such as transport, telecommunication, water and drainage, and financial transactions need electric power in order to function. In this case, the electric power distribution had just been privatized, and there was a poor preparedness to act in a crisis like this one. In the future it is essential to be better prepared to handle such situations. There must be a better cooperation between the public administration and industry. Both the Canadian and the Auckland case demonstrate how important it is to have reserve capacity available (MED, 1998).

A Sabotage Attempt against the Electric Power Distribution System of London
In 1996, one IRA (Irish Republican Army) unit had made advanced plans to blow up six important transformer stations in London and Southern England. A joint operation between the British police and the Security Service (MI5) succeeded in preventing the sabotage. The investigations that followed showed that only a small amount of explosives was needed; 2.5 kg Semtex per station would have been enough. This example shows that infrastructures that are critical for the society can be attractive targets for terrorist organizations. The result could have been a power outage for several weeks, achieved with only small resources for the terrorists (Bennetto, 1997).

The Cascading Blackout in United States and Canada in 2003
The outage was caused by a combination of electrical, computer, and human failures. Uncorrected problems in northern Ohio developed into a cascading blackout. This included an ineffective system-monitoring tool, loss of generation, no efforts to reduce load, lines tripping because of contact with overgrown trees, and finally, overloading. The blackout affected some 50 million people in the Midwest and northeast United States and Ontario in Canada. Power was not restored for two days in some parts of the United States, and parts of Ontario suffered from rolling blackouts for more than a week (US–Canada Task Force, 2004).

11.4 The Vulnerability Concept

We now consider more precisely the concept of *vulnerability*. Let us first recall what is meant by a *risk*. It is defined as a combination of random or uncertain events with negative consequences for human life, health, and the environment, together with the corresponding probabilities. When we consider the vulnerability of infrastructures, that definition has to be altered. By vulnerability we mean the probability of a total or limited collapse of an infrastructure and the loss of its ability to uphold important social functions during a certain time period. Effects on human health may not necessarily be noted.

When an infrastructure collapses, it goes through several stages. In some cases, the function can be upheld by a reserve system, e.g., by buses replacing trains. The electric power production can use diesel-driven generating stations. If the preparedness for repair is good, a fast return to a normal function will be possible.

Vulnerability is a relative concept. Power systems can, for example, be less or more sensitive to extreme weather. The vulnerability can be measured as the probability of a large outage during a certain time. The probability of extreme weather should also be included. If we consider the vulnerability to a terrorist attack, we can compare different net configurations with regard to their vulnerability given a certain attack; so-called conditional vulnerability.

11.5 The Nature of Infrastructures

A common feature of most infrastructures is that they are stable in time. (There are also infrastructures which develop with time, so called dynamic infrastructures.) Generally speaking, an infrastructure can be regarded as a *combination of units with some kind of connection between the units.* There can be purely technical structures, and also structures consisting of individuals and their connections. Nearly all infrastructures are a combination of these types; they are socio-technical, consisting of technical systems, individuals, and organizations. Organizations and individuals are crucial in the control and management of the infrastructures.

One can have different perspectives in the modeling of a system. Here we shall consider a system from the perspective of sustaining important societal functions. It could also have been possible to study them from an industry perspective. An example of the latter is when the problem is to produce service and transportation in a cost-efficient way. In contrast an interest in quality, low prices, and availability is characteristic of a customer-related perspective. Regardless of the perspective, it is natural to use a *graph model,* consisting of a number of points (nodes) which are connected by (one-directional or two-directional) links, see Figure 11.3.

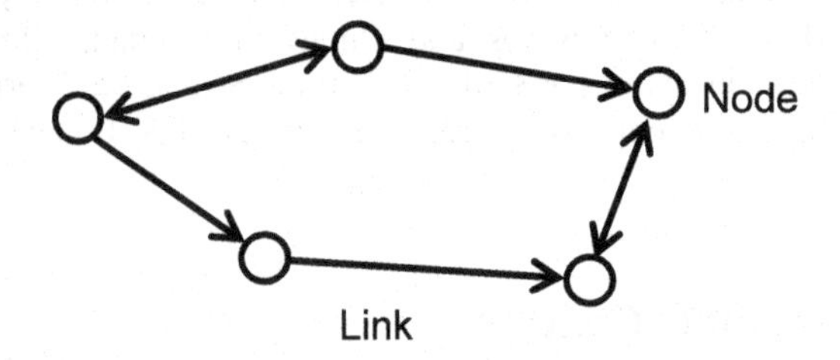

Figure 11.3. Graph model of infrastructures

A railway network consists of stations (nodes) and rails (links), along which the trains transport passengers and goods, and this traffic is managed from control desks.

Road traffic has fixed links (the roads), and the nodes depend on the scale of the system. At a national level the nodes are the villages and towns. At a local level the houses, the parking lots, and the working places are the nodes. The traffic flows are determined by the decisions of the individual drivers, but they are also to some extent controlled by, e.g., signals.

Air and sea transportation systems have fixed nodes: airports and ports, respectively. The links are fixed close to the nodes but are less determined as routes in the air or at sea between these nodes. There is a strict traffic control close to airports and ports but the control is freer elsewhere.

The fact that most infrastructures are interdependent has both positive and negative aspects. Infrastructures can complement each other, e.g., in a collapse of railway traffic, trains can be replaced by buses. On the other hand, most infrastructures are highly dependent on electric power and on control systems based on computers, see Figure 11.4.

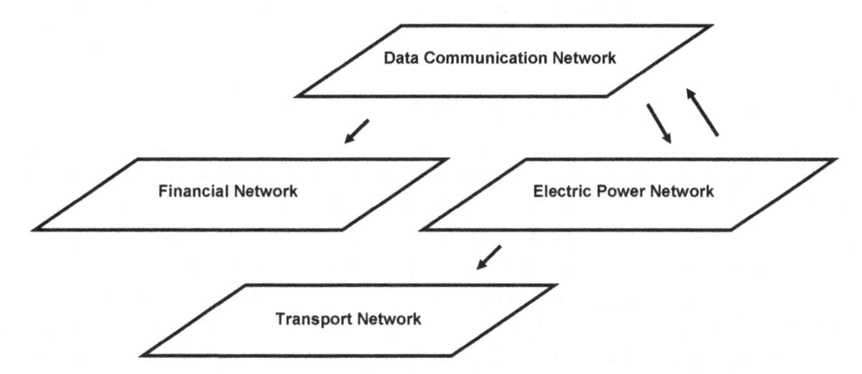

Figure 11.4. Some critical infrastructures and their interdependences

We shall now introduce some important graph concepts. The *link distribution* is the distribution of the number of links with an end point at a random node. The corresponding mean value is called the *degree.* The *mean distance* is the average of the least number of links connecting two random nodes. There are often restrictions on the flow in the links, i.e., capacity limits. A road can, e.g., just take a maximum flow, i.e., the number of vehicles per hour. An infrastructure can then be regarded as a graph, with flows in the links bounded by capacity limits in a network with human or technical control systems. See West (2001) for an introduction to graph theory.

We have defined what we mean by critical infrastructures and their vulnerabilities and have introduced graphs as models. Let us now consider a very simplified example of an electric network modeled by a connected graph, see Figure 11.5. The network consists of two generating stations *A* and *B*. From these there are links to transformers *a* and *b*, i.e., *Aa, Ab*, and *Bb*. The transformers are connected by the link *ab*. The transformers distribute power at a lower voltage to consumer groups *1* and *2* by the links *a1* and *b2*.

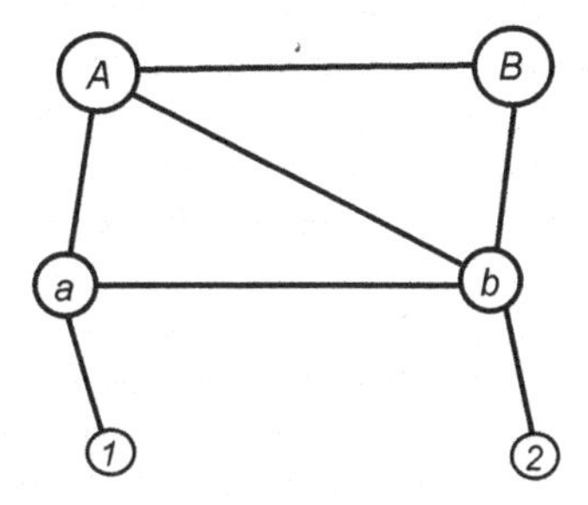

Figure 11.5. A simple electric power network

It is evident that a break of both *Aa* and *ab* will put consumer group *1* out of power, whereas a break of *Bb* will cause no outage. If *Ab* also breaks, in addition to the breakage of *ab* and *Bb* this will cause an outage for group *2*. It is also possible to estimate the possible outage for group *a* and *b* if we assume failure probabilities for respective links. The same reasoning can be applied to real distribution grids, but we then have to use simulation methods.

11.6 Threats Against Infrastructures

An initiating event can result in a partial or total collapse of a network during a certain time period. Such an event may be of different kinds, from purely random to a hostile attack. Let us consider a railway system. A heavy snowfall can cause a derailment, and this in turn can stop the regional railway traffic for many days – a severe consequence for society. The initiating event (snowfall) is of a random character, with a certain distribution. The same holds for other natural events. Another initiating event can be a material failure, e.g., a rail break, or a failure of the computer system. Such failures are usually random. In these cases it is possible to manage the outcome by the choice of material, construction, and control. Human failure and organizational weaknesses are only partly of a random character and they have unknown distributions (see Chapter 15).

Sabotage and terrorist attacks are of a completely different type, not being of a random character. The aim of a terrorist attack can be not only to destroy a critical infrastructure but also to create chaos in society.

The vulnerability of an infrastructure is considered to be conditioned by a given attack, a scenario. Similar problems have long been considered in defense analyses using so-called game theory.

11.7 Approaches to Vulnerability Analysis

To assess the vulnerability and to compare measures to control it, we need methods for its analysis. Since the collapse of an infrastructure is a rare event, and since the structures have individual characteristics, we cannot rely solely on empirical experience and statistical analysis. Instead we have to combine the empirical data with mathematical models of the infrastructures (compare with Chapter 13).

Robustness is a concept that is the opposite of vulnerability. One can ask how robust a certain infrastructure is against attack, when modeled as a graph. The structure of the graph, its topology, may influence the robustness. It is possible to study the effect of different attacks against the infrastructure. If the threat is associated with extreme weather, the failures can be seen as random, i.e., links are broken with a probability p. This influences the flow through the network.

Since there are not many data available of collapses it is in principle, difficult to carry out a post-analysis. Instead we have to rely on simulations of models to study the vulnerability and the effects of various possible measures to ameliorate the effects. For instance, one can simulate the effects of threats against the links and nodes, assuming that links are randomly broken with a probability p due to bad weather, and we can assess the effects of doubling the links – redundancy – or of strengthening some links or nodes. The important goal of any vulnerability analysis is to estimate the effects of changes in the infrastructure structure. As is the case with other technical systems, we have to use a proactive approach, i.e., we have to take measures before a collapse has occurred.

11.8 Practical Examples

11.8.1 Vulnerability Analysis of Electric Power Systems

An analysis of empirical output data can provide a good understanding of everyday disturbances and of the ability of power systems to withstand them. There are few available data of disturbances with severe consequences (a low probability, severe consequences). Holmgren and Molin (2006) analyzed the vulnerability of power delivery systems by a statistical analysis of empirical disturbance data, including data from the Swedish national transmission grid and the Stockholm distribution grid. They demonstrated that the size of large disturbances follows a power law (a distribution with a heavy tail).

Holmgren (2006) presented a graph-based analysis of the vulnerability of the Nordic Interconnected grid and of the Western States (US) transmission grid, in which detailed data on the structure of the two transmission grids were classified, and therefore not available. Various different indirect graph measures were used to estimate the consequences of removing nodes from the network and the networks were shown to have a lower attack tolerance than failure tolerance (compare with Albert et al. (2000) and Holme et al. (2002)). The electric power networks have similar topological properties, and they exhibit similar disintegration patterns. An important field of application for vulnerability analysis is to evaluate the effect of introducing changes in an existing system, see Figure 11.6.

As an experiment, the graph of the Nordic power transmission grid is modified by incorporating two new links (power lines) between Sweden and each country in the region, i.e., six new links. The new power lines are located as in an internal study proposal from the utility that operates the Swedish national transmission grid. A comparison of the augmented Nordic grid with the present Nordic grid yields, however, small, if any changes in the graph-based vulnerability analysis. A generic graph analysis, based on open-source data of the structure of the networks, is then too simplistic for practical purposes.

In Holmgren et al. (2007), the interaction between an attacker and a defender of an electric power system is modeled as a game. In the model, the defender can only spend resources on increasing the component protection (e.g., fortification) and/or on decreasing the recovery time after an attack (e.g., repair teams). The protection of an element is a function of the resources spent on protecting that element. The defender distributes the resources for protection between the elements in the network. The repair time of an element depends on the resources spent on recovery, as well as on the type of the disabled element and on the attack method. In the model, it is assumed that the defender has a basic recovery capacity for maintenance and for repairing minor failures. Thus, the relative effect of spending extra resources on recovery can be studied.

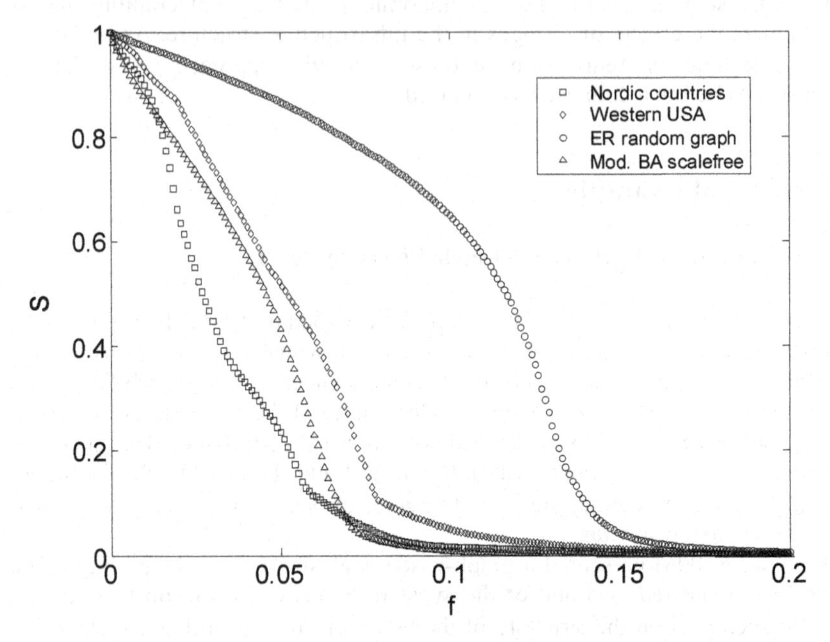

Figure 11.6. Defragmentation of four different networks of approximately the same size. Two electric power networks ("Nordic countries" and "Western USA") are compared with two theoretical network models, i.e., a modified version of the Barabási and Albert scale-free network ("Mod. BA scalefree"), and the random graph of Erdös and Réyni ("ER random graph"). The nodes (fraction *f*) are removed in the order of decreasing degree (i.e., the node with most connected links is removed first). After each node that has been removed, the degree is recalculated. The relative size of the largest connected subgraph (component) *S* is used as a measure of the consequence of removing nodes, i.e., a measure of the attack tolerance of the network (Holmgren, 2006)

The attack model considers only qualified antagonists, i.e. determined, well-informed, and competent antagonists with sufficient resources to be able to make a successful attack against an electric power system. Three different classes of attack strategies are considered:

- Worst-case attack (the antagonist chooses the target that maximizes the expected negative consequences of the attack)
- Probability-based attack (the antagonist tries to maximize the probability that the outcome of an attack is above a certain magnitude)
- Random attack (the antagonist chooses the attack target at random, and each target is attacked with equal probability)

A defense optimized against the worst-case attack strategy will not necessarily provide an optimal defense against other attack scenarios. It is not possible to find a dominant defense strategy in the numerical example, i.e., a defense strategy with lower expected negative consequences than every other defense strategy against every attack scenario. Holmgren et al. (2007) show, however, that it is possible to use a number of statistical methods to give a ranking of the different defense strategies.

11.8.2 Vulnerability Analysis of Road Transportation Systems

Transportation networks such as railways, roads, and air traffic can naturally be modeled as connected graphs with flows in the links and one can study the effect on the total travel time of removing a link in the network. This has been done for the road network of northern Sweden by Jenelius et al. (2006) following a similar study of the Stockholm road network by Berdica (2002). First a simplified graph model was constructed and the real traffic – travel demand, travel time in equilibrium, and the daily traffic load – was given, see Figure 11.7.

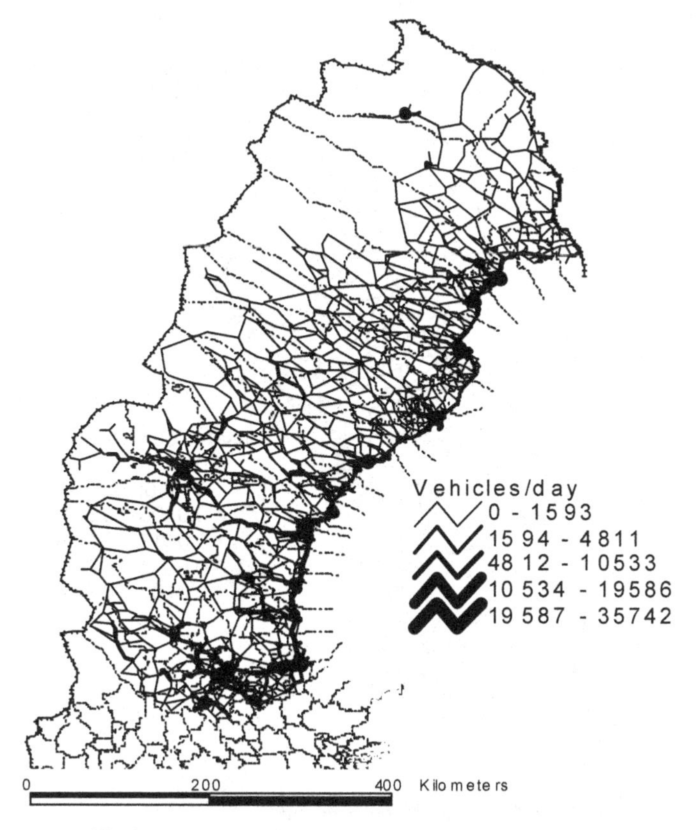

Figure 11.7. Traffic load on the links in a simplified version of the road network of Northern Sweden (Jenelius et al. 2006)

If some specified link is broken, so that no traffic can flow in that link, the total traffic will be reallocated and each driver tries to minimize his travel time. If we disregard the transition of the traffic to a new equilibrium, it is possible to simulate the new traffic situation. In particular, the unsatisfied demand and the total travel time represent the conditional vulnerability corresponding to the breakage of the specific link. If all links are studied in a similar manner it is possible to estimate the relative importance of the various links, see Figure 11.8.

Such an analysis may serve as a basis for deciding where to invest in new roads, if we assume that all links have the same probability to be out of order. It can be of great value in underpinning decisions concerning how to reduce the vulnerability of a specific road network, but it gives little help regarding how to construct new less vulnerable road networks.

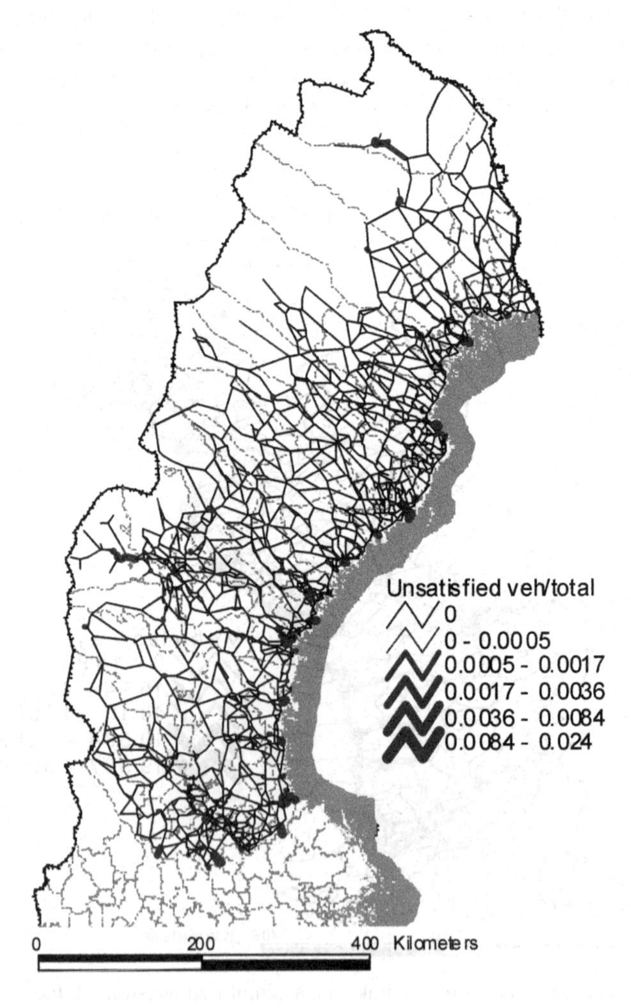

Figure 11.8. Total link importance for the road whole network of Northern Sweden (Jenelius et al. 2006)

11.9 Control and Management

It is evident that the infrastructures of any modern society are vulnerable to random incidents as well as to antagonistic attacks. How can one counteract these threats?

It is important to collect data and to study incidents in order to get an idea about what can happen. The description and analysis of critical infrastructures make it easier to plan defense measures. In principle one can act in the following way:

- Ensure that the nodes and the links of the network are strong against disturbances (robust) so that the vulnerability is reduced. For example the power lines can be placed underground to reduce the effect of extreme weather.
- Ensure that the network has redundancy, i.e., reserve nodes or links, that will diminish the vulnerability. The reserves may always be present, e.g., a small-meshed power net. Another example is reserve generators for a hospital.
- Consider a complementary infrastructure. If a part of a railway net is out of order one can temporarily replace the trains with buses.
- Limit the time to return to a normal state in order to mitigate the consequences of a collapse (resilience). This should include planning of repair and restart.

References

Albert R, Jeong H, Barabási A-L (2000) Error and attack tolerance of complex networks. Nature 406:378–381

Berdica K (2002) An introduction to road vulnerability: what has been done, is done and should be done. Transport Policy 9:117–127

Bennetto J (1997) How IRA plotted to switch off London. The Independent, April 12

Holme P, Kim BJ, Yoon CN, Han SK (2002) Attack vulnerability of complex networks. Physical Review E 65:056109-1–14

Holmgren ÅJ (2006) Using graph models to analyze the vulnerability of electric power networks. Risk Analysis 26:955–969

Holmgren ÅJ, Molin S (2006) Using disturbance data to assess vulnerability of electric power delivery systems. Journal of Infrastructure Systems 12:243–251

Holmgren ÅJ, Jenelius E, Westin J (2007) Evaluating strategies for defending electric power networks against antagonistic attacks. IEEE Transactions on Power Systems 22:76-84

Jenelius E, Petersen T, Mattsson L-G (2006) Importance and exposure in road network. Transportation Research A:517–560

MED (1998) Auckland power supply failure 1998: The report of the ministerial inquiry into the Auckland power supply failure. Ministry of Economic Development (MED), Wellington. www.med.govt.nz/templates/Page____12136.aspx

NPCC (1998) January 1998 Ice storm. Final Report. Northeast Power Coordinating Council (NPCC)

US–Canada Task Force (2004) Final report on the August 14th 2003 blackout in the United States and Canada: causes and recommendations. US–Canada Power System Outage Task Force. https://reports.energy.gov/BlackoutFinal-Web.pdf

West DB (2001) Introduction to graph theory. 2nd ed, Prentice Hall, Upper Saddle River

12

Risk Management

Terje Aven

12.1 Introduction

This chapter reviews and discusses the basic issues and principles of risk management, including:

- Risk acceptability (tolerability)
- Risk reduction and the ALARP principle
- Cautionary and precautionary principles

and presents a case study showing the importance of these issues and principles in a practical management context. Before we take a closer look, let us briefly address some basic features of risk management.

12.2 Overview of Risk Management

The purpose of risk management is to ensure that adequate measures are taken to protect people, the environment, and assets from possible harmful consequences of the activities being undertaken, as well as to balance different concerns, in particular risks and costs. Risk management includes measures both to avoid the hazards and to reduce their potential harm. Traditionally, in industries such as nuclear, oil, and gas, risk management was based on a prescriptive regulating regime, in which detailed requirements were set with regard to the design and operation of the arrangements. This regime has gradually been replaced by a more goal-oriented regime, putting emphasis on what to achieve rather than on the means of achieving it.

Risk management is an integral aspect of a goal-oriented regime (Figure 12.1). It is acknowledged that risk cannot be eliminated but must be managed. There is nowadays an enormous drive and enthusiasm in various industries and in society as a whole to implement risk management in organizations. There are high expectations that risk management is the proper framework through which to achieve high levels of performance.

Risk management involves achieving an appropriate balance between realizing opportunities for gain and minimizing losses. It is an integral part of good management practice and an essential element of good corporate governance. It is an iterative process consisting of steps that, when undertaken in sequence, can lead to a continuous improvement in decision-making and facilitate a continuous improvement in performance.

To support decision-making regarding design and operation, risk analyses are carried out. They include the identification of hazards and threats, cause analyses, consequence analyses, and risk descriptions. The results are then evaluated. The totality of the analyses and the evaluations are referred to as risk assessments. Risk assessment is followed by risk treatment, which is a process involving the development and implementation of measures to modify the risk, including measures designed to avoid, reduce ("optimize"), transfer, or retain the risk. Risk transfer means sharing with another party the benefit or loss associated with a risk. It is typically effected through insurance. Risk management covers all coordinated activities in the direction and control of an organization with regard to risk. This terminology is in line with ISO/IEC Guide 73 (2002).

Figure 12.1. Complex industrial activities require systematic Risk Management. Here, a view of the Mexican Oil Refinery Francisco I Madero. (Photo: Adalberto Ríos Szalay/AGE/Scanpix)

In many enterprises, the risk management tasks are divided into three main categories: strategic risk, financial risk, and operational risk. Strategic risk includes aspects and factors that are important for the enterprise's long-term strategy and plans, for example mergers and acquisitions, technology, competition, political conditions, legislation and regulations, and labor market. Financial risk includes the enterprise's financial situation, and includes:

- Market risk, associated with the costs of goods and services, foreign exchange rates and securities (shares, bonds, etc.)
- Credit risk, associated with a debtor's failure to meet its obligations in accordance with agreed terms
- Liquidity risk, reflecting lack of access to cash; the difficulty of selling an asset in a timely manner

Operational risk is related to conditions affecting the normal operating situation:

- Accidental events, including failures and defects, quality deviations, natural disasters
- Intended acts; sabotage, disgruntled employees, etc.
- Loss of competence, key personnel
- Legal circumstances, associated for instance, with defective contracts and liability insurance

For an enterprise to become successful in its implementation of risk management, top management needs to be involved, and activities must be put into effect on many levels. Some important points to ensure success are (Aven 2008a):

- The establishment of a strategy for risk management, i.e., the principles of how the enterprise defines and implements risk management. Should one simply follow the regulatory requirements (minimal requirements), or should one be the "best in the class"?
- The establishment of a risk management process for the enterprise, i.e., formal processes and routines that the enterprise is to follow.
- The establishment of management structures, with roles and responsibilities, such that the risk analysis process becomes integrated into the organization.
- The implementation of analyses and support systems, such as risk analysis tools, recording systems for occurrences of various types of events, etc.
- The communication, training, and development of a risk management culture, so that the competence, understanding, and motivation level within the organisation is enhanced.

Given the above fundamentals of risk management, the next step is to develop principles and a methodology that can be used in practical decision-making. This is not, however, straightforward. There are a number of challenges and here we address some of these:

- Establishing an informative risk picture for the various decision alternatives
- Using this risk picture in a decision-making context

Establishing an informative risk picture means identifying appropriate risk indices and assessments of uncertainties. Using the risk picture in a decision-making context means the definition and application of risk acceptance criteria, cost benefit analyses and the ALARP principle, which states that risk should be reduced to a level which is as low as is reasonably practicable.

It is common to define and describe risks in terms of probabilities and expected values. This has, however, been challenged, since the probabilities and expected values can camouflage uncertainties; see, e.g., Rosa (1998) and Aven (2003, 2008b). The assigned probabilities are conditional on a number of assumptions and suppositions, and they depend on the background knowledge. Uncertainties are often hidden in this background knowledge, and restricting attention to the assigned probabilities can camouflage factors that could produce surprising outcomes. By jumping directly into probabilities, important uncertainty aspects are easily truncated, and potential surprises may be left unconsidered.

Let us, as an example, consider the risks, seen through the eyes of a risk analyst in the 1970s, associated with future health problems for divers working on offshore petroleum projects. The analyst assigns a value to the probability that a diver would experience health problems (properly defined) during the coming 30 years due to the diving activities. Let us assume that a value of 1 % was assigned, a number based on the knowledge available at that time. There are no strong indications that the divers will experience health problems, but we know today that these probabilities led to poor predictions. Many divers have experienced severe health problems (Aven and Vinnem, 2007). By restricting risk to the probability assignments alone, important aspects of uncertainty and risk are hidden. There is a lack of understanding about the underlying phenomena, but the probability assignments alone are not able to fully describe this status.

Several risk perspectives and definitions have been proposed in line with this realization. For example, Aven (2007a, 2008a) defines risk as the two-dimensional combination of events/consequences and associated uncertainties (will the events occur, what will the consequences be). A closely related perspective is suggested by Aven and Renn (2008a), who define risk associated with an activity as *uncertainty about and severity of the consequences of the activity*, where severity refers to intensity, size, extension, scope and other potential measures of magnitude with respect to something that humans value (lives, the environment, money, etc.). Losses and gains, expressed for example in monetary terms or as the number of fatalities, are ways of defining the severity of the consequences. See also Aven and Kristensen (2005).

In the case of large uncertainties, risk assessments can support decision-making, but other principles, measures, and instruments are also required, such as the cautionary/precautionary principles (HES, 2001; Aven and Vinnem 2007) as well as robustness and resilience strategies (IRGC, 2005); see Sections 2.4 and 12.6. An informative decision basis is needed, but it should be far more nuanced than can be obtained by a probabilistic analysis alone. This has been stressed by many researchers, e.g., Apostolakis (1990) and Apostolakis and Lemon (2005): qualitative risk analysis (QRA) results are never the sole basis for decision-making. Safety- and security-related decision-making is *risk-informed*, not *risk-based.* This conclusion is not, however, justified merely by referring to the need for addressing uncertainties beyond probabilities and expected values. The main issue here is the fact that risks need to be balanced with other concerns.

12.3 Risk Management and the Risk Analysis Process

The risk analysis process is a central part of risk management, and has a basic structure that is independent of its area of application. There are several ways of presenting the risk analysis process, but most structures contain the following three key elements (compare with Figure 12.2):

- Planning
- Risk assessment (execution)
- Risk treatment (use)

Risk analyses, cost–benefit analyses, and related types of analyses provide support for decision-making, leaving the decision-makers to apply decision processes outside the direct applications of the analyses. We speak of managerial review and judgment (Figure 12.2). It is not desirable to develop tools that prescribe or dictate the decision. That would mean too mechanical an approach to decision-making and would fail to recognize the important role of management in performing difficult value judgments involving uncertainty.

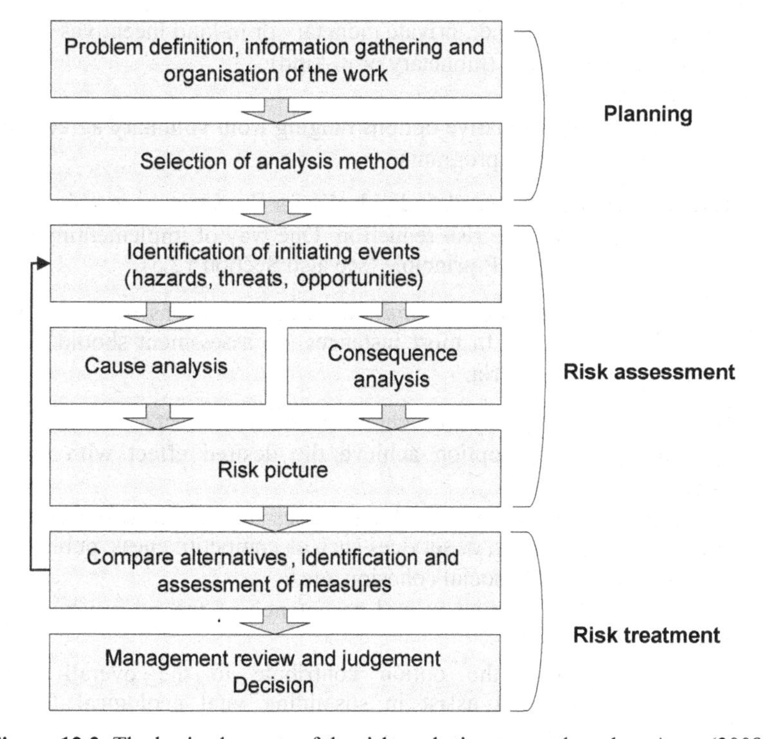

Figure 12.2. The basic elements of the risk analysis process, based on Aven (2008b)

Nonetheless, there is a need for guidance and a structure for decision-making in situations involving high risks and large uncertainties. The aim must be to achieve a certain level of consistency in decision-making and of confidence in obtaining desirable outcomes. Such guidance and structure exist to some degree, and the challenge is to find the right level. This will be discussed in more detail in the following sections.

More generally, the risk management process can be divided into the following steps (IRGC, 2005) according to the basic model used by decision theory (Lindley, 1985; Morgan, 1990; Keeney, 1992; Hammond et al., 1999):

1. Identification and generation of risk management options. Risk reduction can be accomplished by many various means, including:

- Technical standards and limits that prescribe the permissible threshold of concentrations, emissions, take-up or other measures of exposure
- Performance standards for technological and chemical processes such as minimum temperatures in waste incinerators
- Technical prescriptions referring to the blockage of exposure (e.g., via protective clothing) or the improvement of resilience (e.g., via immunization or earthquake-tolerant structures)
- Governmental economic incentives including taxation, duties, subsidies, and certification schemes
- Third party incentives, i.e., private monetary or in-kind incentives
- Compensation schemes (monetary or in-kind)
- Insurance and liability
- Cooperative and informative options ranging from voluntary agreements to labelling and education programs

All these options can be used individually or in various combinations to accomplish even more effective risk reduction. One way of implementing the risk reduction is to apply the ALARP-principle (see also Section 12.5).

2. Assessment of risk management options. Each of the options will have desired and unintended consequences. In most instances, an assessment should be made according to the following criteria:

- Effectiveness. Does the option achieve the desired effect?
- Efficiency. Does the option achieve the desired effect with the least resource consumption?
- Minimization of external side-effects. Does the option infringe on other valuable goods, benefits, or services such as competitiveness, public health, environmental quality, social cohesion, etc.?
- Does it impair the efficiency and acceptance of the governance system itself?
- Sustainability. Does the option contribute to the overall goal of sustainability? Does it assist in sustaining vital ecological functions, economic prosperity, and social cohesion?
- Fairness. Does the option burden the subjects of regulation in a fair and equitable manner?
- Political and legal acceptability. Is the option compatible with legal requirements and political programs?

- Ethical acceptability. Is the option morally acceptable?
- Public acceptance. Will those individuals who are affected by it accept the option? Are there cultural preferences or symbolic connotations that have a strong influence on how the risks are perceived?

Measuring management options against these criteria may give rise to conflicting messages and results. Many measures that prove to be effective may turn out to be inefficient or unfair to those who will be burdened. Other measures may be sustainable but not acceptable to the public or important stakeholders.

3. A broad evaluation and judgment of risk management options: This step integrates the evidence as to how the options perform with regard to the evaluation criteria with a value judgment concerning the relative weight which should be assigned to each criterion. Ideally, the evidence should come from experts and the relative weights from decision-makers. In practical risk management, options are evaluated in close cooperation between both experts and decision-makers.

4. Selection of risk management options: Once the different options have been evaluated, a decision has to be made as to which options are selected and which are rejected. In most cases trade-offs have to be made between different concerns.

12.4 Risk Management Strategies

We introduce four categories of risk problems characterized by simplicity, complexity, uncertainty, and ambiguity (IRGC, 2005; Aven and Renn, 2008b; Renn, 2008). Using this classification, it is possible to design generic strategies of risk management to be applied to classes of risks, thus simplifying the risk management process, see Table 12.1.

Simplicity

Simplicity is characterized by situations and problems with low complexity, and few uncertainties or ambiguities. Examples include car accidents, smoking, regularly recurring natural disasters, or safety devices for high buildings. Note that simplicity does not mean that the risks are low. The possible negative consequences may indeed be very large. The point is that the values that are exposed are non-controversial and the uncertainties low. It is possible to predict fairly accurately the occurrence of events and/or their consequences.

Complexity

Complexity refers to the difficulty of identifying and quantifying causal links between a multitude of potential causal agents and specific observed effects. The nature of this difficulty may be traced back to interactive effects among these agents (synergisms and antagonisms), long delay periods between cause and effect, inter-individual variation, intervening variables, and others.

Complex risk problems are often associated with major scientific dissent about dose–effect relationships or the alleged effectiveness of measures to decrease vulnerabilities. Examples of activities and systems with high complexity include sophisticated chemical facilities, synergistic effects of potentially toxic substances, failure of large interconnected infrastructures, and critical loads to sensitive ecosystems.

Uncertainty

Uncertainty refers to the difficulty of predicting the occurrence of events and/or their consequences based on incomplete or invalid databases, possible changes in the causal chains and their context conditions, extrapolation methods when making inferences from experimental results, modeling inaccuracies or variations in expert judgments. Uncertainty may result from an incomplete, or inadequate reduction of complexity, and it often leads to expert dissent about the risk characterization.

Examples of high uncertainty include many natural disasters (such as earthquakes), possible health effects of mass pollutants, acts of violence such as terrorism and sabotage, and long-term effects of introducing genetically modified species into the natural environment. In the case of a terrorism risk, the consequences of an attack can be fairly accurately predicted, but the time and type of attack are subject to large uncertainties. The uncertainty may be a result of "known uncertainties" – *we know what we do not know*, and "unknown uncertainties" (ignorance or non-knowledge) – *we do not know what we do not know*.

Ambiguity

Ambiguity refers to different views relating to:

- The relevance, meaning and implications of the basis for the decision-making (interpretative ambiguity); or
- The values to be protected and the priorities to be made (normative ambiguity)

What does it mean, for example, if neuronal activities in the human brain are intensified when subjects are exposed to electromagnetic radiation? Can this be interpreted as an adverse effect or is it just a bodily response without any health implication? Examples of high interpretative ambiguity include low-dose radiation (ionizing and non-ionizing), low concentrations of genotoxic substances, food supplements, and the hormone treatment of cattle. Normative ambiguities can be associated, for example, with passive smoking, nuclear power, pre-natal genetic screening, and genetically modified foodstuffs.

Discussion

Let us look somewhat closer into the uncertainty-induced risk problems. If there is a high degree of uncertainty, risk management needs to create robustness, to incorporate criteria including aspects such as reversibility, persistence, and ubiquity, and to select management options also dealing with worst-case scenarios (such as containment of hazardous activities, close monitoring of risk-bearing activities, securing reversibility of decisions in case the consequences turn out to be greater than expected).

Table 12.1. Risk problem categorisations and their implications for risk management

Risk problem category	Management strategy	Appropriate instruments
Simple risk problem	Risk informed	Statistical analysis Risk assessments Cost–benefit analyses
	Routine-based risk treatment (risk reduction)	Trial and error Technical standards Economic incentives Education, labeling, information Voluntary agreements
Complexity-induced risk problems	Risk informed (risk agent)	Risk assessments Cost–benefit analyses Characterising the available evidence: • Expert consensus seeking tools (e.g., Delphi) • Results fed into routine operation
	Risk informed	Risk assessments Cost–benefit analyses
	Robustness focused (risk-absorbing system)	Improving buffer capacity of risk target through: • Additional safety factors • Redundancy and diversity in designing safety devices • Improving coping capacity • Establishing high-reliability organizations
Uncertainty-induced risk problems	Risk informed and caution/ precaution based (risk agent)	Risk assessments. Broad risk characterizations, highlighting uncertainties and features like persistence, ubiquity, etc. Tools include: • Containment • ALARP (as low as reasonably practicable) • BACT (best available control technology), etc.
	Risk informed	Risk assessments. Broad risk characterizations.
	Robustness and resilience focused (risk-absorbing system)	Improving capability to cope with surprises • Diversity of means to accomplish desired benefits • Avoiding high vulnerabilities • Allowing for flexible responses • Preparedness for adaptation
Ambiguity-induced risk problems	Risk informed and Discourse based	Political processes. Application of conflict resolution methods for reaching consensus or tolerance for risk evaluation results and management option selection: • Integration of stakeholder involvement in reaching closure • Emphasis on communication and social discourse

Modified from IRGC (2005) and Aven and Renn (2008b).

Management of risks characterized by large uncertainties should be guided by the cautionary and precautionary approach (see Section 12.6). One should pursue a cautious strategy that allows learning by restricted errors. The main management philosophy for this risk class is to allow small steps in implementation (containment approach) that enable risk managers to stop or even reverse the process as new knowledge is produced or negative side-effects become visible.

With respect to risk-absorbing systems, the main objective is to make these systems resilient so they can withstand or even tolerate surprises. In contrast to robustness, where potential threats are known in advance and the absorbing system needs to be prepared to face these threats, resilience is a protective strategy against unknown or highly uncertain events. Instruments for resilience include the strengthening of the immune system, diversification of the means for approaching identical or similar ends, reduction of the overall catastrophic potential or vulnerability even in the absence of a concrete threat, design of systems with flexible response options, and the improvement of conditions for emergency management and system adaptation. Robustness and resilience are closely linked, but they are not identical. To some extent they require different types of actions and instruments.

12.5 Risk Acceptability (Tolerability) and Risk Reduction (ALARP)

Risk regulation in industry is nowadays to a large extent goal-oriented, i.e., high-level performance measures need to be specified and various type of analyses have to be conducted to identify the best possible arrangements and measures according to these performance measures. There has been a significant trend internationally in this direction in recent years. On the other hand, different approaches have been adopted in order to implement this common objective, if worldwide regulatory regimes are considered.

Whereas the objective may seem simple as a principle, there are certainly some challenges to be faced in the implementation of the principle. One of the main challenges is related to the use of pre-determined quantitative risk-acceptance criteria (tolerability limits), expressed as upper limits of acceptable (tolerable) risk. See also Aven (2007b). Some examples of risk-acceptance criteria used for an offshore installation are:

- The FAR value should be less than 10 for all personnel on the installation, where the FAR value is defined as the expected number of fatalities per 100 million hours of exposure.
- The individual probability that a person is killed in an accident during one year should not exceed 0.1 %.

Note that in the following, when the term "risk-acceptance criteria" is used, we always have in mind such upper limits. Should we define such criteria before any analysis of the systems has been carried out? The traditional textbook answer is yes. First come the criteria and then the analysis to see if these criteria are met, after which, and according to the assessment results, the need for risk-reducing measures is determined. Such an approach is intuitively appealing, but a closer look reveals some challenges:

- The introduction of pre-determined criteria may give the wrong focus, meeting these particular criteria rather than achieving overall good and cost-effective solutions and measures. As stressed by Fischhoff et al. (1981), one does not accept risk, but one accepts options that entail some level of risk among their consequences.
- The risk analyses – the tools used to check whether the criteria are met – are not generally sufficiently accurate to permit such a mechanical use of the criteria.

The first item above is the main point. The adherence to a mechanistic use of risk-acceptance criteria does not provide a good structure for the management of risk to personnel, environment, or assets. This is clearly demonstrated in the case of environmental risk. The acceptability of operations with respect to environmental risk is typically decided on the results of a political process and, following this process, risk acceptance is not an issue and risk-acceptance criteria have not an important role to play. Norwegian petroleum authorities have required risk-acceptance criteria for more than 10 years, but such criteria have almost never led to any improvement from an environmental point of view. This is further discussed by Aven and Vinnem (2005). If a high level of safety or security is to be achieved, mechanisms other than risk acceptance (tolerability) criteria need to be implemented. Such criteria give focus to obtaining a minimum safety standard instead of continuous improvement and risk reduction.

The ALARP principle represents such a mechanism. The ALARP principle expresses that the risk should be reduced to a level that is as low as is reasonably practicable. A risk-reducing measure should be implemented provided it cannot be demonstrated that the costs are grossly disproportional to the gains obtained (HSE, 2001). Risk assessments play an important role in ALARP processes, as risk reduction needs to be based on risk assessment. Risk must be described and the effect of risk-reducing measures determined. Although the ALARP principle has been in use for many years, its interpretation and implementation procedures are still being discussed. The ALARP principle is normally applied together with a limit for intolerable risk and a limit for negligible (acceptable) risk. The interval between these two limits is often called the ALARP region.

In most cases in practice, the risk is found to be within the ALARP region, and the ALARP principle is adopted and an ALARP assessment process is required. This will include a dedicated search for possible risk-reducing measures and a subsequent assessment of these in order to determine which to be implemented.

To verify ALARP, it is usual to use procedures based mainly on engineering judgments and codes, but traditional cost–benefit analyses and cost–effectiveness analyses are also utilized. In such analyses, guidance values are often used to specify the values that define "gross disproportion." Such values may vary from substantially less than USD 0.1 million, up to more than USD 10 million.

A typical figure for the value of statistical life used in cost–benefit analysis in the transport sector is USD 1–3 million (HSE, 2006; Aven and Vinnem, 2005). For other sectors the figure is much higher, for example in the offshore UK industry it is common to use GBP 6 million (about USD 10 million) (HSE, 2006). The higher figure takes into account the potential for multiple fatalities and uncertainty.

The practice of using traditional cost-benefit analyses and cost-effectiveness analyses to verify ALARP has been questioned (Aven and Abrahamsen, 2007). The ALARP principle gives weights to risks. However, cost–benefit analyses calculating expected net present values ignore the risks (uncertainties), and the use of this approach to weight the risk is therefore problematic. Aven and Flage (2009) provide a detailed discussion of this issue.

In our view it is necessary to acknowledge that there is no simple and mechanistic method or procedure for balancing different concerns. When using analyses we have to adopt a pragmatic perspective. We have to acknowledge the limitations of the tools and use them in a broader process where the results of the analyses are seen as just one part of the information supporting the decision-making. The results need to be subjected to extensive sensitivity analysis.

One way of assessing "gross disproportion" is outlined below (Aven and Vinnem, 2005, 2007):

- Perform a crude analysis of the benefits and burdens of the various alternatives addressing attributes related to feasibility, conformance with good practice, economy, strategy considerations, risk, and social responsibility.
- The analysis should typically be qualitative and its conclusions summarized in a matrix with performance shown by a simple categorization system such as Very positive, Positive, Neutral, Negative, Very negative. From this crude analysis a decision can be made to eliminate some alternatives and include new ones for further detailed analysis. Frequently, such crude analyses give the necessary platform for choosing one appropriate alternative. When considering a set of possible risk-reducing measures, a qualitative analysis in many cases provides a sufficient basis for identifying which measures to implement, as these measures are in accordance with good engineering or with good operational practice. Also many measures can quickly be eliminated as soon as the qualitative analysis reveals that the burdens are much greater than the benefits.
- From this crude analysis the need for further analysis is determined, to give a better basis for deciding on which alternative(s) to choose. This may include various types of risk analyses.
- Other types of analyses may be conducted, for example, costs, and indices such as expected cost per expected number of saved lives could be computed to provide information about the effectiveness of a risk-reducing measure or in order to compare various alternatives. The expected net present value may also be computed when appropriate. Sensitivity analyses should be performed to see the effects of varying the values of a statistical life and other key parameters. The conclusions are often fairly straightforward when calculating indices such as the expected cost per expected number of saved lives over the field life and the expected cost per

expected averted ton of oil spill over the field life. If a conclusion regarding gross disproportion is not clear (the costs are not so large), then these measures and alternatives are clear candidates for implementation. If a risk-reducing measure has a positive expected net present value it should be implemented.

- An assessment of uncertainties in the underlying phenomena and processes is carried out. Which factors can yield unexpected outcomes with respect to the calculated probabilities and expected values? Where are the gaps in knowledge? What critical assumptions have been made? Are there areas where there is substantial disagreement among experts? What are the vulnerabilities of the system?
- An analysis of manageability takes place. To what extent is it possible to control and reduce the uncertainties and thereby arrive at the desired outcome? Some risks are more manageable than others in the sense that there is a greater potential to reduce risk. An alternative can have a relatively large calculated risk under certain conditions, but the manageability could be good and could result in a far better outcome than expected.
- An analysis of other factors such as risk perception and reputation, should be carried out whenever relevant, although it may be difficult to describe how these factors would affect the standard indices used in economy and risk analysis to measure performance.
- A total evaluation of the results of the analyses should be performed, to summarize the pros and cons of the various alternatives, where considerations of the constraints and limitations of the analyses are also taken into account.

Note that such assessments are not necessarily limited to the ALARP processes. The above process can be used also in other contexts where decisions are to be made under uncertainty.

ALARP assessments require that appropriate measures be generated. If the aim is to satisfy the risk-acceptance criteria or tolerability limits, there may be little incentive for identifying risk-reducing measures if the criteria and limits are relatively easy to meet. Risk acceptance (tolerability) can in such cases be reached without implementing any specific measures.

As a rule, suggestions for measures always arise in a risk analysis context, but a systematic approach for the generation of these suggestions is often lacking. In many cases, the measures also lack ambition. They lead to only small changes in the risk picture. A possible way to approach this problem is to apply the following principles:

- On the basis of existing solutions (base case), identify measures that can reduce the risk by, for example, 10 %, 50 % and 90 %.
- Specify solutions and measures that can contribute to reaching these levels.

The solutions and measures must then be assessed before any decision is made regarding possible implementation.

A potential strategy for the assessment of a measure, if the analysis based on expected present value or expected cost per expected number of lives saved has not produced any clear recommendation, can be that the measure be implemented if several of the following questions are assessed in the affirmative (Aven and Vinnem, 2007):

- Is there a relatively high personnel risk or environmental risk?
- Is there considerable uncertainty (related to phenomena, consequences, conditions) and will the measure reduce the uncertainty?
- Does the measure significantly increase manageability? High competence among the personnel can give increased assurance that satisfactory outcomes will be reached, for example fewer leakages.
- Does the measure contribute to a more robust solution?
- Is the measure based on best available technology (BAT)?
- Are there unsolved problem areas related to personnel safety and/or the work environment?
- Are there possible areas where there is conflict between these two aspects?
- Are there strategic considerations?

12.6 The Cautionary and Precautionary Principles

The cautionary principle is a basic principle in risk management, expressing the idea that, in the face of uncertainty, caution should be a guiding concern. This principle is being implemented in all industries through regulations and requirements.

For example, in the Norwegian petroleum industry it is a regulatory requirement that fireproof panels of a certain quality in all walls facing process and drilling areas shall protect the living quarters on an installation. This is a Standard adopted to achieve a minimum safety level, based on established practice of many years of operation of process plants. If a fire occurs it represents a hazard for the personnel. In the case of such an event, the personnel in the living quarters must be protected. The assigned probability that the living quarters on a specific installation will be exposed to fire may be judged to be low, but we know that fires occur from time to time in such plants. The justification is experience from similar plants and sound judgment. A fire is not an unlikely event, and we should be prepared. We need no references to risk analysis or cost–benefit analysis. The requirement is based on cautionary thinking. See also PSA (2001).

Risk analyses, cost–benefit analyses, and similar types of analyses are tools providing insight into risks and the trade-offs involved. But they are merely tools, with strong limitations. Their results are dependent on a number of assumptions and suppositions. The analyses do not express objective results. Being cautious also reflects this fact. We should not put more trust in the predictions and assessments of the analyses than can be justified by the methods used.

In the face of uncertainties related to the possible occurrence of hazardous situations and accidents, we should be cautious and adopt principles of safety management, such as:

- Robust design solutions, so that deviations from normal conditions do not lead to hazardous situations and accidents
- Design for flexibility, meaning that it is possible to utilise a new situation and adapt to changes in the frame conditions
- Implementation of safety barriers, to reduce the negative consequences of hazardous situations if they should occur, for example a fire
- Improvement of the performance of barriers by employing redundancy, maintenance and testing, etc.
- Quality control/quality assurance
- The precautionary principle, saying that if there is a lack of scientific certainty on the possible consequences of an activity, we should not carry out the activity, or measures should be implemented
- The ALARP-principle, saying that the risk should be reduced to a level, which is as low as is reasonably practicable

The level of caution adopted will of course have to be balanced against other concerns such as costs. However, all industries should introduce some minimum requirements to protect people and the environment. These requirements can be considered justified by reference to the cautionary principle.

The precautionary principle has many definitions, see, e.g., Löfstedt (2003) and Sandin (1999). The most commonly used definition is probably that given in the 1992 Rio Declaration (UNCED, 1992):

> *In order to protect the environment, the precautionary approach shall be widely applied by States according to their capabilities. Where there are threats of serious or irreversible damage, lack of full scientific certainty shall not be used as a reason for postponing cost-effective measures to prevent environmental degradation.*

Seeing beyond environmental protection, a definition such as the following reflects what we believe to be a typical way of understanding this principle:

> *The precautionary principle is the ethical principle that, if the consequences of an action, especially the use of technology, are subject to scientific uncertainty, then it is better not to carry out the activity rather than to risk the uncertain, but possibly very negative, consequences.*

The key aspect is that if there is a lack of scientific certainty as to the consequences of an action, then that activity should not be carried out, or measures should be implemented. The problem in this statement is that the meaning of the term "scientific certainty" is not at all clear. As the focus is on the future consequences of the action, there can be no (or at least very few) cases with known outcomes. Hence scientific certainty must mean something else. Three natural candidates are:

- Knowing the type of consequences that could occur
- Being able to predict the consequences with sufficient accuracy

- Having accurate descriptions or estimates of the real risks, interpreting the real risk as the consequences of the action

If one of these interpretations is adopted, the precautionary principle could be applied either when we do not know the type of consequences that might occur, or when we have poor predictions of the consequences, risk descriptions, or estimates. Aven (2006) and Aven and Vinnem (2007) argue that the second point above should be adopted, restricting the precautionary principle to situations where there is a lack of understanding of how the consequences of the activity may be influenced by the underlying factors. In the case of an oil spill, there is a state of scientific uncertainty if biologists cannot explain with sufficient confidence how the oil exposure will affect the environment. It must be acknowledged that it is not possible to establish science-based criteria for when the precautionary principle should apply. Judging when there is a lack of scientific certainty is a value judgment.

12.7 Process Plant Safety – A Case Study

12.7.1 Introduction

This case study is based on Aven (2008b). A risk analysis is to be carried out for a process plant. The task involves a significant modification of the plant, adding new production equipment which will have an impact on the risk level. New equipment units mean additional potential leak sources with respect to gas and/or oil, which may lead to fire and/or explosion. The decision to be made is whether or not to install additional fire protection for the personnel in order to reduce expected consequences in the event of a critical fire. The design of the plant means that there is limited protection for the personnel using escape routes against fire and explosion effects.

Following a review of the problem, it was soon evident that there were three alternatives:

- Making minor improvements in order to compensate for the increased risk due to new equipment, but no further risk reduction
- Installing protective shielding on existing escape routes together with over-pressure protection in order to prevent smoke from entering into the enclosed escape routes
- Doing nothing, accepting the situation as it is

The purpose of the risk analysis is to provide a basis for selecting an alternative. This basis requires a risk description and an associated evaluation. The analysis emphasizes both qualitative and quantitative aspects, forming part of an ALARP process.

Following a traditional approach to risk analysis, the ambition now would have been to calculate risk using risk indices such as IR (individual risk, FAR (fatal accident rate) and PLL (potential loss of lives). If the expected cost in relation to the cost of a statistical life is high, the safety investments cannot be justified. A typical result of such an analysis would have been that protective shielding costs too much relative to the calculated risk reduction.

In the following, we present a broader risk analysis process, acknowledging that risk involves more than expected values and probabilities. The process highlights the uncertainties and phenomena and processes that could lead to surprises relative to the expected values, as well as manageability factors. Manageability is related to the extent to which it is possible to control and reduce the uncertainties, and achieve the desired outcome. The expected values and the probabilistic assessments performed in the risk analyses provide predictions for the future, but some risks are more manageable than others, in the sense that the potential for reducing the risk is greater for some risks than for others. Proper management seeks to obtain desirable consequences. This leads to considerations, for example as to how to run processes to reduce the risks (uncertainties) and how to deal with human and organizational factors and develop a good safety culture.

12.7.2 Risk Assessment

In this case, we have only one type of hazard, hydrocarbon leakages, and the main barrier functions are:

- Prevent loss of containment (i.e., prevent leakage)
- Prevent ignition
- Reduce cloud/emissions
- Prevent escalation
- Prevent fatalities

For each barrier function, a qualitative analysis is conducted as in a standard quantitative risk analysis (QRA) using fault trees and events. The analysis also highlights operational aspects, such as a possible hydrocarbon release starting from the initiating event "valve in wrong position after maintenance." In order to analyze the performance of the barriers, fault trees can be constructed. In addition, influence diagram, are used to show how the risk-influencing factors (such as training, time, pressure, competences, etc.) influence the various events; see Aven (2008b).

A qualitative analysis was performed, based on the following information:

- Historical records for hazardous situations and accidental events, such as leakages
- Historical records for the performance of barrier elements/systems, such as the reliability of the gas and fire detection systems
- Measurements and evaluations of the state of the technical conditions of the various systems in the installation
- Investigation reports relating to near misses and accidents
- Evaluations of the performance of the main barrier systems based on the historical records and expert judgments
- Interviews with key personnel regarding the operation and maintenance of the installation

The aim of this part of the analysis is to establish a picture of the state of the system (plant), and to identify critical safety issues and factors, as well as uncertainties in phenomena and processes. It constitutes an important part of the basis (background information) for the risk assignments.

From this analysis, it is concluded that the main safety challenges are:

- The increased accident risk caused by the modification
- Poor performance on a set of barrier elements
- Deterioration of critical equipment, making substantial maintenance necessary

In the presentation of the risk picture, these challenges are an integrated part of the results, as is explained below.

In the risk quantification, the analysis group assigns probabilities for a number of events, using a rather crude risk matrix approach. For example, the hazardous situations identified are gas releases of different magnitudes, which may divide into three categories, r_1, r_2, and r_3. For each of these, the number of fatalities C is assessed. The uncertainty distribution has an expected value (center of gravity) $E[C|r_i]$ which, together with the assigned probabilities $p_i = P$(release of category r_i during the coming year of operation), forms the risk matrix. If p_i is large (typically larger than 0.5) the probability should be replaced by the frequency, the expected number of leakages. Combining the probabilities p_i and the expected values $E[C|r_i]$ yields the expected number of fatalities caused by this scenario. The total expected number of fatalities is obtained as the summation of these values.

Similarly, risk matrices are established for ignited releases (fires) and the expected consequences. If f_i denotes the assigned probability for the ith fire scenario (severe fire) and $E[C|\text{fire}_i]$ denote the expected number of fatalities given this scenario, the pair (f_i, $E[C|\text{fire}_i]$) forms the risk matrix.

A set of scenarios is defined, and relevant consequence calculations are carried out providing insight into, e.g., initial release rates and development of the discharge concentration (when and where a combustible mixture may arise). From these studies and the analysis group's general risk analysis experience, the uncertainties related to releases and consequences are assessed. The assessments are based on all relevant information, including the identified poor performance of some of the safety barriers and the equipment deterioration problem. The consecutive assigned probabilities and expected values represent the analysis group's best judgments, based on the available information and knowledge.

To simplify the assessments, some categories of probabilities and expected values are defined. For the probabilities p_i the following categories are used: >50 %, 10–50 %, 1–10 %, 0.01–1 % and < 0.01 %.

Hence for the second category 1–5 events are predicted for a 10-year period. To specify a probability, a default value is assigned, by general comparison with similar types of systems, and based on studies performed as a part of the QRA. All key factors influencing the probabilities are then assessed and, based on this assessment, the probabilities are adjusted. In the present case, the equipment deterioration was identified as a critical issue. However, instead of adjusting the probabilities an assumption was made that extended maintenance activities will be carried out.

The effect of the modification was assessed and the probabilities and expected values adjusted, and the effect of the implementation of protective shielding on the existing escape routes was also assessed. The following changes were quantified:

1. Modification:

- Increased p_i, 5 %
- Increased f_i, 5 %
- Increased PLL, 5 %
- Increased IR, 10 % for a specific personnel group

2. Effect of the implementation of the protective shielding on existing escape ways (after modification):

- Reduced PLL, 30 %
- Reduced IR, 50 % for a specific personnel group

These figures represent the best judgments of the analysis group. This risk picture is supplemented with uncertainty considerations and assessments of vulnerability.

Equipment Deterioration and Maintenance
It was assumed that the deterioration of critical equipment would not cause safety problems if a special maintenance program were implemented. However, experience of such plants shows that unexpected problems occur with time. Production of oil leads to changes in operating conditions and problems that often need to be solved by the addition of chemicals. These are all factors increasing the probability of corrosion, material brittleness, and other conditions that may cause leakage. The quantitative analysis has not taken into account the fact that surprises may occur. The analysis group was concerned about this uncertainty factor, and it was reported together with the quantitative assessments.

Barrier Performance
The historical records show a poor performance for a set of critical safety barrier elements, particularly with some types of safety valves. The assignments of the expected number of fatalities $E[C]$ given a leakage or a fire scenario were based on average conditions for the safety barriers, and adjustments were made to reflect the historical records. However, the changes made were small. The poor performance of the barrier elements would not necessarily result in any significant reduction in the probabilities of barrier system failures, as most of the barrier elements were not safety-critical.

The barrier systems were designed with a lot of redundancies. Nevertheless, this also causes concern, as the poor performance may indicate that there is an underlying problem of an operational and maintenance character, which results in reduced availability of the safety barriers in a hazardous situation. There are a number of dependencies among elements for these systems, and the risk analysis methods for studying these are simplified with strong limitations. Hence there is also an uncertainty aspect related to the barrier performance.

These issues are related to the quantitative risk result in the following way. A probability assignment is a value $P(A|K)$, a probability of an event given the background information K. Assumptions are a part of K. In the ageing case, for example, K includes the assumption that an effective maintenance program is implemented. The probability assignments do not therefore reflect all types of uncertainties. We may ask whether the equipment deterioration may cause surprises, resulting in leakages. The numbers do not explicitly express this issue. This is always the case – the quantitative analyses have to be placed in a broader context of the constraints and limitations of the analyses. Aspects to consider relate to hidden uncertainties in the assumptions, the problem of expressing uncertainties using probabilities, and the selection of appropriate risk indices.

12.7.3 Summary of the Risk Picture

The results of the analysis in the case study are summarized in Tables 12.2 and 12.3.

Table 12.2. Overall assessment of modification and measures

Modification	The resulting fire risk is not deemed to represent any great problem, the risk increases by 5 %
Fire protection implemented	Reduced individual risk by 50 % for specific personnel group

Given this information, what is the conclusion? Should the protective shielding be implemented? The decision depends on the attitude of the management towards risks and uncertainties and, of course, the costs. The analysis does not direct the decision-makers but it provides a basis for making appropriate decisions. If (say) the investment cost for this measure is of the order of USD 5 million, can this extra cost be justified?

Table 12.3. Uncertainty factors

Uncertainty factors	Minor problem	Significant problem	Severe problem
Deterioration of equipment		×	
Barrier performance	×		

The expected reduced number of fatalities is rather small, and hence the expected cost per expected saved life (the implied value of a statistical life) would be rather high, and a traditional cost–benefit (cost-effectiveness) criterion would not justify the measure. Suppose the assigned expected reduced number of fatalities is 0.1. The implied value of a statistical life, i.e., the expected cost per expected number of lives saved, is then given by a factor 5/0.1, i.e., USD 50 million. If this number is considered in isolation, in a quantitative context, it would normally be considered too high to justify the implementation of the measure.

This number is strongly dependent on key assumptions. Sensitivity analyses show that the implied value of a statistical life is very sensitive to changes in the probabilities of leakage and ignition.

It is necessary to look beyond the results of the cost-effectiveness analysis. The purpose of the safety investment is to reduce risks and uncertainties, and the traditional cost-effectiveness (cost–benefit) analyses do not reflect these concerns in an appropriate way. These analyses are based on the use of expected values, and hence give little weight to risks and uncertainties. Investments in safety are not only justified by reference to expected values. Management review and judgment is required to balance the different concerns and give the proper weights to the risks and uncertainties. If management places an emphasis on the cautionary principle, then the investment in fire protection can be justified, despite the fact that a traditional cost–benefit consideration indicates that the measure cannot be justified.

12.8 Discussion

When various solutions and measures are to be compared and a decision is to be made, the analysis and assessments that have been conducted provide a basis for such a decision. In many cases, established design principles and standards provide clear guidance. Compliance with such principles and standards must be among the first reference points when assessing risks. It is common thinking that risk management processes, and especially ALARP processes, require formal guidelines or criteria (e.g., risk acceptance criteria and cost-effectiveness indices) to simplify the decision-making. Care must, however, be shown when using this type of formal decision-making criteria, as they easily result in a mechanization of the decision-making process. Such a mechanization is unfortunate because:

- Decision-making criteria based on risk-related numbers alone (probabilities and expected values) do not capture all the aspects of risk, costs, and benefits
- No method has a precision that justifies a mechanical decision based on whether the result is over or below a numerical criterion
- It is a managerial responsibility to make decisions under uncertainty, and management should be aware of the relevant risks and uncertainties

Apostolakis and Lemon (2005) adopt a pragmatic approach to risk analysis and risk management, acknowledging the difficulties of determining the probabilities of an attack. Ideally, they would like to implement a risk-informed procedure, based on expected values. However, since such an approach would require the use of probabilities that have not been "rigorously derived," they see themselves forced to resort to a more pragmatic approach.

This is one possible approach when facing problems of large uncertainties. The risk analyses simply do not provide a sufficiently solid basis for the decision-making process. We argue along the same lines. There is a need for a management review and judgment process. It is necessary to see beyond the computed risk picture in the form of the probabilities and expected values. Traditional quantitative risk analyses fail in this respect. We acknowledge the need for analyzing risk, but question the value added by performing traditional quantitative risk analyses in the case of large uncertainties. The arbitrariness in the numbers produced can be significant, due to the uncertainties in the estimates or as a result of the uncertainty assessments being strongly dependent on the analysts.

It should be acknowledged that risk cannot be accurately expressed using probabilities and expected values. A quantitative risk analysis is in many cases better replaced by a more qualitative approach, as shown in the examples above; an approach which may be referred to as a semi-quantitative approach. Quantifying risk using risk indices such as the expected number of fatalities gives an impression that risk can be expressed in a very precise way. However, in most cases, the arbitrariness is large. In a semi-quantitative approach this is acknowledged by providing a more nuanced risk picture, which includes factors that can cause "surprises" relative to the probabilities and the expected values. Quantification often requires strong simplifications and assumptions and, as a result, important factors could be ignored or given too little (or too much) weight. In a qualitative or semi-quantitative analysis, a more comprehensive risk picture can be established, taking into account underlying factors influencing risk. In contrast to the prevailing use of quantitative risk analyses, the precision level of the risk description is in line with the accuracy of the risk analysis tools. In addition, risk quantification is very resource demanding. One needs to ask whether the resources are used in the best way. We conclude that in many cases more is gained by opening up the way to a broader, more qualitative approach, which allows for considerations beyond the probabilities and expected values.

The traditional quantitative risk assessments as seen for example in the nuclear and the oil & gas industries provide a rather narrow risk picture, through calculated probabilities and expected values, and we conclude that this approach should be used with care for problems with large uncertainties. Alternative approaches highlighting the qualitative aspects are more appropriate in such cases. A broad risk description is required. This is also the case in the normative ambiguity situations, as the risk characterizations provide a basis for the risk evaluation processes. The main concern is the value judgments, but they should be supported by solid scientific assessments, showing a broad risk picture. If one tries to demonstrate that it is rational to accept risk, on a scientific basis, too narrow an approach to risk has been adopted. Recognizing uncertainty as a main component of risk is essential to successfully implement risk management, for cases of large uncertainties and normative ambiguity.

A risk description should cover computed probabilities and expected values, as well as:

- Sensitivities showing how the risk indices depend on the background knowledge (assumptions and suppositions)
- Uncertainty assessments
- Description of the background knowledge, including models and data used

The uncertainty assessments should not be restricted to standard probabilistic analysis, as this analysis could hide important uncertainty factors. The search for quantitative, explicit approaches for expressing the uncertainties, even beyond the subjective probabilities, may seem to be a possible way forward. However, such an approach is not recommended. Trying to be precise and to accurately express what is extremely uncertain does not make sense. Instead we recommend a more open qualitative approach to reveal such uncertainties. Some might consider this to be less attractive from a methodological and scientific point of view. Perhaps it is, but it would be more suited for solving the problem at hand, which is about the analysis and management of risk and uncertainties.

References

Apostolakis G (1990) The concept of probability in safety assessments of technological systems. Science 250:1359–1364

Apostolakis GE, Lemon DM (2005) A screening methodology for the identification and ranking of infrastructure vulnerabilities due to terrorism. Risk Analysis 24(2):361–376

AS/NZS 4360 (2004) Australian/New Zealand Standard: Risk management

Aven T (2003) Foundations of risk analysis – A knowledge and decision oriented perspective. Wiley, New York

Aven T (2006) On the precautionary principle, in the context of different perspectives on risk. Risk Management: an International Journal 8:192–205

Aven T (2007a) A unified framework for risk and vulnerability analysis and management covering both safety and security. Reliability Engineering & System Safety 92:745–754

Aven T (2007b) On the ethical justification for the use of risk acceptance criteria. Risk Analysis 27:303–312

Aven, T (2008a) Risk Analysis. Wiley, N.J.

Aven T (2008b) A semi-quantitative approach to risk analysis, as an alternative to QRAs. Reliability Engineering & System Safety 93:768 –775

Aven T, Abrahamsen EB (2007) On the use of cost–benefit analysis in ALARP processes. International Journal of Performability Engineering 3:345–353

Aven T, Flage, R (2009) Use of decision criteria based on expected values to support decision-making in a production assurance and safety setting. Reliability Engineering and System Safety 94:1491–1498

Aven T, Kristensen V (2005) Perspectives on risk – Review and discussion of the basis for establishing a unified and holistic approach. Reliability Engineering & System Safety 90:1–14

Aven T, Renn O (2008a) On risk defined as an event where the outcome is uncertain. J. Risk Research. To appear

Aven, T, Renn, O (2008b) The role of quantitative risk assessments for characterizing risk and uncertainty and delineating appropriate risk management options, with special emphasis on terrorism risk. Risk Analysis. To appear

Aven T, Vinnem, JE (2007) Risk management, with applications from the offshore oil and gas industry. Springer-Verlag, New York

Aven T, Vinnem JE (2005) On the use of risk acceptance criteria in the offshore oil and gas industry. Reliability Engineering & System Safety 90:15–24

Fischhoff B, Lichtenstein S, Slovic P, et al. (1981) Acceptable risk. Cambridge University Press, Cambridge, UK

Hammond J, Keeney R, Raiffa H (1999) Smart choices: A practical guide to making better decisions. Havard Business School Press, Cambridge, MA

HES (2001) Reducing risk, protecting people. HES Books, ISBN 0 71762151 0

HSE (2006) Offshore installations (safety case) regulations 2005 regulation 12 demonstrating compliance with the relevant statutory provisions

IRGC (International Risk Governance Council) (2005) Risk governance – Towards an integrative approach, White Paper no 1, Renn O with an Annex by Graham P, IRGC, Geneva

ISO (2002) Risk management vocabulary. ISO/IEC Guide 73

Keeney R (1992) Value-focused thinking. A path to creative decision making. Harvard University Press, Cambridge, MA

Lindley D (1985) Making decisions. Wiley, New York

Löfstedt RE (2003) The precautionary principle: Risk, regulation and politics. Trans IchemE 81:36–43

PSA (2001) Risk management regulations. Petroleum Safety Authority Norway (PSA), Stavanger

Renn O (2008) Risk governance. Earthscan, London

Rosa EA (1998) Metatheoretical foundations for post-normal risk Journal of Risk Research 1:15–44

Rosa EA (2003) The logical structure of the social amplification of risk framework (SARF); Metatheoretical foundations and policy implications. In: Pidgeon N, Kasperson RE, Slovic P (eds) The social amplification of risk, pp 47–79. Cambridge University Press, Cambridge

Sandin P (1999) Dimensions of the precautionary principle. Human and Ecological Risk Assessment 5:889–907

UNCED (1992) Rio declaration on environment and development. The United Nations Environment Programme (UNEP). www.unep.org/Documents/Default.asp?DocumentID=78&ArticleID=1163

13

Risk Analysis

Åke J. Holmgren and Torbjörn Thedéen

13.1 Introduction

In this chapter, risk is regarded as a combination of a random (uncertain) event with negative consequences for human life and health and the environment and the probability (likelihood) that this event may occur.

The rapid technical development involves risks that we, somehow, want to manage. During recent decades, there has been an increasing concern about risks in connection with public policy and decision-making. Today, *risk analysis* is an important part of the formalized risk management process that often precedes the launch of a new technical system (or activity). This *proactive* approach of dealing with risks can be compared with the traditional, *reactive* approach in the conduct of accident investigations.

Risk analysis, sometimes used synonymously with *safety analysis*, is about systematically using all available information in order to be able to describe and calculate the risks associated with a given system. The following questions usually summarize the major steps in the traditional risk analysis of technical systems:

- What can go wrong?
- How likely is it to happen?
- What are the consequences?

Here, we emphasize the use of risk analysis in the context of (formal) decision-making. Ideally, the risk analysis is conducted by experts and the result is then fed into some policy and decision-making process. The analysis helps the decision-makers to make informed choices by providing, e.g., a numerical decision-support. Risk analysis, as we treat the subject here, aims at creating a starting point for the process of risk valuation, i.e., the subsequent step in the risk management process. However, the risk analysis methodology we describe in this chapter is merely of a principal nature, and does not consider the level of risk that should be acceptable in any given situation.

Consequently, the most important contribution from a risk analysis of a technical system may be that it can be used to make comparisons between options for avoiding, controlling, reducing, transferring, or in some other way, managing risk. Risk analysis does not, however, only help us to understand the risks associated with a particular activity or system. The work process of conducting a risk analysis is of great value in itself, i.e., the analysis creates a tangible result that can facilitate the thought process, bring together different stakeholders in the planning process, and raise the awareness of these issues in an organization.

The predecessors of contemporary engineering risk analysis can be traced far back, and nowadays a multitude of risk analysis methods are found in the literature. The predominant quantitative risk analysis methods have mainly been developed within areas where there is a potential for major accidents, for example in military and civil air and space travel, as well as in the nuclear process, and offshore industries. Standardized risk analysis methods are presently used in many situations. This is not only a result of legal demands, such as mandatory instructions for the handling of pressure vessels or hazardous substances, but also for commercial requirements. An example of the latter is that risk analysis is now often used as a quality tool in the process of procuring goods and services.

13.2 Risk and Decision

In most situations, risks are associated with decisions – one has to choose between alternatives. It can, for example, be a question of investing in a road transportation system which may lead to fewer traffic accidents, shorter travel times, and higher costs. In this particular situation, it is usually a public body which makes the final decision. An individual may also have to choose between different means of transport such as car, bus, train, or aircraft. Each of these alternatives is associated with risk, travel time, and cost. Even if we can identify the same kind of risk in both examples (e.g., traffic accidents), we are in fact considering two completely different *systems*.

Risk often involves externalities (costs or benefits that spill over onto third parties). Societal decision-making on these issues may result in governmental intervention. The practical meaning of the risk concept depends on the decision-situation involved. In general decision theory, it may be natural to treat risk in a strict utility perspective. Here, however, we emphasize the risk dimension. Consequently, a simple way to describe the decision-situation is to consider the following three ideal stakeholders: the *decision-maker*, the *cost-bearer/benefit-receiver*, and the *risk-taker* (i.e., those who are affected by the inconveniences caused by the activity).

In the first example above, the decision-maker is perhaps the agency or political body in charge of transport policy, the cost-bearers/benefit-receivers may indirectly all be citizens (depending on how the road system is funded), but in any case, all users of the road system, and the risk-takers are the drivers and passengers using the road system. In societal decision-making regarding transport policy, many groups and stakeholders are involved, and the decision-maker may never be affected by the negative outcomes of the decision nor have to bear the direct costs of the decision. In the second example, the individual will make the decision, bear the costs and receive the benefits, and also take the risk (we disregard the fact that the choices can indirectly affect the risks of, e.g., other travelers). Thus, depending on the specific decision-situation (the perspective), the three parties can be overlapping or separated (Figure 13.1).

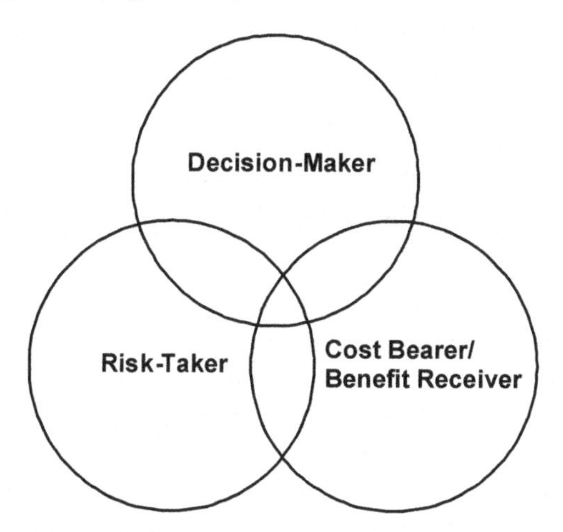

Figure 13.1. Risk and decision – a simple model of the stakeholders involved. The model can be used as a basis for defining the system in a risk analysis

The simple system model described here can be applied to most risk decisions and it serves to help delimit the system in *time* and *space*. A technical system has a natural life cycle, schematically described by the following phases: planning, construction, operation, and decommissioning. Risk analysis can serve different purposes in the different phases of the life cycle. In the *planning* and *construction* phases, risk analysis can be employed as a design tool, i.e., to specify requirements, and also for following-up of the construction work risk levels. In the *operational* phase, risk analysis may be used to analyze the cause of accidents or other problems and also for studying the effects of operational changes, optimizing test intervals, scheduling maintenance and repairs, analyzing processes and operational procedures, prioritizing between measures for risk management, etc. This can also require a continuous updating of previous risk analyses to verify that the level of risk is not increasing. When a technical system finally enters the *decommissioning* phase, risk analysis can be utilized for preserving the safety culture, i.e., to control the risk level during the whole de-escalation process.

The nature of the system determines what *risk measure* is suitable. We have already noted that risk has two dimensions – probability (likelihood) and consequence – but so far we have given no operational definition of the concept. Probability is, by definition, measured by a number between 0 and 1 (see Appendix A), but there are many different value systems that may be used for measuring consequences. A practical decision situation can involve a mix of different value systems such as the number of deceased persons, the number of severely injured persons, or simply the cost in monetary terms. Consequently, in the most straightforward situation, risk is represented by a pair of numbers (probability, consequence). In practice, however, the risk concept has numerous meanings and, since it is seldom defined, its meaning is often taken for granted. Below, we give three examples:

- Risk is taken to mean the *likelihood* of the negative consequences. This is perhaps the most common way of using the risk concept in an everyday situation.
- Risk is represented by the *maximal negative consequences* (disregarding the likelihood). This way of representing risk has been used in the public debate on nuclear power, where the focus has been on possible, but unlikely, catastrophic scenarios.
- Risk is equal to the *expected consequences*, i.e., the sum of the products of consequence and probability. In this case, risk is often considered in relation to some appropriate unit of exposure (a size measure), such as per person, per kilometer, per working hour, or per unit of energy. It should, however, be kept in mind that the risk measure used affects how risks are considered in decision-making.

As we have already pointed out, the risk measure must be related to the actual decision situation. If an individual, for example, is to decide between different means of transport for a particular journey, all the different means of transport which may be used to reach the final destination should be taken into account. A flight may for instance start with a car drive to the airport, and end with a car drive from another airport. Also, different means of travel may not be directly comparable using the same risk measure. A classical example is how to compare the risk of traveling by train and by air – using the number of fatalities per person-kilometer as a risk measure will give air travel an advantage since take-off and landing in fact are the most critical moments (i.e., the total traveled distance may be less important).

A particular risk may have different degrees of randomness depending on the systems level at which the risk is observed. To further explain what we mean by this, we schematically classify risks into the following types:

(a) Risks with Small Random Variations ("Deterministic Risks")

The term "deterministic risk" is in itself something of a contradiction, since risk always involves chance or uncertainty. However, let us take the number of fatalities due to traffic accidents in Sweden as an example. In 2006, this number was 445. The natural random variation predicts that it will fall between 400 and 500 people the following year, even if traffic safety in reality has not changed at all, i.e., here we have rather small random variations, and hence we use the expression "deterministic." Similar results can be obtained if one observes the number of occupational injuries on a national level. The law of large numbers shows that many observations (accidents) will give a stable mean (see Appendix A). In a situation where traffic safety on a national level is considered, it would be suitable to use the mean number of fatalities per year as a risk measure.

(b) Risks With Relatively Large Random Variations

In many situations, the random variations are rather large. The number of fatalities due to traffic accidents in a small region or a city may be taken as an example.

Here, there are very large variations compared with the result for the national level. This means that the risk measure also needs to include the probability of the different consequences.

(c) Catastrophes
A catastrophe is an event that is concentrated in time and space. From the perspective of an individual (let us again consider a road user) a traffic accident resulting in a severe injury may be regarded as a catastrophe. However, dealing with traffic safety on the policy level, a single traffic accident is not considered to be a disaster. In general, catastrophes can be defined as events with low probability and high negative consequences. This leads us to the major problem, and the main reason why formalized risk analysis methods were developed, i.e., the fact that it is extremely difficult to estimate the probability of a very unlikely event.

(d) Uncertainty
In (c), it may be possible to calculate the negative consequences, but the probability is often very difficult to estimate. Sometimes, for example when considering biological or environmental risks, we may face the opposite situation. Let us take pollution as an example. We know that the effects on human health are negative, and it is often possible to estimate the collective dose in affected areas. On the other hand, it might be very hard to estimate the type and the extent of the consequences. We may also have difficulties in estimating the probability. Uncertainty can also be classified in more elaborate ways; see Morgan and Henrion (1990).

The classification above can also be compared with the conventional way of understanding risk in decision analysis, i.e., decision under risk is present when it is possible to assign probabilities to events. We shall now discuss risk estimation in more detail, and will show that the way we understand risk is closely associated with how we treat the probability concept.

13.3 General Principles of Risk Analysis

13.3.1 Systems Definition and Identification of Threats and Hazards

The first step in the risk analysis is the *systems definition* – i.e., to describe and delimit the system under consideration. The second step is the *threat and hazard identification* – i.e., to find the sources of risk.

In the previous section, we presented a simple stakeholder classification that can be used as a starting point for the system definition (Figure 13.1). Here we give no formal definition of the concept of a system. Many appropriate definitions can be found in the literature, but the following themes are often included: *elements* (or components), an *interaction* between the elements, a *systems environment* (and hence boundaries of the system), and the notion that the parts of the system perform to accomplish a *given task*. A fundamental problem in studies of large technical systems is that we may not understand the response of the system to all possible stimuli. Thus, describing and delimiting a system is often a daunting task in itself.

An essential feature of the risk analysis is finding, organizing, and categorizing the set of *risk scenarios*, i.e., the threat and hazard identification. Threats and hazards are the sources of *potential harm* or situations with a potential for harm. Hazards often relate to accidental events, whereas threats often relate to deliberate events. Threats and hazards can be classified in several ways. Firstly, a classification can be based on the *cause* of the threats and hazards. Secondly, the classification can be based on the potential *consequences*. Thirdly, the classification can be based on the *resources* needed to manage the consequences of the threats and hazards.

In Figure 13.2, we present a cause-related classification of threats and hazards. Here, a separation is made between endogenous and exogenous causes. The system boundaries can of course, from a theoretical point of view, be chosen rather arbitrarily. However, a practical choice of system boundaries is to make sure that the considered system coincides with the system where a specific actor, such as an infrastructure operator, has authority. The main reason for doing this is that the organizational and technical prerequisites for managing risks will depend upon whether or not the sources of risk are within the operator's own system. The type of threat or hazard affects both the potential range of a disturbance and the degree of difficulty of managing a disturbance.

An engineering risk analysis has historically been focused on technical failures, natural hazards, and human error. Today, we often apply an "all threats all hazards approach," i.e., we include both threats from *antagonistic attacks* and the conventional hazards in the risk analysis. An important difference between technical failures and attacks by antagonists is that the latter are to a higher degree *planned*, while the former are usually *random*. This has implications for the risk estimations, and is discussed briefly in the next subsection, and further in Section 13.7.

Non-human actions	Human actions		
	Intentional	Unintentional	
Natural disasters Adverse weather Technical failures	Labor conflicts Sabotage Terrorism Acts of war	Human factors (human errors)	**External causes**
Technical failures	Insiders Infiltration	Human factors (human errors)	**Internal causes**

Figure 13.2. A cause-related classification of threats and hazards

The threat and hazard identification can be performed with more or less formalized methods such as checklists, expert workshops, case studies, data analyses, and mathematical modeling.

13.3.2 Risk Estimation

The third, and final step in the risk analysis is the *risk estimation* – i.e., describing the negative consequences from the initiating events, and estimating the corresponding probabilities.

In many situations it is not difficult to describe conceivable *events with negative consequences*. Mathematical (numerical) modeling is usually employed to conduct a more detailed investigation of the consequences resulting from different initiating events. For example, there are engineering models for calculating the course of events resulting from the release of hazardous chemicals (gas dispersion), explosions, and fire (see examples in Chapter 14).

As we have already pointed out, the resulting damage to the environment, and to humans, from hazardous substances may not be fully understood, and we may even face a genuine uncertainty. A particular chemical substance can be proven to cause damage above a certain dose. On the other hand, the effects of low doses may be impossible to assess (the so-called dose–response problem). Experts still hesitate as to whether or not very low radioactive radiation doses have a negative effect on humans. In the case of uncertain consequences, the precautionary principle is frequently referred to. However, the avoidance of all activities that may result in uncertain consequences is not necessarily a wise choice. This subject involves both ethical and technical issues, and the principles for setting limit values are treated more extensively in Chapter 8.

The other part of the risk concept, i.e., the *probability*, is often more difficult to assess. For some *initiating events* (e.g., failure of technical components), it may be possible to estimate their frequency. For other events (e.g., antagonistic attacks), we may have to settle with the conditional probability, i.e., given that a specific attack occurs, how likely are certain negative consequences? Attacks involve a strategic interaction between the attacker and the defender of a system. A more sophisticated way of modeling this is to use game theory (Section 13.7).

Let us now consider the probability of a specific consequence, e.g., a particular fatal occupational accident. The ideal situation, from a statistical point of view, is when stable, and extended, time series accident data are available. It is then possible to use *classical statistical methods* to estimate the corresponding probabilities and also to obtain statistical distributions of, e.g., injuries. The actual decision situation will determine the kind of basic information needed in order to make the decision.

If, for example, the responsible political body were to make a decision regarding traffic safety measures, it would be natural to use official traffic accident data as a starting point. Such a data set may comprise the number of deceased, severely injured, and injured people. It is also possible to obtain information about the total traffic volume. In all situations where one collects accident data there are unrecorded observations. For instance, the number of unrecorded observations is relatively large in the case of minor traffic injuries, and only a fraction of the actual injuries may be included in the data set. The number of fatalities is a more reliable statistics. One then chooses the number of fatalities in relation to the traffic volume as a risk measure, i.e., as an *indicator* of the road traffic risk (Chapter 9).

The situation described above corresponds to the first step in the risk analysis information hierarchy described in Figure 13.3. Traffic accidents and occupational accidents on nation-wide levels are examples of areas where this ideal situation is often present. For the individual car driver, nation-wide accident data may not be very useful for decision-making about how to drive. Accident data will, however, indirectly affect the choices of the individual car driver through regulation and information from road traffic agencies.

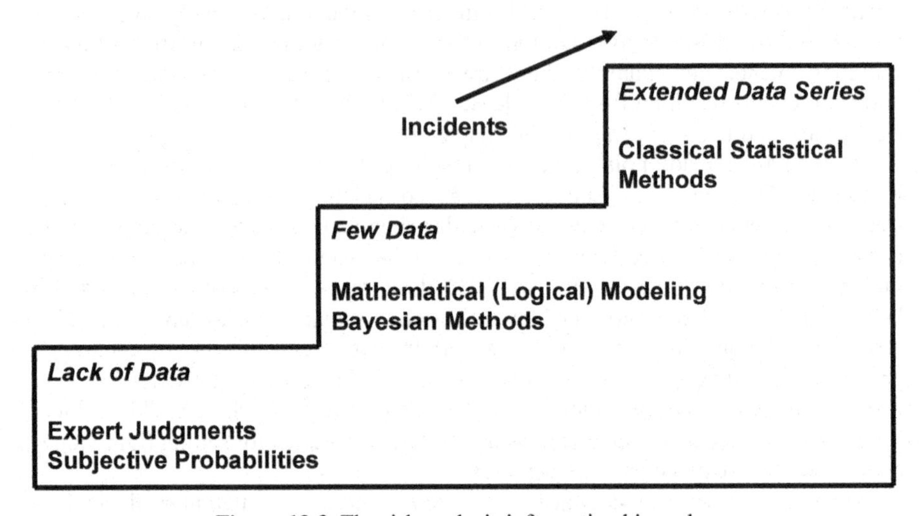

Figure 13.3. The risk analysis information hierarchy

Today many technical systems are large and have an individual design – i.e., there are few identical systems for comparison. Further, accidents in the form of complete system collapses are rare and we therefore seldom have access to extended, stable, accident data sets as a starting point for the risk estimation. Sometimes we can obtain valuable information from *incidents* (precursors or near-misses). The number of incidents is larger than the number of accidents, and we may obtain enough observations to use statistical analyses to estimate (indirectly) the risk, and also to study how the level of risk changes over time. Thus, we are back at the first step in Figure 13.3. In the case of air traffic, it is not possible to use accident data as a basis for risk estimation (although world-wide accident data can sometimes be useful). Instead, one usually relies on systems for the continuous, and systematic, collection of incident and pilot reports to study changes in the risk level.

Another way of estimating the risk when we have access to few accident data is to combine a knowledge of a *mathematical (logical) model of the system* with *component failure data*. We are then at the middle step in the risk analysis information hierarchy depicted in Figure 13.3. During the Second World War, reliability theory was developed (as a part of operations analysis) and used in military applications. For the last few decades, reliability theory has been an important part (as the mathematical framework) for quantitative risk analysis in, e.g., nuclear and process industries. We shall discuss quantitative risk analysis later, but to illustrate the general principles, let us give a simple example (see also Figure 13.4).

In the 1950s, automobiles were equipped with a simple footbrake system which consisted of a hydraulic system filled with oil. If there was a leakage in one of the pipes in the system, the braking effect was lost very quickly. Assuming that the probability of component failure is 0.1 % ($p = 0.001$), the probability of success (i.e., that it is possible to brake the car) is then 99.9 % ($1 - p = 0.999$). The footbrake system in modern cars is constructed using at least two independent brake circuits (i.e., a parallel system). Assuming that the probability of component failure is still 0.1 %, then the probability that at least one of the brake circuits is working will be 99.9999 % ($1 - p^2 = 0.999999$). This means that we have reduced the probability of system failure from 0.001 to 0.000001.

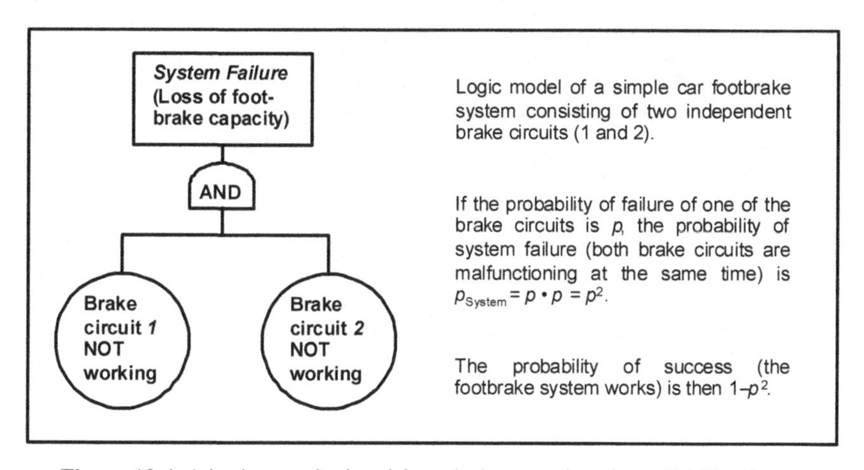

Figure 13.4. A basic quantitative risk analysis example using reliability theory

We cannot always base the risk estimation on mathematical modeling and empirical data, and for this reason we have placed *expert assessments* on the lowest step of the risk analysis information hierarchy. It may appear unscientific to resort to what can be thought of as almost guessing as a method for risk estimation, but even on the other two steps in our flight of stairs in Figure 13.3 expert assessments play an important role. For example, expert judgments come into play when choosing a mathematical or statistical model, estimating parameters in models, and assessing the quality of data sets. The quality of a data set may for example be assessed through the revision of qualities such as definitions and classifications, the data collection procedure, and the number of unrecorded observations.

Especially in the planning phase of a technical system, we must often use expert assessments, and even more so when we are dealing with new technology or new system designs. Later, in the operational phase, empirical observations can support the analysis. A typical situation in the risk analysis of a technical system is that we have access to some expert knowledge, and a few empirical observations (accident data), and it is then possible to use *Bayesian statistical methods* to combine expert knowledge with empirical data.

Before considering how to combine expert assessments with empirical observations in practice, let us discuss the probability concept (see also Appendix A). The *frequency* interpretation of the probability concept is the most widely used, i.e., the probability $P(A)$ can be thought of as the proportion of times the event A would occur in a long run of repeated trials (sometimes referred to as objective probability). The probability $P(A)$ can, however, also be interpreted as a *degree of belief* in the event A (sometimes referred to as subjective probability). In the Bayesian paradigm, the latter probability concept plays a fundamental role.

The quality of a quantitative risk analysis depends heavily on the quality of the probability estimates. Lind and Thedéen (2007) argue that the dichotomy between subjective and objective probabilities is inadequate for decision-making, that the "true probability" of an event does not exist, and that there is thus no "true risk." Estimating probability should, therefore, be understood as an expression that means "assigning probability," and probability is best viewed as a "quantified expert judgment." In engineering risk and safety analysis, aggregated probability is judgmental, and should be subject to competent peer review. As a result, Lind and Thedéen introduce two other concepts that are more relevant to engineering decision-making: *judgmental probability* and *consent probability*.

Let us return to the question of how to combine expert knowledge with data. First one must choose a so-called *prior* probability estimate. It is an estimate of the probability of the consequence based on expert judgment, i.e., experts express their degree of belief in the probability of the event occurring by selecting, e.g., a probability distribution. Next, the prior probability estimate is combined with a *likelihood*, which is obtained from empirical observations. This leads to the *posterior* probability:

$$\textit{prior} \times \textit{likelihood} = \textit{posterior}. \qquad \textit{(Bayes theorem)}$$

As more and more empirical observations are obtained, the prior distribution becomes less important. From a practical point of view, the Bayesian framework does not introduce any limitations into the risk assessment methodology, and it does not prevent us from using empirical observations in a traditional way. Bayesian methods are used not only to estimate the risk but also to obtain component failure probabilities in the quantitative risk analysis when there is a shortage of reliable empirical data. (See Appendix A.)

If the subjective probability concept is accepted, it is always possible to obtain a probability estimate (since an expert according to the Bayesian paradigm is always prepared to express his or her degree of belief in an event as a probability). The Bayesian approach is in contrast to the notion of uncertainty in classical decision theory, i.e., a situation where it is not possible to assign probabilities to events. Whereas in the classical statistical paradigm one makes an estimate (of, e.g., some parameter in a statistical distribution) and is then faced with the difficult task of estimating the uncertainty (e.g., deriving a confidence interval for the parameter), the estimate in the Bayesian setting is in fact already an uncertainty estimate in itself. For a more thorough discussion of the Bayesian risk analysis approach; see Aven (2003) and Singpurwalla (2006).

13.4 Risk Analysis Methods

13.4.1 Introductory Remarks on Risk Analysis Methods

The traditional engineering risk analysis offers a toolbox of established quantitative and semi-quantitative methods for the safety analysis of well-defined systems. Quantitative risk analysis (QRA), also called *probabilistic safety analysis (PSA)* or *probabilistic risk analysis (PRA)*, is based on well-known mathematical and statistical models and methods, collectively often taken to form what is called reliability theory.

In this section, we briefly discuss some of the most common risk analysis methods currently in use, by giving a few practical examples. The first two subsections discuss the three main methods in the traditional PSA framework, namely *fault tree analysis (FTA)*, *event tree analysis (ETA)*, and *failure modes, effects and criticality analysis (FMECA)*. We then treat a modeling technique that describes the walk between different system states as a random process, so-called *Markov modeling*. Finally, the fourth subsection focuses on methods for qualitative risk analysis.

For a stringent mathematical treatment of the subject of quantitative risk analysis the interested reader is recommended to consult more technical literature for proofs and detailed instructions on how to use these methods. Some useful textbooks are: Andrews and Moss (2002), Aven (2008), Ayyub (2003), Bedford and Cooke (2001), Modarres (2006), NASA (2002), and Singpurwalla (2006). A well-written introduction to system reliability theory is provided by Høyland and Rausand (1994). Several of the engineering risk analysis methods are also described in various standards (Section 13.5.3).

13.4.2 Risk Analysis of a Fuel-cell-driven Electric Vehicle

FEVER (Fuel-Cell-Driven Electric Vehicle for Efficiency and Range) was a prototype vehicle developed by a consortium of several European companies in the late 1990s. The vehicle was powered by a fuel cell that converts hydrogen into DC power by a process called electrolysis. Bigün (1999) presents a risk analysis of the vehicle itself and of the vehicle as a part of a traffic system.

Hydrogen has been used as an energy carrier for a long time in many industrial applications. A fuel-cell-driven vehicle is not a novel idea, but the technology has not been subject to a full-scale introduction in any traffic system. Hydrogen is a non-toxic, colorless, and odorless gas that disperses quickly into the atmosphere. It is a very clean fuel in the sense that there are no toxic compounds or greenhouse gases as byproducts of its combustion. Hydrogen has a higher power density than gasoline, methanol, or liquid methane, but it requires a larger storage volume. However, hydrogen has very low ignition energy (e.g., compared to gasoline), and is explosive when mixed with air in certain proportions. The gas detonates very quickly. This property, amplified by events such as the Hindenburg airship accident in the 1930s, has created distrust in hydrogen as an energy carrier. The acceptance of hydrogen as a conventional vehicular fuel will probably depend largely on whether or not people are convinced that the safety issues can be managed properly.

In the *first phase* of the FEVER risk analysis, only the vehicle itself was considered. The system that was studied involved components that, if they where malfunctioning, had the potential to jeopardize the safety of the entire vehicle. After an initial analysis, the following components were selected for a more detailed analysis: liquid hydrogen tank, fuel cell, catalytic burner, high-voltage battery, supervisory unit, hydrogen detection system, battery management unit, and contactors. In general, the choice of the appropriate risk analysis method depends on the *purpose* of the analysis, and also on what *information* is available. In the FEVER case, the *level of detail* in the analysis was affected by the fact that the vehicle was still in its development phase, and that only a single prototype existed. Since there was very little data from the actual use of the vehicle (including experimental life testing of the main components), a qualitative analysis was carried out.

Fault tree and event tree analyses are perhaps the two best-known detailed engineering risk analysis methods. In the *fault tree analysis (FTA)*, one starts with an undesired event, i.e., the *top event*, and then tries to find all the paths through which the top event could occur. FTA is thus a top-down approach, where one tries to identify the underlying chains of events by a stepwise and systematic process. The result of the analysis is displayed in a tree diagram, and there are more or less standardized graphical FTA symbols. The diagram (the fault tree) has a clear and logical structure represented by gates in the branching points of the tree. The most common gates are the AND-gate (the event occurs if all underlying events occur), and the OR-gate (the event occurs if any underlying event, or combination of underlying events, occurs). In the *event tree analysis (ETA)*, we also start with an undesired event, which is often denoted the *initiating event*, and try to follow the chain of events forward in order to determine the possible outcomes of this event. The event tree has a simple logic structure; either a particular event in the sequence occurs (success), or it does not occur (failure). There are, however, no standardized graphical ETA symbols, or conventions for representing an event tree.

Both FTA and ETA can be carried out to identify hazards and to estimate the risk, i.e., the methods can be used for either qualitative or quantitative analysis. In a quantitative analysis (e.g., PSA) the two techniques are often used together. The FTA is then employed to estimate the probabilities in the branching points of the ETA. More thorough descriptions of both methods can be found in many textbooks, and they are also referred to in Standards. In Figure 13.4, we presented a short fault tree example. The FEVER risk analysis involved a large number of qualitative FTAs, where the failure of the main components of the vehicle constituted the different top events. In order to study the consequences of a number of risk scenarios, e.g., leakage of liquid hydrogen due to collision, ETA was carried out, as shown as, in Figure 13.5.

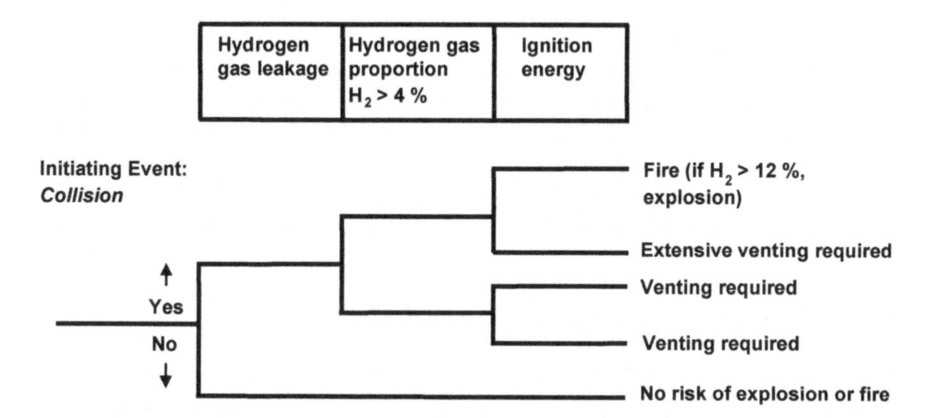

Figure 13.5. Example of a qualitative event tree analysis (ETA) from Bigün (1999). If the initiating event "Collision" leads to a leakage of hydrogen, the resulting consequences may be an explosion, or a fire, depending on the hydrogen–air mix composition

The *second phase* of the FEVER risk analysis concerned the vehicle as a part of a traffic system. This meant a new systems description, and at the same time the boundary of the system had to be expanded. For example, the consequences of an accident with the FEVER vehicle will depend extensively on the traffic environment being considered. The release of liquid hydrogen in a confined or semi-confined space, e.g., a tunnel or a city with narrow roads surrounded by high-rise buildings, may result in a so-called vapor cloud explosion (Chapter 14), whereas a hydrogen leakage in gaseous form may create a jet fire or a flash fire.

As the system is expanded, the number of new issues that must be dealt with increases rapidly. For example, what operational aspects of FEVER should be considered? What kind of training does the driver have? Must we study the use of the vehicle during different seasons, i.e., in different weather conditions? Naturally, there are always resource and information constraints, and a lot of practical simplifications must be introduced in order to carry out any risk analysis. To choose what to exclude is often the most difficult part of the analysis. This is also a process where one must rely mainly on practical risk analysis experience and tacit knowledge. The FEVER risk analysis did not include aspects such as refueling and maintenance or vehicle degradation, i.e., components did not wear down or start to malfunction. Regarding the human factor aspects, the driver was assumed to have a perfect knowledge of how to operate the vehicle, and to understand the safety systems correctly. Since hydrogen disperses quickly into the atmosphere, it became clear that the meteorological conditions had an important effect on the likelihood of fire or explosion, particularly the wind and how confined the space was.

The scope of the FEVER risk analysis was rather technical. If hydrogen is introduced on a large scale as a vehicular fuel, many societal changes are necessary. Each of these changes would be associated with different risks and benefits. Thus we could have included a *third phase* in the FEVER risk analysis. A risk analysis on a *societal level* will require an even more forward-looking perspective, and the system boundaries in time and space will have to be expanded extensively. The transition from cars that run on gasoline, or similar fuels, to hydrogen-fueled vehicles will not take place immediately. For some time, both types of vehicles will coexist and require maintenance, fuel, and specific considerations. Further, hydrogen has to be produced and distributed to the refueling stations. This will create completely new infrastructure systems, and the way in which many present systems are managed or used will also change. Not only will there be new socio-technical systems to consider in the analysis – new threats and hazards may also appear.

13.4.3 Risk Analysis of the Spillway in a Hydropower Dam

In this subsection we discuss the risk analysis of a spillway function of a hydropower dam presented by Berntsson (2001). The system considered is composed of the following parts: the reservoir, the water-retaining structures, the discharge facility, and the immediate downstream area. Here we discuss primarily the discharge facility, and especially the spillway gate function.

Water from the reservoir is discharged through the spillway, which is closed by a gate, as illustrated in Figure 13.6. If the water level rises above a certain level, e.g., because of heavy rain or thaw, the gate body is automatically lowered in order to avoid overtopping of the dam and eventually a complete collapse of the dam body (a dam breach). The mechanism that lowers the gate (i.e., discharges water) consists of a float, which runs in a pipe and is attached to the gate body via a wire and a drum. The gate, which runs in a guide, has the same weight as the float. Discharged water flows down the ski board, and the energy in the fast-flowing water is dissipated as the water flows up the ski jump (block).

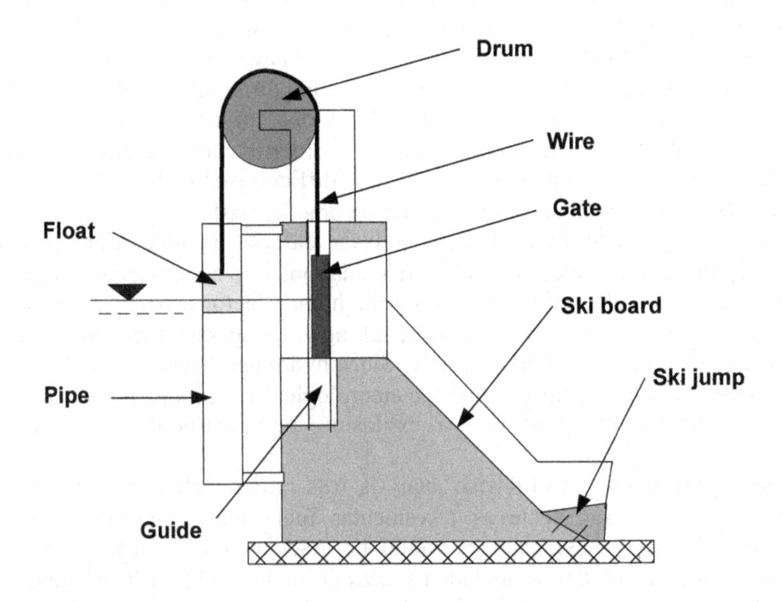

Figure 13.6. Cross-section of water discharge system (spillway) (Berntsson, 2001)

A risk analysis method that is often used to generate input data to the FTA is *failure modes, effects and criticality analysis (FMECA),* a qualitative, or semi-quantitative, technique to identify and analyze the following:

1. The failure modes of each part of the system, i.e., how each component fails
2. The causes of failure
3. The effects these failures may have on the system or on subsystems
4. How to avoid failures or mitigate the effects of failures

If it is decided not to consider and rank the criticality of each failure, the analysis is called a failure mode and effects analysis (FMEA). Nowadays, the distinction between FMECA and FMEA has become blurred. FMECA is a popular risk analysis method in early stages of system development, and procedures and worksheets are described in several Standards (see below). Table 13.1 shows a simple example of a FMEA that has been adapted from the spillway gate risk analysis presented by Berntsson (2001)

Table 13.1. Schematic FMEA example from a spillway gate risk analysis (Figure 13.6). Adapted from Berntsson (2001).

Component	Failure mode	Failure cause (example)	Failure effect (subsystem)	Failure effect (system)
Gate guide	Gate gets caught	Gate guide blocked or damaged	Gate is not lowered	Possible overtopping of dam (dam breach)
Pipe	Float gets caught	Pipe blocked	Gate is not lowered	Possible overtopping of dam (dam breach)
Drum	Drum does not rotate	Lack of maintenance	Gate is not lowered	Possible overtopping of dam (dam breach)
Cable	Cable breaks	Mechanical wear and tear	Gate is lowered	Water is discharged (reservoir is emptied)

There are, of course, many initiating events that may cause a dam breach, such as sabotage, landslide or avalanche, earthquake, heavy rain or thaw (large floods), and various forms of technical failure (e.g., a faulty operation of an upstream dam release gate that may provoke a surge wave). The *scenario* that was treated in the spillway gate risk analysis was a major inflow of water into the reservoir, in combination with a malfunction of the discharge system (inadequate spillway capacity or a spillway gate failure). As a worst-case consequence, the dam is overtopped and eventually there is a collapse of the water-retaining structure. The load on the retaining structures (the response of the dam), and thus the likelihood of overtopping and dam breach, depends on the reservoir level. For the hydropower system considered, there was a direct relationship between inflow and the resulting water level in the reservoir. Thus, a fundamental question in the analysis was what magnitude of reservoir inflow should be considered. Particularly important are extreme inflows.

To determine the so-called *design flood*, there are several possibilities (compare with Figure 13.3). Firstly, if time series of observed flood data are available, we can use statistical techniques (e.g., based on *extreme value theory*) and calculate an inflow with a particular theoretical *return period* (frequency), e.g., 100, 1 000 or 10 000 years. Secondly, physical modeling can be used and a theoretical flood based on different hydrological situations (e.g., levels of precipitation and infiltration) can be calculated. Thirdly, some form of expert assessment can be used. Finally, all these approaches can be combined. Regardless of how a design flood value is obtained, it must be recognized that the calculations will always contain large uncertainties, which means that some form of *sensitivity analysis* must be included in the risk analysis.

There are few new major hydropower dams under construction in the industrialized countries and we are here considering a technical system that has been operational for a long period, and where the design engineers are no longer on active duty. This situation is completely opposite to that of the hydrogen-fueled vehicle described in the previous section.

The traditional quantitative risk analysis methods, especially FTA and ETA, are appropriate for studies of mechanical systems such as the spillway gate but it is much more difficult to carry out a quantitative risk analysis for other parts of a hydropower system. It is in practice impossible to construct a logical model of the water-retaining structure, i.e., the dam body. Even if incident data is used in combination with an engineering dam response model, it is difficult to predict the behavior of a specific dam.

For most technical systems there are, however, indirect ways of following up the risk level. One important cause of damage to embankment dams (rock and earth-fill dams) is internal erosion (e.g., erosion in the body of the dam or the foundation, as well as seepage or erosion at the crest of the dam). Erosion in the dam body leads to an increase in the flow of water through the dam. Thus the distribution and rate of water flow may be a suitable *indicator* for measuring and supervising the condition of the dam body. The water flow can, for example, be detected and quantified by measuring the temperature and the electric resistance at various points in the dam body.

13.4.4 Markov Modeling of a Simple Parallel System

This subsection presents a brief description of a quantitative risk analysis technique based on a knowledge of the various *states* of a system and of the *transitions* between these states. The transition between different system states can be described mathematically as a random process (e.g., a Markov process) in time, which in turn may be depicted in a *state-space diagram* (Figure 13.7).

Let us, as an example, consider a system consisting of two identical parallel components (subsystems), such as an emergency cooling system in a nuclear power plant supported by two identical pumps (i.e., a redundant system). Assuming that the system is operational if at least one of the components is operational, we can define the different system states (Table 13.2). Further, we assume that only one component can be repaired at a time and that, when a component fails, repairs will start immediately provided the other component is not then being repaired. Finally, we assume that both components have *failure intensity* z and *repair intensity* μ. Figure 13.7 shows the state-space diagram of the system.

Table 13.2. System states for a simple parallel system.

System State	Comment
0	Both components are operational (system operational)
1	One of the components is operational (system operational)
2	Both components are in a failure state (system failure)

The concept of failure intensity may loosely be perceived as a measure of how prone a component is to fail at a given point in time. More formally, the failure intensity $z(t)$ is the limit value of the quotient of the mean number of failures of a repairable unit in the time interval $(t, t+\Delta t)$, and the length of the interval (Δt), when the length of the interval tends to zero. Thus, the probability that a failure will occur in the time interval $(t, t+\Delta t)$ is approximately equal to the product $z(t) \cdot \Delta t$.

It is evident in Figure 13.7 that the transition from State 0 to State 1 has the intensity $2z$, i.e., one of the two components fails. When the states of the system have been defined and the corresponding intensities are known, the quantitative analyses can be carried out and the average time to *system failure* (System State 2) can be calculated. In a repairable system, the aim is usually to calculate the availability of the system.

In this schematic example, a number of simplifications have been made. For example, it has been assumed that a component can only be either fully operational or in a failed state, i.e., the component cannot have a reduced functionality. It has also been assumed that the components can fail independently of each other, and that only one component can fail at a given time. However, a particularly hazard of a technical system is so-called *common cause failures*, i.e., multiple failures that are the direct result of a common or shared root cause, e.g., a lightning strike or a fire.

In addition, it has been assumed that the components have *constant failure intensities*. Many technical systems go through an initial phase where the system is adjusted and also may face "teething troubles," but where the failure intensity gradually decreases and stabilizes as the system enters its operational phase. However, such technical systems (especially systems involving mechanical components), often eventually enter an aging phase where the failure intensity again increases. This type of failure intensity behavior is sometimes referred to as the bath tub curve.

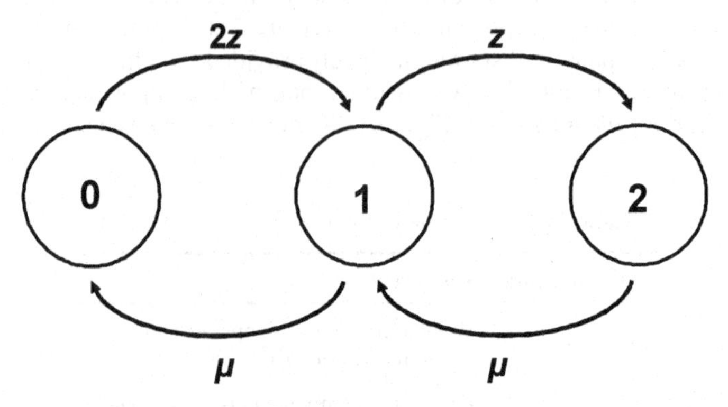

Figure 13.7. State-space diagram for a simple parallel system. The figure shows that the transition from System State 1 (only one of the components is operational) to System State 2 (both components have failed) takes place with intensity z

State-space modeling (Markov modeling) is appropriate for the detailed analysis of small systems. It requires an experienced risk analyst with a knowledge of probability theory (particularly mathematical techniques for analyzing Markov processes), and is usually computerized. More advanced techniques for modeling dependent and common cause failures are also available; see, e.g., Høyland and Rausand (1994). For many components, it may be very difficult to get the data necessary to be able to perform the analysis. For other components it may be feasible to perform controlled experiments, so-called life tests, to estimate the failure rates of components. An important advantage of the method is that it is possible to analyze the effects of a safety system based on so-called active reserves (standby systems), and that repair and maintenance aspects can be considered.

13.4.5 More on Qualitative Risk Analysis

A large number of semi-qualitative, or qualitative, risk analysis methods are described in the literature. The purpose of these methods is usually to give a broad picture of the risk situation. They are often used in early design stages or as an initial step in a detailed risk study. Typically, a semi-qualitative risk analysis relies on expert assessments. There are formalized ways of collecting experts' judgments, e.g., interviews, surveys, and group discussion (ranging from "brain storming" to so-called Delphi techniques). The outcome of the *expert-opinion elicitation process* can be either a consensus among the experts or a disagreement. The important questions of how to select experts, extract knowledge from experts, collect experts' opinions, and (if we face a non-consensus process) weigh together different experts' opinions are, however, outside the scope of this text. See Ayyub (2003) for a more detailed discussion of formal expert-opinion elicitation.

From an academic point of view, most of the semi-qualitative risk analysis methods have a similar set-up. There are, of course, more or less elaborate worksheets and process support, and different steps in the risk analysis process, e.g., the risk ranking or the hazard identification, may be emphasized.

The result of a *preliminary hazard analysis (PHA)*, sometimes called *hazard identification (HAZID)* or *rapid risk ranking (RRR)*, is a risk mapping without any detailed considerations of technical details, and an estimate or categorization of the likelihood and consequences of different hazards. The analysis is often presented in a *risk matrix* where the frequency of occurrence of the negative event (e.g., low, medium, high) is given on one axis, and the consequence (e.g., insignificant, moderate, severe) is noted on the other axis.

Hazard and operability studies (HAZOP) is a hazard identification technique that can also be used to study operational processes and procedures. HAZOP was originally developed for the chemical process industry and is an example of a so-called *deviation analysis.* By applying a set of standardized guide words to process variables in each study node (e.g., major item of equipment), a systematic hazard analysis of the process is obtained. For example, if the guide word "more" is applied to the process variable "flow" at various study nodes, the deviation "more flow" is studied at these parts of the system. The next step in the analysis is to determine the possible causes of deviation, and the potential negative consequences as well as the required actions.

Finally, many risk analysis methods seek to address the interplay between humans and technology. The subject of human factors is studied extensively in the nuclear and process industries, as well as in the aviation industry. Chapter 15 contains a more detailed discussion of the subject, and provides references to literature on human reliability and human failure analysis.

13.5 Practical Aspects of a Risk Analysis

13.5.1 Conducting a Risk Analysis

Risk analysis is foremost a *craftsmanship* that requires the ability to integrate knowledge from many different areas and scientific disciplines. Analytical tools, e.g., mathematical and statistical modeling, often play an important role in quantitative risk analysis. To be able properly to use these means, a good overview of theories and methods, their applicability and limitations, is needed. A *forward-looking perspective* is usually required since the analyses can deal with systems that do not yet exist, have only recently have become operational, may be operated under different circumstances, or face a different set of threats in the future. Thus, fantasy and *creativity* are always important elements in any risk analysis, as well as an understanding of the problems involved, and the context in which the results are to be used.

Expert knowledge is a necessary prerequisite for conducting a thorough analysis. A risk analysis requires a broad range of expertise, e.g., from engineering disciplines, mathematics and statistics, health sciences, psychology, and economics. Any risk analysis team needs at least to include people with *practical knowledge* of the system being studied, and an *experienced risk analyst* (facilitator) who can guide the team through the practical work. Further, if one really wants the risk analysis to have a major impact, the senior management's commitment to the task and the support of the organization are crucial. Experience shows that this also makes it easier to elicit knowledge from the participating experts and less complicated to treat sensitive matters in the analysis.

Often, *third parties* such as consultants carry out the risk analysis. This can be a practical solution if the necessary risk analysis expertise is not available within the organization. An external party without ties to the organization may also be able to conduct a more independent analysis. However, people with practical knowledge of the system being studied still need to be included in the analysis process. Furthermore, when external parties carry out the risk analysis, *confidentiality and trust* must be maintained to ensure that information of strategic importance cannot be spread outside the organization.

The use of computers in the risk analysis helps to increase the size and complexity of the analysis. A wide range of *computer programs* are available for both quantitative and qualitative risk analysis. Properly used, a risk analysis computer program is a valuable tool. The choice of software package depends on the scope of the analysis, and different lines of business have their own preferences. For some applications, e.g., a probabilistic safety analysis in the nuclear power industry, a software package with advanced mathematical (logical) modeling and statistical data analysis capabilities is required. Consequence analyses, such as numerical simulations of gas dispersion and effects of explosions, require specific software packages and sometimes extensive computational capacity. In other situations, it may be sufficient to use a simple software program that provides a framework for the analysis, e.g., interactive work sheets or templates.

Finally, *documenting and reporting* are important practical parts of the risk analysis. Documentation should be carried out continuously, throughout the course of the analysis process. The purpose of the final documentation is to present the analysis in a clear and systematic way in order to help experts, decision-makers (commissioners), and others to interpret and evaluate the results. The final documentation will also serve as a starting point for risk communication, and may, therefore, be distributed in some form to politicians, the general public, and the media.

13.5.2 Quality Requirements on Risk Analyses

Risk analysis is not in itself an exact science, and two independent analyses of the same subject can lead to different results, depending on, e.g., the assumptions on which the analyses are based. Consequently, a risk analysis must always be exposed to a thorough *critical examination* and quality assessment.

A risk analysis of good quality must be carried out in accordance with basic scientific principles, combined with personal integrity and common sense. The risk analysis should be problem-oriented, and it is very important that the objectives are clearly defined before the analysis is carried out. A fundamental quality requirement is that the analysis is properly documented. From this documentation, at least the following must be evident: the *scope* of the analysis, the *methods* and *data sources*, the most important *results*, and, if this is a requirement, the recommended *measures* and *actions*. In summary, the ideal risk analysis is characterized by *traceability* and *transparency*.

Assumptions and delimitations are necessary parts of any risk analysis. In general, it is important that uncertainties and assumptions are not concealed from decision-makers and others. The *underlying assumptions* on which the consequence calculations, and the corresponding probability estimations, are based should be clearly stated, but it should be remembered that it can be difficult to illustrate the risk concept for decision-makers and laymen. The probability concept, and particularly low probabilities, may be especially difficult to convey (compare with Chapter 16).

From experience, we know that there is often a very strong belief in quantitative data, and that an advanced risk analysis method may appear accurate and reassuring. Although there are many advantages in a quantitative risk analysis – especially that it yields a very precise risk measure and often a detailed understanding of the system studied – there are situations where this approach should be avoided. If the *data quality* is poor, and the data sources are questionable, a qualitative analysis may be preferable. Thus, a basic requirement of any quantitative risk analysis is that it shall contain a discussion on the quality of data and the data sources, as well as a sensitivity analysis.

Accidents occur as a result of the interaction between humans, organizations, technology, and nature. A description of the system and the identification of the sources of risk is the point of departure for all forms of risk analysis. Thus, when critically examining a risk analysis, we should ask whether the *system delimitations* are reasonable and whether the *identification of threats and hazards* is complete. Since human beings are often an important source of risk – or, more correctly, since technology is often not designed to appropriately handle human deficiencies and weaknesses – we must ensure that a technical perspective alone is not applied in the analysis.

13.5.3 Risk Analysis Standards

A risk analysis Standard can be of great help. The use of *generally accepted frameworks* for the analysis, i.e., well-known definitions, methods, templates, checklists, and symbols, facilitates external quality reviews, and helps to avoid misunderstandings. It can also make it easier for external parties to relate to the analysis.

Nevertheless, a risk analysis Standard cannot replace a creative risk analyst or practical knowledge of the system studied. Compliance with a Standard does not ensure that one has identified the critical hazards and risk scenarios. Nor can it guarantee that the assumptions are reasonable or that the estimates are correct. However, following a Standard will help to ensure that the analysis has been carried out systematically and in accordance with good practice. Also, references to Standards and established procedures may help to avoid contractual disputes.

Table 13.3 presents a few examples of risk and reliability Standards. More information is available via the Internet, and the various standardization bodies.

Table 13.3. Risk-related Standards (examples)

Subject	Standards
Risk concepts and terminology	IEC 60050-191 International Electrotechnical Vocabulary – Chapter 191: Dependability and Quality of Service (IEC, 1990)
	ISO/IEC Guide 73:2002 Risk management – Vocabulary – Guidelines for use in standards (ISO/IEC, 2002)
Risk and safety analysis	IEC 60300-3-9 Dependability management – Part 3: application guide – section 9: risk analysis of technological systems (IEC, 1995)
	Z-013 Risk and emergency preparedness analysis (NORSOK, 2001)
	MIL-STD-882D Standard Practice for System Safety (DoD, 2000)
Risk analysis methods	IEC 61025 Fault Tree Analysis (IEC, 2006a)
	IEC 60812 Analysis techniques for system reliability – Procedure for failure mode and effects analysis (FMEA) (IEC, 2006b)
	IEC 61165 Application of Markov Techniques (IEC, 2006c)
	IEC 61078 Analysis techniques for dependability – Reliability block diagram and boolean methods (IEC, 2006d)
	IEC 61882 Hazard and Operability Studies (HAZOP Studies) – Application Guide (IEC, 2001)

13.6 The Role of Risk Analysis in the Risk Management Process

In general, there are several strategic options for the management of risks:

- *Risk avoidance*, e.g., not allowing a certain kind of activity
- *Risk reduction*, i.e., by prevention or mitigation
- *Risk retention*, i.e., an intentional or unintentional choice not to do anything about a specific risk, e.g., the organization itself pays claims
- *Risk distribution* (or transfer) to another individual or organization, e.g., insurance
- *Risk sharing*, i.e., a combination of risk retention and risk distribution, e.g., joint ventures

The formalized risk management process is described in detail in Chapter 12. Here, we briefly discuss the role of risk analysis in the risk management process, and more specifically the connection between risk analysis and how risks are valued. AS/NZS (2004)[1] describes the main elements in the risk management process as:

(a) Communicate and consult
(b) Establish the context
(c) Identify risks
(d) Analyze risks
(e) Evaluate risks
(f) Treat risks
(g) Monitor and review

Risk analysis, as described in this chapter, deals mainly with (c) and (d). We have emphasized the relation between risk and decision. In the formalized risk management process, the risk evaluation, activity (e), the task is to compare the level of risk against pre-established criteria and consider the balance between potential benefits and adverse outcome (AS/NZS, 2004). The process of establishing risk criteria and evaluating risks is not as straightforward as it might seem. As has already been emphasized, there are several ways of interpreting the risk concept. Even if it is agreed that risk is a pair of numbers, i.e., probability and consequence, there are many consistent ways of weighing together the two parts. Consequently, the preferences of the involved parties (the risk valuation) play an important role, especially in societal policy and decision-making regarding risks. This will have implications for the risk analysis, but the risk valuation also affects how risks are perceived and can be communicated.

Let us take an example to illustrate this point. The intense debate on nuclear power in the 1970s often concerned the risk of a nuclear meltdown (severe damage to the reactor core). The advocates of nuclear power emphasized the extremely low number of expected fatalities per generated unit of electric energy, whereas the opponents of nuclear power described the disastrous potential consequences of a meltdown. These are two completely different ways of communicating risk, which unfortunately twisted the debate. The differences in opinion on this matter depend in fact on the differences in the fundamental idea of the grounds on which decisions are made or should be made.

To elaborate on this, let us first present a few examples of seemingly obvious and natural values that can be perceived to be attainable:

- A longer average lifetime
- Fewer fatalities per working hour, per produced unit, etc
- The same risk levels for different professional groups
- As effective risk reduction efforts as possible

A few examples are, however, sufficient to show that these criteria can easily come into conflict with each other. The last criterion, effectiveness, might lead to an unequal risk distribution which would violate the third criterion. Thus it is very

[1] The forthcoming ISO 31 000 Standard ("Risk Management – Principles and guidance on implementation") will, to a large extent, be based on AS/NZS (2004).

important that the risk analysis is neutral in this sense, i.e., it does not provide the decision-maker with information that is connected only with a particular set of values or criteria.

In summary, valuation for decision-making regarding risks lies on a scale with two extremes. The first is the purely economically rational approach, i.e., to save as many lives as possible given a certain allocation of resources. The second seeks to describe, and to some extent explain, how we as human beings perceive risks. These two perspectives are described in Chapters 16 and 17.

13.7 The Future of Risk Analysis

Increasingly advanced computer-based *supervisory, control and data acquisition (SCADA) systems* are presently being employed to optimize industrial processes. Programmable electronic systems and computer control of safety-related functions continue to replace the traditional electromechanical implementations in commercial products (see also Chapter 10). When software is combined with hardware to create programmable systems, the ability to ensure conformity assessment through analysis, testing, and certification becomes more difficult. There are several reasons why it is complicated to use the conventional engineering risk analysis framework to study IT-related threats; the number of potential consequences resulting from a threat can be very large and, since the consequences can be abstract, e.g., loss of data integrity, it may be difficult to describe them in a way that fits the framework. Further, we have the same problem described below, namely the estimation of the probability of initiating events.

The rapid proliferation and integration of telecommunications and computer systems connects previously isolated technical systems. Here, the interactions between the large technical systems in society, often referred to as *interdependencies*, are particularly important because several systems must often act together to provide a service. This creates an increasing level of complexity, i.e., more complicated, interdependent, and all-embracing technical systems.

As already indicated, the traditional risk analysis of technical systems has focused on technical failures, natural hazards, and human factors, i.e., a safety perspective has usually been employed. There is, however, a growing international concern about antagonistic attacks, e.g., sabotage and terrorism, against technical systems (a security perspective). This chapter has not focused on antagonistic attacks, although the general risk analysis principles presented here are still valid. For some events it can, however, be problematic to use the probability concept as such. In an analysis of planned attacks against technical systems, there is an interaction between the attacker and the defender, and the measures applied to protect a system will affect the antagonist's course of action. This means that we face a game situation and not a decision situation (see also Chapter 11). Several attempts have been made to provide risk and vulnerability frameworks that can also deal with security problems; see, e.g., Aven (2007), Garrick et al. (2004), Holmgren (2007), and McGill et al. (2007). In addition, game theory has been applied to study the allocation of resources to defend systems against attacks; see, e.g., Bier (2007) and Holmgren et al. (2007).

In summary, existing quantitative risk analysis methods can to some extent be adjusted to embrace interdependencies and security problems. The major challenge we face today is, however, to develop risk analysis methods that can help us to manage the rapidly increasing level of complexity in technical systems.

References

Andrews J, Moss T (2002) Reliability and risk assessment. Professional Engineering Publishing, Suffolk

AS/NZS (2004) Risk management. AS/NZS 4360:2004, Australian/New Zealand Standard (AS/NZS), Sydney and Wellington

Aven T (2003) Foundations of risk analysis: a knowledge and decision-oriented perspective. John Wiley & Sons, Chichester

Aven T (2007) A unified framework for risk and vulnerability analysis and management covering both safety and security. Reliability Engineering & System Safety 92:745–754

Aven T. (2008) Risk analysis: assessing uncertainties beyond expected values and probabilities. John Wiley & Sons, Chichester

Ayyub BM (2003) Risk analysis in engineering and economics. Chapman & Hall/CRC, Boca Raton

Bedford T, Cooke R (2001) Probabilistic risk analysis: foundations and methods. Cambridge University Press, Cambridge

Berntsson S (2001) Dam safety and risk management. Case study: risk analysis of spillway gate function. Licentiate thesis, Center for Safety Research and Division of Hydraulic Engineering, Royal Institute of Technology (KTH), Stockholm

Bier VM (2007) Choosing what to protect. Risk Analysis 27:607–620

Bigün H (1999) Risk analysis of hydrogen as a vehicular fuel. Licentiate thesis, Center for Safety Research, Royal Institute of Technology (KTH), Stockholm

DoD (2000) Standard practice for system safety. MIL-STD-882D, Department of Defense (DoD). www.safetycenter.navy.mil/instructions/osh/milstd882d.pdf

Garrick BJ, Hall M, Kilger JC, McDonald T, O'Tool T, Probst PS, Parker ER, Rosenthal R, Trivelpiece AW, Arsdale LAV, Zebroski EL (2004) Confronting the risks of terrorism: making the right decisions. Reliability Engineering & System Safety 86:129–176

Holmgren ÅJ (2007) A framework for vulnerability assessment of electric power delivery systems. In: Murray A, Grubesic T (eds.). Critical infrastructure: reliability and vulnerability. Springer-Verlag, Berlin

Holmgren ÅJ, Jenelius E, Westin J (2007) Evaluating strategies for defending electric power networks against antagonistic attacks. IEEE Transactions on Power Systems 22:76-84

Høyland A, Rausand M (1994) System reliability theory: models and statistical methods. Wiley, New York

IEC (1990) International electrotechnical vocabulary – chapter 191: dependability and quality of service. IEC 60050-191, International Electrotechnical Commission (IEC), Geneva

IEC (1995) Dependability management – Part 3: application guide – section 9: risk analysis of technological systems. IEC 60300-3-9, International Electrotechnical Commission (IEC), Geneva

IEC (2001) Hazard and operability studies (HAZOP studies) – application guide. IEC 61882, International Electrotechnical Commission (IEC), Geneva

IEC (2006a) Fault tree analysis. IEC 61025, International Electrotechnical Commission (IEC), Geneva

IEC (2006b) Analysis techniques for system reliability – procedure for failure mode and effects analysis (FMEA). IEC 60812, International Electrotechnical Commission (IEC), Geneva

IEC (2006c) Application of markov techniques. IEC 61165, International Electrotechnical Commission (IEC), Geneva

IEC (2006d) Analysis techniques for dependability – reliability block diagram and boolean methods. IEC 61078, International Electrotechnical Commission (IEC), Geneva

ISO/IEC (2002) Risk management – vocabulary – guidelines for use in standards. ISO/IEC Guide 73:2002, International Organization for Standardization (ISO), Geneva

Lind N, Thedéen T (2007) Consent probability. International Journal of Risk Assessment and Risk Management 7:804–812

McGill WL, Ayyub BM, Kaminskiy M (2007) Risk analysis for critical asset protection. Risk Analysis 27:1265–1281

Modarres M (2006) Risk analysis in engineering: techniques, tools, and trends. CRC Press, Boca Raton

Morgan MG, Henrion M (1990) Uncertainty. A guide to dealing with uncertainty in quantitative risk and policy analysis. Cambridge University Press, New York

NASA (2002) Probabilistic risk assessment procedures guide for NASA managers and practitioners. Office of Safety and Mission Assurance, NASA Headquarters, Washington D.C. www.hq.nasa.gov/office/codeq/doctree/praguide.pdf

NORSOK (2001) Risk and emergency preparedness analysis. NORSOK Z-013, Standards Norway. www.standard.no/pronorm-3/data/f/0/01/50/3_10704_0/Z-013.pdf

Singpurwalla N (2006) Reliability and risk: a bayesian perspective. John Wiley & Sons, Chichester

14

Fire and Explosion

Håkan Frantzich and Göran Holmstedt

14.1 Introduction

This chapter deals with the consequences and risks associated with fires and explosions. The objective is to give a brief insight into the development of a fire scenario, the methodology of risk analysis for fire safety engineering and the calculation of the consequences of a fire or explosion. At the end of this chapter some precautionary activities relating to fires and explosions are presented. The content is mainly focused on the protection of people in a building. There are, however, some other aspects that must be taken into consideration. For example, the property damage that affects a company or the loss of production capacity that emerges as a result of a fire. Such analyses are rather extensive and the risk analysis is often greatly dependent on the company and its particular circumstances.

Unfortunately people get killed in fires. The number of fatalities varies between countries but the number is generally low compared to other risks in society. The fatality rate is quite constant over the years. The fatality rate in Western countries ranges between from 0.5 to approximately 2 fatalities per 100 000 persons per year, Figure 14.1. A majority of those killed in fires are elderly people. It is therefore probable that we shall see an increase in the number of people killed by fires, since the numbers of elderly people are constantly growing.

The fatality rate is even higher in developing countries. It should, however, be noted that the variation within a certain country or region may be significant, since the fatality rate is dependent on for example age group, income, and geographical location within a region.

14.2 Fire Development

The development of a fire is normally divided into three phases; the early stage, the fully developed fire, and the decay phase. An evacuation has to be completed within the early stage of the fire, when there is a significant stratification with a layer of hot smoke below the ceiling and cooler air below this hot layer. Evacuation must be completed before the hot gas temperature reaches 175–200°C, which corresponds to a radiation level of 2.5 kW/m^2 at floor level. A person can survive this exposure for a short period of time, although it may be painful. The time available for evacuation can be counted in minutes. In a normal type of apartment the time available is usually only a few minutes. In larger premises like warehouses, malls, and fairs, the time available for evacuation is usually somewhat longer.

The transition between the early and the fully developed fire is called the flashover. It is characterized by a rapid temperature increase within the room and is very dramatic. A flashover is normally defined as the phase in the fire development when the hot gases present in the smoke reach a temperature of 500–600°C and the radiation level towards the floor is approximately 15–20 kW/m^2. In the fully developed fire, the temperature is more or less constant, approx. 800–900°C. This phase is important when the stability of the building and fire separation between different fire compartments are considered. The fully developed fire can last for hours.

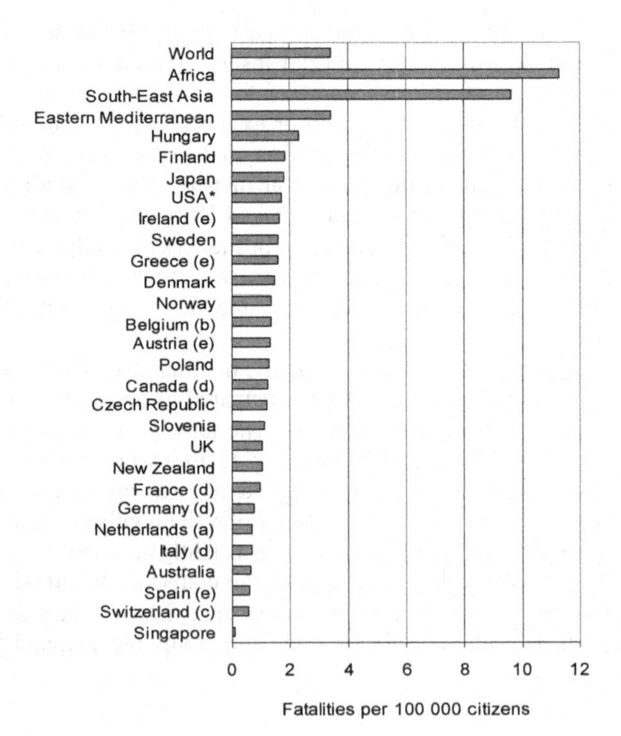

Figure 14.1. A compilation of fire death frequencies (Geneva Association, 2006; WHO, 1999). The numbers are given as the number of persons killed per 100 000 citizens and are presented for the period 2001–2003, with the following exceptions: (a) 1994–1996, (b) 1995-1997, (c) 1998-2000, (d) 1999-2001, (e) 2000-2002. * The US numbers also include 2791 deaths in 2001 arising from the attack on the World Trade Center in New York on September 11

14.3 Factors that Affect the Enclosure Fire Development

To estimate the dynamics of a fire, it is necessary to know what factors influence the fire development. Physical factors like the type of structure, the room geometry, and the furnishing of the room are such factors. Fire development is usually described in terms of the heat release rate (measured in kW or MW) from the combustible material. Typical heat release rates of fires vary from approx. 50 kW (single items like a small wastepaper basket) up to 10 megawatts and higher (liquid fires, enclosure fires, etc). A sofa develops approx. 2–4 MW depending on the material of which it is made.

Since a fire scenario depends on several factors, some of which are specific to the particular building, it is not feasible to specify representative fire developments for certain types of buildings, for example industrial or public buildings. As fires often have a dynamic character it is also important to put the development in a time perspective, in which the spread of fire and smoke are considered as functions of time. Fire safety is in many cases characterized in a time perspective, e.g., evacuation time and time to building collapse.

The layout and size of a building also affect the fire development. The locations of walls and ceiling, openings between spaces within the structure and to the outside, combustion and thermal properties of materials, mechanical ventilation, etc. – all are of importance. Active systems, like sprinkler installations and fire smoke extraction systems, influence the development. Some of these factors are described below. This chapter is focused mainly on the enclosure fire dynamics, i.e., the room fire. The description is relevant for typical rooms in a building and for tunnel fires which show a similar fire development. The dynamics of a tunnel fire are, however, slightly different, as the geometry is very important for the development, i.e., the smoke cannot usually be evacuated through other exits than the tunnel entrances.

The geometry of the room is an important factor. Hot gases rising from a fire eventually reach the ceiling and form a heated layer beneath it. The hot gases and flames radiate back towards the burning object and therefore increase the rate of combustion. This effect is greater in a room with a low ceiling height and small floor area. If the floor area is limited, the accumulation of fire gases increases and a thicker layer is formed. As the layer grows thicker it also increases the radiating effect. In premises with high ceilings the re-radiation is less, due to a more diluted upper layer with a lower temperature. The fire develops more slowly when the flames do not reach the ceiling (high ceilings and large area floors).

The fire can spread from one object to another mainly due to radiation heating from the flame and from the hot fire gases. The distance between the objects is, therefore, of great importance for the fire development in a room.

Oxygen has to be supplied to achieve continuous flame combustion. If the oxygen supply is limited, the fire will self-extinguish after some time. When this occurs, there will probably be a large amount of unburned gases in the room. If a door or window is opened during this phase, a backdraft may occur with a total flashover as a consequence.

The materials in the surrounding constructions, like walls and ceilings, affect the temperature development of the hot fire gases and therefore also the propagation of the fire. In a fire situation, materials normally used to reduce the energy consumption, such as insulating mineral wool, limit the heat flow through a wall. As a result, the heat loss of the upper hot gases to the walls and the ceiling is small. Materials with a high thermal inertia, for example brick and concrete, usually give higher initial conduction losses, and this leads to a lower temperature in the hot gas layer.

Combustible materials in the furnishings are, of course, an important aspect. Furnishings of porous plastic materials and wood contribute to a rapid development. If the furnishings also have a large area compared to the size of the premises, this will facilitate spreading of the fire. Plastic materials such as expanded polyurethane and polystyrene that are nowadays very common in furnishings not only emit heat but also large amounts of soot and toxic gases, and this complicates evacuation. The amount of combustible materials in the furnishings is probably the most crucial fire parameter.

After a period of sustained fire, spreading to neighboring rooms is likely to occur. Initially it is primarily fire gases that spread through openings like doors, and, gaps around pipes, ducts, and electrical wires. This means that a small fire can cause a lot of damage due to a large spread of fire gases. If the spreading of fire gases is not confined, for example by closing an open door, it is possible for the fire to spread. In this scenario, heat transfer through the walls and windows may also spread the fire. Different actions that can be taken to reduce the consequences and the probability of a fire are discussed later in this chapter.

14.4 Explosions

14.4.1 Introduction

An explosion can be caused by many different physical phenomena, as discussed by Eckhoff (1991) and Harris (1983). The word "explosion" consists of two parts, *ex* that describes a direction (from the origin) and *plosion* that describes a swift current. Correspondingly, in an implosion, *im* describes a direction towards the origin. In an explosion, gases flow out from an area because of a generated over-pressure, and in an implosion gases flow in towards an area where an under-pressure has been generated. In the course of an explosion there is always an increase of pressure. In many cases the pressure increase is followed by an increase in temperature, which substantially contributes to the damage caused by the explosion. The word explosion is also used to describe something that happens rapidly. Expressions such as "the fire developed like an explosion" are often used. The course of an explosion is very rapid, generally occurring in a fraction of a second up to some seconds, i.e., too fast for anyone to escape.

14.4.2 Different Types of Explosions

In connection with strong electrical discharges, such as a flash of lightning or a flash-over in a transformer station, the temperature increases in the flash-channel, and this is followed by a large increase in pressure and a subsequent explosion.

An explosion of a pressure vessel can occur when a gas container is filled to an excess pressure or if the container is subjected to a mechanical, chemical or thermal impact until it breaks. The energy from the pressure stored in the pressure vessel is quite large. In a normal 20–50 liter (5–15 gallon) container, the energy from the pressure varies between 0.5 and 2.5 MJ at 15 MPa (150 atm). If the gas in the container is also combustible and perhaps condensed under pressure, e.g., LPG (liquefied petroleum gas), the consequences can be severe since the stored combustion energy is more than 100 times greater than the energy associated with the pressure. Gas containers that have been affected by heat in a fire must be handled with great caution.

Figure 14.2. An explosion and the resulting fire destroyed the Piper Alpha oil platform in the North Sea on July 6, 1988. The accident is believed to be the world's worst offshore oil disaster with a total death toll of 167 people, and the insured losses were about GBP 1.7 billion (USD 3.4 billion). Piper Alpha began production in 1976 as an oil platform, and was later converted to gas production. The immediate cause of the accident was a leaking gas condensate that ignited. (Photo: David Giles/PA Wire/PA Photos/Scanpix)

Vapor explosions are similar to explosions of pressure vessels. When a liquid is heated in a closed vessel, a high pressure can arise if the liquid starts to boil. When the liquid vaporizes, the volume increases 500–2 000 times and this increase of volume can lead to a high pressure, which may cause the vessel to break.

Various kinds of *chemical reactions* can result in an explosion. The most common types are:

- *Gas explosion*: The reaction between gaseous or vaporized fuel and oxygen in the air. Combustible gases, e.g., hydrogen, methane, natural gas, LPG, petrol fumes, etc. belong to this category. Leaking LPG systems in boats and caravans have caused many explosions (Figure 14.2).

- *Mist explosion*: The reaction between a mixture of vaporized liquid (mist consists of small droplets suspended in air) and the oxygen in the air. The vapor can be generated in different ways, e.g., by condensation when a combustible gas is cooled or by a spray formation from a leak in a high-pressure hydraulic system.
- *Dust explosion*: The reaction between a mixture of pulverized solid material (dust consists of minute solid particles) and oxygen in the air. Stirred-up dust of many organic and inorganic materials can lead to an explosion. The dust is usually stirred up when material is transported or when solid material is exposed to a strong air flow. Examples of materials that can cause dust explosions are flour, sugar, coal, sawdust, peat, and metal powder. In most cases, dust explosions occur in connection with transportation (transporters), storage (silos, storage premises), and processing of materials (sawmills).
- *Explosions caused by chemical reactions to which the oxygen in the air does not contribute*: Examples are unstable substances, e.g., acetylene gas and explosives that are present both as liquid and solid materials.

14.4.3 Limits of Flammability

For an explosion to occur in a mixture of fuel (gas, vapor, dust) and air, the mixture must have a certain composition and a source of ignition must be present. The upper and lower limits of the fuel–oxygen mixture that are combustible define the flammability range. Below the lower flammability limit there is insufficient fuel and above the upper limit there is too much fuel for the mixture to ignite and maintain a sustained reaction. The risk of an explosion generally arises when the lower flammability limit is exceeded.

Many combustible gases have a lower flammability limit corresponding to approximately 40 grams of fuel per cubic meter of air. Hydrogen is an extreme exception, as it is combustible down to 4 g/m^3.

Liquid mists have a more complicated flammability range. The range depends on both the fuel and the size of the drops. The drops must be small (less than approx. 0.05 mm in diameter) for ignition to lead to an explosion. For very small drops, the lower flammability range for vaporized liquids is approximately the same as for vaporized fuels, i.e., 25–100 g/m^3.

Dust has an even more complicated flammability range, depending on several parameters such as combustible substance, size distribution, geometric form, and moisture content. The more dry and pulverized the dust, the higher is the risk. The lower flammability limit therefore varies broadly between 50 and 1 500 g/m^3.

The second necessary condition for an explosion to occur is ignition. The self-ignition temperature for most fuel–air–mixes is high, which means that ignition is usually caused by a local ignition source. If the ignition source has sufficient energy it can ignite the mixture. The amount of energy needed depends on the composition of the fuel–air–mixture, the size of the spark distance, etc. An ideal (stoichiometric) mixture, in which all fuel and oxygen is consumed when combusted, requires less energy to trigger an ignition than a mixture close to the flammability limits. Gas fuels require very little energy, usually of the order of microjoules to millijoules.

14.4.4 Ignition Sources

A hot surface in contact with the fuel-air mixture, e.g., a cigarette or a welding, cutting, and grinding operation, can be an ignition source. A welding spark can have an energy of more than 1 mJ and grinding fragments 10 mJ. The surface of the hot fragments must have a temperature above the self-ignition temperature of the organic materials, usually 300–400 °C.

An open flame from gas welding, a candle, etc, in contact with the mixture can be an ignition source. Flames often have a very high temperature and sufficient energy to ignite most mixtures. A match, fully consumed, emits approximately 5 kJ.

Electric sparks and arcs can be generated when power is turned on or shut off, or in a static discharge. The energy when power is turned on or shut off can amount to several joules. In a static discharge, the energy depends on the object's capacitance and on the voltage. On a dry day a spark from a person can release up to 60 mJ, a 50 l bucket up to 5 mJ, an oil barrel up to 15 mJ, and a large cistern more than 100 mJ.

14.5 Fire Risk Identification

The overall risk assessment starts with risk identification (compare with Chapter 13). The risk identification determines the hazards threatening certain objects. Such identification must be made prior to the risk analysis or the design of the object. The risk identification is in most cases a qualitative assessment because there are normally no standardized methods to evaluate which parts of a building are associated with high risks. Instead, the identification usually emanates from a list of relevant questions. On the basis of these questions, probable fire scenarios can be identified. The following are examples of information that is important:

- What activities are carried out in the building/premises?
- How many persons reside in the building/premises?
- Do people sleep in the building? Do they require assistance?
- Are the members of staff trained?
- Is there an accumulation of material that can lead to a rapid fire development?
- Is there an accumulation of material with a high fire load?
- Which probable fire scenarios can be identified?
- What technical systems are present for fire detection, information to persons, and for the limitation/extinguishing of a fire?
- How is the building divided into fire compartments?
- What means of egress exist? What are the qualities of those that exist?

- How do the control and maintenance routines work?
- Is a fire safety management system implemented?

14.6 Fire Risk Analysis

14.6.1 Introduction to Fire Risk Analysis Methods

Fire risk analyses differ considerably from each other in their levels of detail. There are a number of simple grading, ranking, or index methods which can be used to rank different alternatives or activities in buildings. Two common ranking methods are:

- The Gretener method
- NFPA schemes (National Fire Protection Association)

Quantitative fire risk analysis methods, originating from other engineering disciplines, may also be used. These methods are often based on an event tree analysis. If quantitative risk analysis methods are used, they can also be linked to more or less complex systems for sensitivity analysis and uncertainty analysis. When a problem is structured with an event tree, uncertainties dealing with the system's reliability are already taken into consideration. Two types of quantitative risk analysis for fire safety engineering purposes are presented later in this chapter.

14.6.2 Ranking Methods

The Gretener method was developed in Switzerland and has its origin in data from insurance companies. The method is described in BVD (1980). The risk is estimated in terms of three main factors; the risk factor, the protection factor, and the activity factor. The first factor indicates what consequences can arise and cause damage. The second factor indicates how such damage can be reduced and the third factor is linked to the frequency of fires. This results in a risk description which is a combination of consequence and frequency. Each of the three main factors is determined through the grading of a large number of parameters. The Gretener method is mainly applicable in the evaluation of the risk of property damage but it has also been used for risk evaluations regarding life safety. It can be used, e.g., in industrial complexes, hotels, shopping malls, and exhibition halls.

A series of ranking schemes has been developed in the USA by the National Fire Protection Association (NFPA, 1987). There are different schemes for different occupancies such as health care institutions, prisons, and homes for the elderly. A major disadvantage is that they are adapted to the North American building tradition, and this is reflected in the results. They can, however, be used as checklists because they take into consideration factors that are important for the fire protection of any object.

14.6.3 Quantitative Fire Risk Analysis Methods

Quantitative methods are usually based on the scenario analysis resulting from the risk identification process. The purpose is usually to investigate occupant safety in the event of fire, but in some cases the purpose may be to study risks for emergency personnel. Occupant safety is normally quantified in terms of the time margin for evacuation, which is the difference between the time to reach untenable conditions and the evacuation time. Depending on how uncertainties are handled, two methods can be distinguished:

- The scenario analysis or deterministic approach
- The more probabilistic approach where scenario probability and consequences are treated explicitly

An event tree is typically used to structure the problem. Much of the input data to an assessment is based on judgment, but in recent years much effort has been devoted to reducing the uncertainty by working towards a standardization of methods and input data; see, e.g., BSI (2001) and BSI (2003). It should be noted that the deterministic scenario analysis can in some sense be regarded as probabilistic, since information regarding uncertainty is implicitly considered in the choice of scenario and parameters.

Scenario Analysis
The scenario analysis describes how the evacuation process proceeds when all the technical and organizational systems are working as intended. The result is a simple comparison between the time available for evacuation and the time it takes to evacuate. The degree of occupant safety is expressed in terms of the safety margin for evacuation for the scenario investigated. The input data to calculate the two time parameters for the scenario under consideration is usually chosen to give conservative estimates, and no explicit scenario probability is considered. This way of representing safety is simple, but it may lead to an over-conservative solution since there is no probability information.

The scenario analysis should always be complemented by an extensive sensitivity analysis. The sensitivity analysis investigates scenarios where, for example, technical systems like an evacuation alarm do not operate as intended. Fundamentally different fire scenarios and other fire sources should also be included in the sensitivity analysis. In some countries, the scenarios for the sensitivity analysis have been standardized (NFPA, 2006). The problem arising during the analysis of the sensitivity analysis results is to decide whether or not the safety is satisfactory in these situations. The calculated consequences should not differ significantly from the original scenario. The disadvantages of the scenario analysis have led to an increasing use of more traditional quantitative risk analysis methods, where both the scenario frequency and the consequences are explicitly considered. This is, however, a more time-consuming method.

Probabilistic Methods
The probabilistic approach has been used more frequently in recent years, mostly because of the problems associated with evaluating the acceptable safety level in

the scenario analysis. The method is based on a traditional technique to describe scenarios with one or more event trees. As in the scenario analysis, the scenarios in the event tree describe the consequences in terms of evacuation safety. The main difference is how the data for the calculation are chosen. If the calculated risk is to be compared with an acceptable risk level, the input data must correspond to how the acceptable risk criterion is derived. The risk criterion is usually expressed as an expected level of risk and the input data should correspond to this. The risk is often presented in the form of a risk profile (FN-curve) or as a value of the expected loss of life. The FN-curve (frequency number) showing the probable frequency of x fatalities or higher, is a cumulative probability distribution curve (compare with Appendix A).

14.6.4 Models for the Calculation of Consequences

To calculate the consequences for occupant safety, a mathematical model is required that includes factors such as the spread and characteristics of the fire gases, the movement of people in the building, and the detection time, usually the time to detect a fire if an automatic fire alarm system is installed. The time of detection can also be determined from the results given by the smoke spread calculations. Other models can be used to estimate property damage due to smoke exposure or explosions.

Evacuation Time

The evacuation time is often described as a sum of three different times:

- Time of detection
- Time of recognition and response
- Time of movement

Of these three stages, the time of movement has been most studied. There are many models that describe how long a time it takes to evacuate a building. Many of these can be used on a regular PC, but flow equations may also be used in simple cases. Most of the computerized evacuation models do not consider how people behave during the evacuation or how they respond to the cue of the fire or the fire alarm. Models that do consider human behavior operate with specific decision rules, which may be probabilistic. Most models consider the movement in the building, where narrow passageways and doors may cause a delay in the evacuation. Human behavior is therefore usually simplified by adding a delay time to the simulated movement time.

The advantage of translating human behavior into a delay time is that the method is transparent. The disadvantage is obvious; the estimation is very crude and in many cases the delay time estimation is subjective, depending on who performs it. The time of movement may be a few minutes. In large or high buildings, the time of movement may be several tens of minutes or more.

The detection time can be calculated from a knowledge of how smoke spreads in the building. When a certain amount of fire smoke has spread in a room, it can be expected that the persons there become aware of the situation. If a building has automatic detectors, these may alert in the case of fire and the time of detection for these automatic detectors can be calculated. The detection time may vary from a few seconds up to several minutes.

In the calculation of the total evacuation time, the time related to human behavior is subject to the greatest uncertainty. The recognition and response time, sometimes referred to as pre-movement time, includes the time a person, or a group, takes before deciding to evacuate, but this decision is not necessarily the only action. Other common actions are to investigate what has happened, to try to save valuables and to call the fire department. The recognition and response time is supposed to consider all those activities which do not bring the person closer to an exit. This time delay usually has to be estimated based on experiments and experiences of fires, since it cannot be calculated. The time is generally not less than one minute but longer times are common. Informative evacuation alarms may shorten the time of recognition and response. A sensitivity analysis is of course also necessary for the evacuation time calculations.

Time Available for Evacuation

A fire produces a lot of smoke and heat which move through openings in the building by buoyancy forces and external wind pressure. Gradually, smoke and heat accumulate in every room of a building as the fire develops.

At a certain time, hazardous levels of combustion gases and the accumulated heat will be reached and it is no longer possible for people to remain inside the room. These conditions are generally known as untenable conditions.

The models used to describe the smoke spread which determines when untenable conditions occur are traditionally divided into three categories:

- Simple analytical expressions
- Zone models
- Field models or CFD models (computational fluid dynamics)

Analytical expressions can be used for simple calculations of, for example, smoke-filling of a single room. These models are, however, intended mainly for designing the positions of smoke vents and the capacity for smoke-extraction fans.

The other two types of model need computer capacity for the calculations. Zone models can be used in many cases. There are several commercial zone model software and freeware versions available. The most common software is FAST (Fire and Smoke Transport Model) which has been developed by the National Institute of Standards and Technology (NIST) in the USA. Zone models divide each room into a warm upper volume and a cold lower volume, Figure 14.3. All transfer of mass and energy between the two volumes occurs through a fire plume that is created above the fire and works as a pump for mass and energy. Combustion gases that flow from one room to the next transfer mass and energy through a new fire plume in each doorway, and in this way the model calculates the spread of combustion gases between two rooms.

The results produced with zone models have been shown to deviate quite significantly from the true course of events, particularly for fires in premises with a large base area and low ceiling height, or in buildings where the fire gases spread through several rooms or vertically between rooms. Spreading through corridors and shafts is also poorly modeled in zone models. Despite these limitations the zone models are frequently used, although with poor results. Much work has been done to describe the limitations of zone models and to provide recommendations on how to use them.

Typical results from zone models are the temperature in the hot smoke layer and the descent of the smoke layer interface as a function of time. Other results are radiation levels and levels of toxic products. The results of zone models show average values, for example in the warm zone, and local variations are not therefore shown. The variation of temperature in the hot layer is one of the reasons why zone models poorly predict the smoke transport in large premises and tunnels.

The third category, CDF models, can be used for more reliable analyses in cases where zone models are not applicable. These models are more refined and divide each room into a large number, often hundreds of thousands, of smaller control volumes. In each of these control volumes, the heat and mass equilibrium equations are solved separately resulting in a credible description of the conditions in each room.

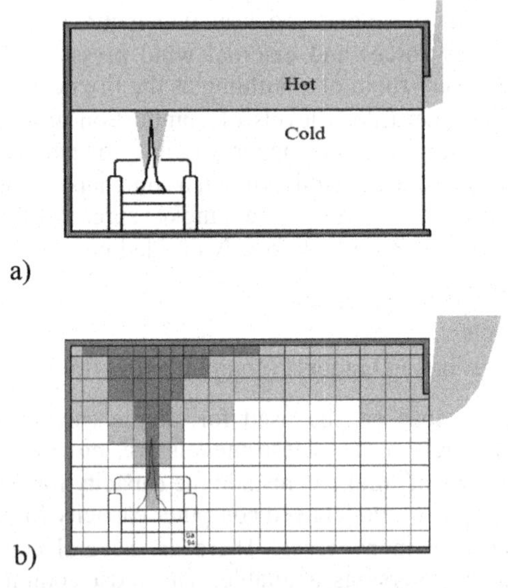

Figure 14.3. Two types of model for the calculation of combustion gas spreading: **a** the zone model, and **b** the CDF model. (Illustration: Stefan Särdqvist)

Both commercial and non-commercial CDF software have been developed in recent years. Some of these software programs have been adjusted to simulate fires where the heat release rate and the fire spread can be calculated. Most CDF models treat the fire as a heat source producing hot air. The disadvantages of CDF models are that they require comparatively large computational resources and expert knowledge on the part of the user, so that only a limited number of persons can perform a CDF simulation. The development is, however, heading towards the production of more user-friendly software, which means that this type of model will eventually replace zone models. These models give a better description of the consequences of a fire scenario, for example in the form of a detailed distribution of the temperature inside a room on fire, Figure 14.4.

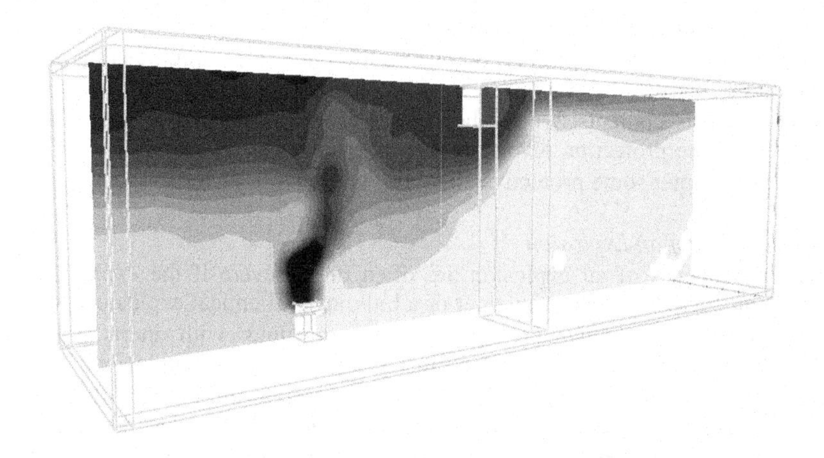

Figure 14.4. The temperature distribution inside a room on fire calculated with a CDF model. Darker shading means higher temperature

A structured sensitivity analysis must be performed, no matter which type of model is chosen. This is especially important because many models emanate from empirical correlations relying on doubtful evidence.

Choice of Level of Untenable Conditions

In a fire safety analysis, the scenario must be described in a quantitative way, indicating when untenable conditions occur. Unfortunately, there is no well-defined description of untenable conditions. What is assumed to be an acceptable level of exposure may vary in the literature and may also depend on the type of risk analysis, i.e., a scenario analysis or a probabilistic analysis. Traditionally, in any scenario analysis a conservative level of tolerable exposure is chosen, representing in a high probability that a person will survive, although the situation may be very uncomfortable.

The desired goal of a risk analysis is, however, to express the consequences in terms of the number of fatalities. The level of tolerable exposure is then higher, resulting in a lower level of the consequences of a given fire scenario. Since the risk analysis includes situations with unfortunate combinations of fire development and evacuation conditions, some scenarios will, however, result in a significant number of fatalities. The important aspect is that the two design strategies are compared to levels of acceptable exposure, relevant for the method chosen. Performing a risk analysis with lethal exposure levels implies that the calculated risk level can be compared with other risks in society. To assess the risk acceptability one must also consider how different people perceive risk.

Consequences on Property
Fires usually damage property significantly. Machines, electrical wires and other sensitive equipment are often affected, even in small fires. A typical kind of damage, even in small fires, is corrosion on surfaces.

Building regulations focus on protecting life rather than on protecting property. The building owner must therefore apply an appropriate protection strategy for the building and its contents, where the continuation of the business also has to be considered. The protection levels set by society are usually not sufficient. At the end of this chapter some protective strategies are presented.

Consequences of an Explosion
The consequences of an explosion are often more severe if the explosion takes place in an enclosed space (container or a building). Chemical explosions are often much more violent than pressure vessel explosions with inert gases. The combustion energy that is released in most gas, mist, and dust explosions lies in the range of 20–50 MJ/kg fuel. In an enclosed explosion, the added volume of gas cannot expand, and this increases pressure until the enclosure bursts. The theoretical maximum pressure in the most common kind of combustion, so-called deflagration, is 0.5–1 MPa, if the pressure is initially 0.1 MPa (1 atm). No building can withstand such a high overpressure, which corresponds to a load of 40–90 metric tons per square meter.

Table 14.1. Human injuries at different overpressures (DoD, 2008)

Overpressure [kPa]	Injuries on people staying in a non-reinforced building
6.2–8.3	Very small risk of death or severe injuries. Mainly splinter injuries
6.9–13.8	No direct deaths or severe injuries are expected. Risk of temporary loss of hearing and splinter injuries
24	Serious bodily injuries from fragments, debris, firebrands, splinter or other objects. Approximately 2 % risk of ruptured eardrums
55	Serious injuries from blast, fragments, debris, and translation. 15 % risk of eardrum rupture
83	Fatal or serious injuries from direct blast, building collapse, or translation. 100 % risk of eardrum rupture
186	Fatal injuries by blast, being stuck by debris, or impact against hard surfaces

In very rare cases and under very special circumstances, a "detonation" can develop, which expands with the speed of sound. In an explosion, the gas also heats up, and this can give rise to thermal damage in addition to pressure injuries. Combustible materials like clothes and movable and fixed furnishing in buildings can ignite and this may result in a fire. As mentioned earlier, an explosion takes place very soon after initiation and normally there is no time to escape. The consequences of an explosion are due to both pressure and heat impact. To be able to evaluate the risks of an explosion and design protection measures, one needs information about tenability limits of both humans and buildings.

Injuries

Information about the effect on humans of a rapid change in pressure is not easily found, for natural reasons. Table 14.1 summarises human injuries at different over-pressures.

Buildings

There are two important variables for pressure-damaged buildings: the maximum overpressure and how long the over-pressure lasts. Different materials have different natural frequencies. The relation between the pressure pulse and the natural frequency is decisive for the impact on the section of the building, particularly in the case of a detonation. The phenomenon can be divided into three parts:

1. If the time for a shockwave is less than the element's natural frequency, the load is smaller than the corresponding static load.
2. If the time is longer than the natural frequency, the load is equal to the static load.
3. If the time for a shockwave is approximately equal to an element's natural frequency, the actual load is up to $\pi/2$ times larger than the static load.

Most explosions are of category (2), i.e., the maximum pressure from a deflagration is equal to its static equivalent.

It is difficult to generalize how much overpressure a building can withstand, since it depends on both the strength of the material and on the constructional design. Many of the pressure damage criteria in the literature are based on shock-waves from short duration explosions, e.g., explosive charges and nuclear weapons, and do not apply to contained deflagrations. However, there are also studies of responses from contained gas explosions in construction parts. Table 14.2 presents a compilation of pressure damage criteria. Figure 14.5 shows an example of the pressure rise in the case of a room explosion and the damage it caused to the building.

Table 14.2. Typical pressures for different damaged construction parts (Harris, 1983)

Construction part	Typical destruction pressure [kPa]
Glass window	3–7
Doors	2–3
Light partition walls	2–5
50 mm "breeze block" walls	4–5
Unloaded brick wall	7–15

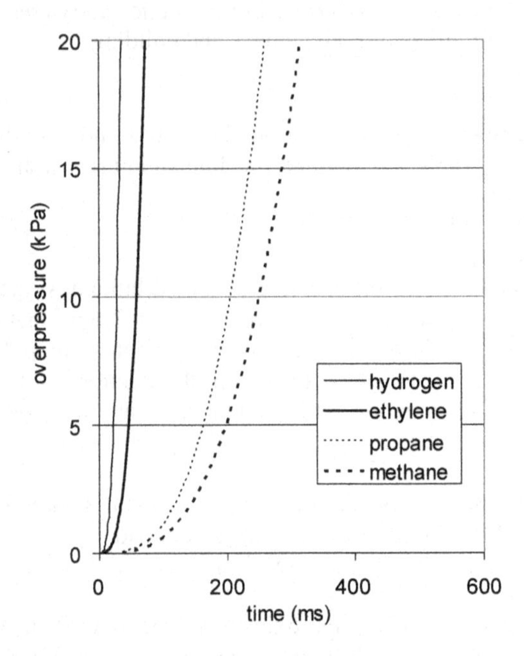

Figure 14.5. The pressure build-up by different fuels in a 70 m^3 room. The fuel–air–mixtures are presumed to be ideal (stoichiometric). Hydrogen gives the fastest build-up, while methane is considerably slower. The difference is caused by variations in flame speeds (Harris, 1983)

14.7 Risk Reduction

14.7.1 Introduction

Traditional fire and explosion risk reduction strategies have focused on mitigating the consequences. Today a broader approach is common, looking also at possibilities of reducing the fire frequency. A pro-active approach is important, and fire safety management procedures are mandatory in some countries.

14.7.2 Fire Risks

The most common way to reduce the risks due to a fire is to install some kind of technical prevention system. This could be an automatic fire extinguishing system (water sprinklers, gaseous extinguishing agents) or alarm systems (fire alarms, smoke detectors, evacuation alarms). Another common technical installation is fire gas ventilation (smoke vents, smoke extraction fans). These systems become active only when a fire occurs.

There are alternative methods which can be used if the fire development is known. Buildings are usually separated into fire compartments to limit the spread of the fire. These fire compartments are separated from each other by specially designed walls, floors, or ceilings. These components are insulated and their purpose is to prevent, for a limited period of time, the fire from spreading outside the fire compartment in which it started. Doors, windows, penetrations for electrical wires, etc, are also specially designed if they are parts of the construction of the fire compartment.

If a fire occurs, its course can be slowed down if combustible materials are not piled on top of each other. The spreading of the fire increases when the combustible material is stored vertically. Combustible surface materials on walls, and especially on the ceiling, should be avoided since they cause the fire to spread more rapidly as the materials are heated. This is the reason for stringent regulations on the surface lining material of walls and ceilings in public premises. Examples of materials to be preferred are brick, concrete, and plasterboard. Untreated wood is less suitable, although wood can be treated in order to improve its fire-resistant characteristics.

The interior design of a room can be designed so as to reduce the fire development rate. Textiles, furniture, cables, etc., can be treated with flame retardants to hinder ignition. The merit of flame retardants is disputed, however, since many of these products have a negative effect on the environment in other respects.

Today it is common to implement organizational fire protection systems, e.g., staff education and the introduction of routines for handling of goods. Routines, documentation, etc., should be put together in a fire safety management system. Many countries have introduced such systems, which correspond well to similar systems for quality assurance (e.g., the ISO 9 000 family of standards), and environmental protection (e.g., the ISO 14 000 family of standards).

When technical installations are included as a part of the overall fire safety system, they must be supplemented with routines for control and maintenance so that the function of the applications can be guaranteed. These routines are a part of the fire safety management system.

Most important is, however, that activities are carried out in a way that reduces the probability of a fire starting in the first place, i.e., the fire frequency is reduced. A fire that never occurs is no threat.

14.7.3 Explosion Risks

To provide protection from explosion accidents, one can take explosion-preventing or explosion-limiting precautions, see Figure 14.6. One explosion-preventing precaution is to reduce the risk of fuel–air–mixes being within the flammable range or to reduce the risk of ignition. Other explosion-limiting precautions could be to reinforce constructions, preventing the spreading of combustion into other spaces and the installation of explosion-relief or explosion-suppressing systems.

Fires and explosions are often controlled by well-known factors. Studying the elementary phenomenon governing the combustion processes makes the problem easier to solve. With knowledge of fire chemistry and dynamics, effects can be predicted and solutions designed to protect against explosions. The main priority is to prevent injuries to humans and to the environment. Humans should not stay in areas where the risk of an explosion is high. The solution is to take explosion-preventing actions or to keep people away from the area. If there is a risk of explosion in places where there are people, for example premises with processing equipment, there are several solutions. If products harmful to the environment are being handled, the consequence of spillage due to an explosion must be kept in mind.

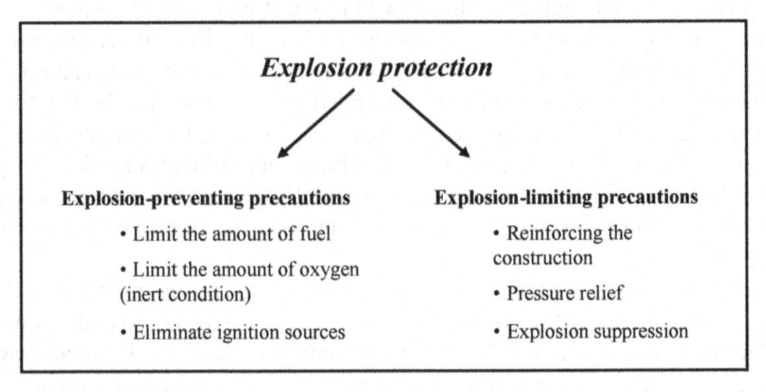

Figure 14.6. Various ways to reduce the effect of an explosion (ignition of a pre-mixed fuel–air–mix)

Examples of explosion-preventing and explosion-limiting precautions:

- *Preventive precautions.* There are a number of preventive precautions which can reduce the risk of an explosion occurring. The fuel–air–mix must be kept outside the flammable range. Spillage of fuel can be reduced with various mechanical constructions. Gas, vapour, and particle detectors can alert long before the concentration of fuel reaches the flammability range. In the case of a leakage, the ventilation could be increased or the room could be made inert. The risk of ignition could be reduced by installation of explosion-safe equipment, discharge of static electricity; precautions when gas welding and handling of open flames, etc.

- *Construction-reinforcing precautions.* As mentioned earlier, the explosion pressure will reach approximately 0.5–1.0 MPa, if the building is intact. Buildings that can withstand such high pressures are costly and such precautions are seldom beneficial other than in small-scale productions.

- *Precautions to prevent spreading.* These are mainly used in pipelines and transporters to prevent explosions from spreading between separated compartments. This can be done with a system that inerts the pipelines before the flame front passes. Other solutions are rapidly closing valves that restrain the deflagration mechanically or a pipe construction that directs a pressure wave to the surroundings.

- *Pressure relief.* Pressure-relieving valves can be installed to prevent harmful pressure. They break at a certain pressure and ventilate gases in order to take power from the explosion. The main power of the combustion often takes place outside the construction. Pressure-relief of a machine or room can in some cases be situated inside a building. The pressure must be relieved through a shaft that leads out from the building. The object in need of protection should be located so that the length of the shaft is as short as possible. The shaft should at least be of the same dimensions as the relief hatches.

- *Explosion suppression.* Another way to protect people and property from explosions is to use an extinguishing system that is fast enough to detect and extinguish a potential explosion before a harmful pressure has been reached. The greatest advantage of using systems for explosion-suppression instead of explosion relief is that there is no discharge from either flames or fuel.

References

BSI (2001) Application of fire safety engineering principles to the design of buildings – code of practice. BS 7974:2001, British Standards Institution (BSI), London

BSI (2003) Application of fire safety engineering principles to the design of buildings – Part 7: Probabilistic risk assessment. PD 7974-7:2003, British Standards Institution (BSI), London

BVD (1980) Fire risk evaluation – Edition B / The Gretener risk quantification method. Draft December 1979, Brand-Verhutungs-Dienst für Industrie und Gewerbe (BDV), Zürich

DoD (2008) DoD ammunition and explosives safety standards. DoD 6055.09-STD, US Department of Defense (DoD), Washington DC www.dtic.mil/whs/directives/corres/html/605509std.htm

Eckhoff RK (1991) Dust explosions in the process industries. Butterworth-Heineman, Oxford

Geneva Association (2006) World fire statistics. Bulletin No 22, Oct 2006. The International Association for the Study of Insurance Economics (“The Geneva Association”), Geneva

Harris RJ (1983) Gas explosions in buildings and heating plant. E & FN Spon, London

NFPA (1987) NFPA 101M: Manual on alternative approaches to life safety. National Fire Protection Association (NFPA), Quincy, MA

NFPA (2006) NFPA 101: Life safety code. National Fire Protection Association (NFPA), Quincy, MA

WHO (1999) World health report. World Health Organization (WHO), Geneva

15

The Human Factor?

Mats Ericson and Lena Mårtensson

15.1 Two Accidents

On March 27, 1977, an aircraft disaster occurred in Tenerife on the Canary Islands that sheds some light on the concept of the human factor. The background to the accident was a terrorist attack on the airport in Las Palmas, which led to the redirection of all air traffic to Tenerife including one Pan AM aircraft and one KLM aircraft. The crews of the two aircraft were very concerned about the redirection. In the case of the Pan Am aircraft, there was a risk that the regulated working hours would be exceeded before they reached their final destination in Las Palmas. The KLM aircraft was to return to the Netherlands, and delays were not accepted by the airline management.

At the time when the KLM flight and the Pan Am flight received clearance to taxi for departure to Las Palmas (where the airport had been opened again) the airport was enveloped in fog. The Pan Am aircraft got too far on the runway and started to turn back.

In the KLM aircraft, the captain started the engines, eager to take off. The first officer said: "Wait – we have not yet got the clearance from ATC." The captain: "I know, go ahead and get it." Without waiting for the clearance, he accelerated the engines for take-off, but the Pan Am aircraft was still on the runway and the two aircraft collided. This was the biggest accident in the history of civil aviation; 583 people died, 58 survived. The Accident Investigation Board found that the accident was caused by the fact that the Dutch captain, who was said to be very authoritarian, did not listen to his first officer but started the engines on his own responsibility.

The accident showed that individuals in the system must cooperate in order to ensure safety. Since this accident occurred most airlines include crew resource management in pilot training, which underlines the importance of taking advantage of all resources of the crew.

Human relations are a part of the concept of human factors. However, it is often in the interface between human beings, technology, and organization that the cause of an accident can be found, despite the fact that human error is given as an explanation. Accidents can be defined as unexpected and undesirable events, especially those resulting in damage or harm. In nearly all accidents there are human actions. It is human beings who make decisions about the design of machines about what materials should be used and about how the work is planned. James Reason (1990) has defined different types of human errors which may lead to an accident. According to the terminology of Reason *active errors* are committed by "front line operators" like pilots, control room crews, ships' officers, and the like. *Latent errors*, on the other hand, are most likely to be triggered by those whose activities are removed in both time and space from the direct control interface: designers, high-level decision-makers, construction workers, managers, and maintenance personnel.

The ship disaster at Zeebrugge in Belgium on March 6, 1987, is here presented as an example of Reason's definitions of active and latent errors. At 18:05 the "Herald of Free Enterprise," a roll-on/roll-off passenger and freight ferry sailed from the inner harbour at Zeebrugge en route to Dover with her bow doors open. As she passed the Outer Mole and accelerated, water came over the bow sill and flooded into the lower car deck. At about 18:27, the Herald capsized rapidly and came to rest in shallow waters with her starboard side above the water (Figure 15.1); 150 passengers and 38 crew lost their lives. Reason gives a description of the chain of events and the latent errors (Reason, 1990).

Figure 15.1. The roll-on/roll-off passenger and freight ferry Herald of Free Enterprise capsized on the evening of March 6, 1987, near the entrance to Zeebrugge Harbor, Belgium. (Photo: Topham Picturepoint/Scanpix)

The most immediate cause of the accident was that the assistant bosun, whose job it was to close the doors, was asleep in his cabin, having just been relieved from maintenance and cleaning duties. The bosun, his immediate superior, was the last man to leave the deck. He noticed that the bow doors were still open, but did not close them, since he did not consider that to be part of his duties: *management failures* according to Reason (1990). The chief officer, responsible for ensuring door closure, was, however, required, by company orders, to be on the bridge 15 minutes before sailing time, which prevented him from inspecting the bow doors: management failure. Because of delays at Dover, there was great pressure on the crew to sail early, again management failure.

Despite repeated requests from the masters to the management, no bow door indicators were available on the bridge, and the master was unaware that he had sailed with bow doors open. The estimated cost of indicators was GBP 400–500: management failure.

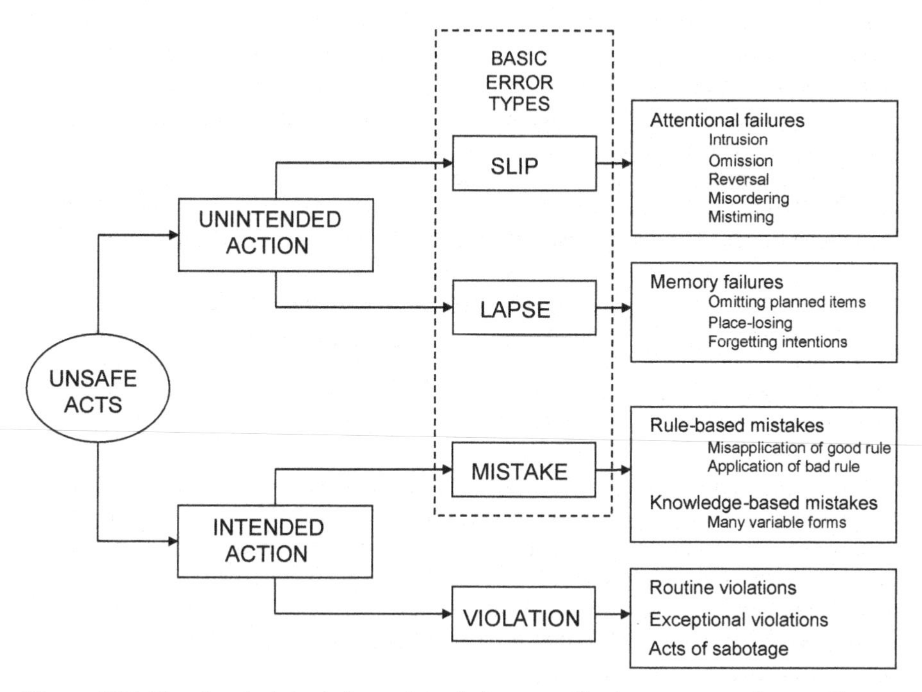

Figure 15.2. Unsafe acts, intended or unintended acts, and basic error types. Source: Reason (1990)

Scuppers were inadequate to void water from the flooded deck, which could be caused by *design and maintenance failure.* Furthermore, the top-heavy design of the Herald and other "ro-ro" ships in its class was inherently unsafe.

With his analysis, Reason wants to show that there are a lot of latent failures, each of which may not be critical, but which in a chain of unfortunate events may lead to a disaster. He wants to take the responsibility from the single operator at the sharp end up to the higher levels of the decision hierarchy.

Reason furthermore analyzes *unsafe acts*; failures or lapses which are committed in the presence of a potential hazard: some mass, energy, or toxicity that, unless properly controlled, could cause injury or damage. He classifies unsafe acts as being *unintended* or *intended* actions (Figure 15.2). The unintended action may be a slip, a lapse, or a mistake. A *slip* may be caused by attentional failures. A *lapse* may be caused by memory failures like omitting planned items. The intended action may lead to a mistake like misapplication of a good rule. Finally *violations,* may be routine violations or acts of sabotage. This analysis is a step towards an objective analysis of an accident, in which the cooperation between man, technology, and organization is vital.

15.2 The Concept of Human Factors – How Is It Used?

To blame the human operator for an accident is quite common in, e.g., tabloid papers searching for a scapegoat. The need to blame somebody is evident for responsible persons within companies and authorities. As part of the vision of a perfect system, technology is good and the human is unreliable.

Media, politicians, and the general public ask for a definition of the concept of human factors and the role of the human operator in the accident, but is there a common definition, upon which researchers can agree? The truth is that there are as many definitions as researchers. Alphonse Chapanis, professor of ergonomics in the United States, defined the research subject of *human factors* (Chapanis, 1991):

> *Human Factors is a body of knowledge about human abilities, human limitations, and other human characteristics that are relevant to design. Human factors engineering is the application of human factors information to the design of tools, machines, systems, tasks, jobs, and environments for safe, comfortable, and effective human use.*

In his book *Human factors in flight* (1987), Hawkins states that, in the world of aviation, the human factor concept deals with humans in their life situation and in their work environment. It is about the relation between humans and machines, equipment and procedures, but also about relations with other people. The concept of human factors is used in the United States. In 1949, the English professor Murrell created the term ergonomics from the Greek *ergon* (work) and *nomos* (law). Ergonomics is the term widely used in Europe.

The International Civil Aviation Organization, ICAO, is a United Nations body within the aviation world. ICAO uses the so-called SHEL model when analyzing accidents. The original SHEL concept, named after the initials of its components, software, hardware, environment, liveware, was initiated by Edwards (1972) and developed by Hawkins (1984). At the center of the model is man, or the *liveware*. This is the most valuable as well as the most flexible component. Linked to the liveware, there are physical factors like body measurements, sight, and hearing, and also physiological ones like state of health, tiredness, lifestyle, and drugs. There may be psychological factors like perception, attention, work load, knowledge, experience, and emotional state. Finally, there are psychosocial factors like mental load, conflicts, and financial problems.

In the Tenerife disaster, it was the tiredness and irritation of the Dutch airline captain of the KLM aircraft which made him take the risk of starting the engines without clearance from air traffic control. Furthermore, his authoritarian leadership style towards subordinates caused the first officer not to insist on abortion of the take-off. As a young first officer, one does not reprimand the most distinguished captain of the airline.

The individual in relation to his colleagues is of vital importance, e.g., communication with other people in the system, as well as in relation to management.

The first of the components which requires matching with the characteristics of man is the *hardware.* This liveware-hardware interface is the one most commonly considered when speaking of man-machine systems. It is about the possibility for man to cooperate with technical equipment like instruments, controls, computers, robots, and word-processing systems. This has been analyzed by Sheridan (2002). The captain of the ferry Herald of Free Enterprise had requested an indicator for the bridge to show whether the bow doors were open or closed, but he had not received this, despite the very low cost of the instrument.

The *liveware–software* interface deals with written information, computers, automation, requirements from authorities, and the like.

The *liveware–environment* interface is about the physical work environment as well as weather conditions. At the airport of Tenerife, the prevailing fog made it difficult for the crews of the two departing aircraft to see each other. If they had done so, they would not have collided. However, there were also problems with the infrastructure in the sense that the dispatch facilities were not sufficient for the air traffic controllers to handle the 40–50 aircraft, which had been redirected from Las Palmas. The technical equipment did not have a sufficiently high standard considering the frequent occurrence of fog due to the surrounding mountains.

The above-mentioned factors must be analyzed in each accident with the aim of finding the critical factors in the chain of events of the accident. The analysis of the technical equipment is of course a vital part of the accident investigation.

Over the past 30 years, it is said that the rate of human-caused accidents within aviation has increased from 30 % to 75 % (Hollnagel, 1998). Since 1960, the number of aircraft accidents has decreased from 60 to 1 per million departures for civil aviation, possibly thanks to the human operator in the system. The increased interest in improving safety has led to the analysis not only of accidents but also of incidents. When this is reported in the press, the ordinary citizen may get the feeling that more accidents occur than earlier. That the incidents do not develop to accidents is due to the judgment of the human operator, his flexibility and competence.

Earl Wiener (1985), a well-known researcher in aviation psychology, has spent a considerable portion of his research time on the human operator in the automated system. He refers to an investigation of 93 aircraft accidents which occurred between 1977 and 1984, which were said to be caused by the crew. Of the causes of the accidents, 33 % could be explained by pilot deviation from basic operational procedures and 26 % could be explained by the second crew member not having performed an appropriate cross-check. The Dutch flight captain deviated from operational procedures by starting the engines before having received clearance from air traffic control. The first officer made an attempt to cross-check, but did not insist that the captain should interrupt the start. On board the ship Herald of

Free Enterprise, the assistant bosun deviated from procedures to close the bow doors by not being present. The chief officer responsible for the operation did not double check that the task had been performed.

15.3 What Is Human?

There is a saying that "to err is human." What is probably meant is that humans are not machines but creatures of flesh and blood with all the shortcomings that we all possess. "To err" is therefore in some sense to be excused. "To be human" has a positive sense of being considerate of these shortcomings, but it is also given the meaning of not being perfect. The human way is to make things in the right way. Most of the things we do are correct in some sense. The "human factor" could be given the meaning of performing things in a correct way, even if the concept is often used in a negative sense.

We try to characterize the special qualities of the human in relation to the environment in terms of machines and technique but also in relation to animals. A great deal of the philosophical, biological, and theological literature is devoted to the theme of the difference between humans and animals and to what could be said to be specifically human. Some of the differences often mentioned are the capacity of the human brain to interpret sensory impressions of symbols and to preserve them in our memory, and putting an emotional value to the impressions. The human being has also via the unique brain an incomparable capacity to clarify timely and logical relations of causes and also to plan his actions in an intelligent manner. Other human characteristics could be said to be the motoric control of thumb and fingers as well as of the vocal cords in our throats. The ability to speak is said to be particularly human.

When being challenged in a crisis situation, a physiological stress reaction is triggered which prepares the body and senses for fight or flight, called the "fight or flight response." The bodily reactions of pulse and breathing are increased, the blood flow is redirected to the brain and the muscles, our attention is increased and our senses are vitalized. The stress makes us focus on the threat which is challenging us. Simultaneously we delete information which at the moment is considered to be unnecessary. This is called a cognitive tunnel view, and is often used in the negative sense that under stress we focus on what is experienced as a threat so that we do not manage to perceive other relevant information. Our capacity to interpret complex information decreases when the stress increases.

We have all experienced that in stressful situations we do not seem to manage simple things which under normal conditions do not cause us any problem. It is highly likely that under extreme stress we behave more like other mammals and their way of reacting instinctively. Several of the unique human characteristics become weakened in extremely threatening situations. In such situations we are forced to trust previously learned strategies in dealing with serious problems. In aviation, the pilots spend part of their training in simulators in order to learn "by heart" items, which are crucial in critical operations. Operators in other safety-critical systems like nuclear power plants have the same kind of simulator training.

Although we frequently make some harmless mistakes, we correct them immediately and do not register the situation as having made a mistake. However, it is sometimes obvious that we have made a mistake. The human being is able to use intelligence to analyze the cause of the mistake in order not to repeat it. Parents often try in their upbringing of their children to teach them not to make the same mistakes that they have made themselves. However, it seems that each generation needs to make its own mistakes in order to learn from them. It may be that it is this personal analysis of one's own mistakes which creates the basis for our personal life experience, knowledge, and skill.

It seems, however, that we have a very high tolerance for the human being with all his mistakes and shortcomings. Consider hypothetically that an airline announced in the daily press that they had reduced the fare by 50 % due to the fact that the aircraft did not need a pilot to control it any more. This modern aircraft is equipped with fully automated systems for departing, cruising, and landing. Would you take this opportunity to fly over the Atlantic in an unmanned aircraft for half the normal price? Probably most of us would say "no." Not until you have seen that it works and that other people have tried it, would you consider it for yourself. So, why do we need pilots, considering the fact that the human being creates so many mistakes and causes 75 % of all aircraft accidents? It may create a feeling of security to hand over to the uniformed pilot and to rely on his or her skill to deal with the disturbances of the technical system. The pilot could be considered as a crisis manager, who can mitigate the effects of the imperfections of the technical system (Figure 15.3).

Figure 15.3. Test pilot Captain Jean-Michel Roy in the cockpit of the world's largest passenger aircraft, the Airbus A380, Pudong Airport in Shanghai, China October 25, 2007. (Photo: Wang Rongjiang/China Foto Press/Scanpix)

It may be true that the human being, due to man's intelligence and flexibility will be the master of big, complex technical systems. It is not possible at the design stage to foresee all the problems that may occur when the technology is used. The designer tries to foresee as much as possible and to provide redundancy in the system for this. Some errors are so unlikely that they are referred to as "10 to the minus nine" errors. The solution is to provide the system with an operator who may interfere when something goes wrong. Sometimes the costs of solving technical problems are not reasonable, so instead an operator will have to deal with the shortcomings of the system. This may be cheaper than to solve the problem technically. The conclusion is that the human with his or her intelligence and flexibility becomes an irreplaceable part of the technical system.

15.4 Technical Errors – Human Errors

Which errors can be considered as technical and which as human? What is actually a technical error? Isn't a technical error a human error, which has been committed in the beginning of the chain of events, physically separate from the place of the accident occurrence? There is a tendency to categorize those errors which have been committed close to the human operators as human errors. All other errors are technical errors, even if they have been committed by humans many years ago during the design or installation of the system. Which are the genuinely technical errors? What remains when all the human decisions and possibilities for incorrect actions have been removed concerning the production of raw material, choice of materials, design, and the installation, and operation of technical systems? Is it possible that the majority of errors are not of a technical nature but are due to human shortcomings? Is a possible design error in the bow doors of a ferry a human or a technical error?

Several researchers within the human factors area are convinced that it is via human knowledge, decisions, and actions that risks are consciously or subconsciously built into the technical systems (Maurino et al., 1998; Sarter and Woods, 1995). Human beings design, produce, install, operate, maintain, and approve machines and production systems, and it is the human beings who make the error analysis.

Legal developments in Europe are heading towards a situation where the responsibility of the designer, producer, and salesman is increased in terms of product responsibility. It is becoming easier to demand legal or economic responsibility from those who could be said to have caused technical disturbances or to have designed an inhuman system where the operator, pilot, or driver will sooner or later commit a fatal human error.

Certain functions could be introduced into automated systems to increase the fault-tolerance of the system to assist the crew to detect errors when they occur. The errors which actually occur should have limits for undesired consequences. Furthermore the transparency of the system should be increased by improving the feedback of the system. The reinforcement of the knowledge of the human beings in the system among crews, designers, management, authorities, and technical researchers is an important contribution to safety.

Consequently, to increase fault-tolerance is a desired action. Systems must give room for incorrect actions carried out by the operator. When, for instance, the pilot is to feed a new parameter into the flight management system during turbulence, mistakes may occur. A validity control of the system may prevent this and ensure that the parameter lies within the tolerance limit.

A simple example of a fault-tolerant system is the word-processing program which may ask: "Do you really want to empty the recycle bin, which contains x number of files?" If a system is fault-tolerant, it has the capacity to recover from an error without stopping, a concept which is used for both hardware and software.

An important issue is the level of knowledge of the operator or of the pilot. What technical knowledge is needed for the operator in relation to the knowledge of the designer? The training for complex systems is carried out in an ordinary classroom context but also in computer-based training, CBT. The student is sitting at a computer screen and is able to work his way through his lessons at his own speed. The next step of the training is carried out in a simulator, the most advanced of which is an aircraft cockpit. A lot of incidents are simulated with the intention of training the pilot in what procedures to follow and what measures to take in a particular situation.

15.5 Vulnerable Professional Groups

Certain groups of professionals are more exposed to the risk of making mistakes which may have more serious human and economic implications than others. Physicians make decisions every day which if they are wrong, may make the illness or injury worse, or may even lead to the death of the patient. The engine-driver may cause serious damage and train accidents. A policeman is sometimes forced under very stressful circumstances to make a decision which may have devastating consequences. These professional groups very often have to stand in the pillory in the tabloid press. Society also demands responsibility from these professionals when they make a mistake. The medical doctor may lose his authorization, the engine-driver may be prosecuted and the policeman may be both prosecuted and fired. Even though we know that mistakes happen every year, society calls for punishment of the individual professional. When choosing your profession you will have to face risks leading to some kind of punishment at least from a statistical point of view. During military service, soldiers will have to take the responsibility for material and human lives.

Sometimes it pays off to take risks both on an individual and on an organizational level. You do something which is dangerous and is sometimes forbidden. Very often this works out well and risk-taking is normally accepted in society, as long as nothing happens. This positive feedback may of course increase the inclination to take risks. When something goes wrong, an earlier accepted action is suddenly condemned.

A mistake which is made under stress, under time pressure in the course of a few seconds, may in the ensuing of an investigation be condemned. Over several years, investigators, experts, and lawyers may take their time to analyze what the operator or pilot should have done.

One way of decreasing the number of errors in a technical system is to have routines of feedback and analyses of all the incidents and mistakes which occur daily. There must be a balance between the penalty for the person who makes a mistake and the need of the organization for information. A "no-blame culture" is necessary for a voluntary reporting system to function, where incidents and near misses are reported.

15.6 Final Remarks

Most accidents or catastrophes are linked to human actions. The further away in time and space one is from the incident, accident, or catastrophe, the greater is the readiness to identify the error as technical errors.

To err is human, but it is even more human to act correctly. The flexibility of the human being is a necessary component of most technical systems.

References

Chapanis A (1991) To communicate the human factors message, you have to know what the message is and how to communicate it. Human Factors Society Bulletin 34:1–4

Edwards E (1972) Man and machine: systems for safety. Proceedings of British Airline Pilots Association Technical Symposium, 21–36

Hawkins FH (1984) Human factors education in European air transport operations. In: Breakdown in human adaptations to stress. Towards a multidisciplinary approach. (Vol 1, for the commission of the European Communities), Martinus Nijhoff, The Hague

Hawkins FH (1987) Human factors in flight. 2nd edn. Orlady HW (ed). Gower Technical Press, Aldershot

Hollnagel E (1998) Cognitive reliability and error analysis method (CREAM). Elsevier Science, Oxford

Maurino DE, Reason J, Johnston N, Lee RB (1998) Beyond aviation human factors: safety in high technology systems. Ashgate Publishing, Hampshire

Reason J (1990) Human Error. Cambridge University Press. New York

Sarter NB, Woods DD (1995) How in the world did we ever get into that mode? Mode error and awareness in supervisory control. Human Factors 37:5–19

Sheridan TB (2002) Humans and automation: system design and research issues. John Wiley & Sons, New York

Wiener EL (1985) Beyond the sterile cockpit. Human Factors 27:75–90

16

The Perception of Risks of Technology

Britt-Marie Drottz-Sjöberg and Lennart Sjöberg

16.1 Introduction

What makes us perceive some risks as large and others as small? Why can some risks be accepted or tolerated and others not? These are questions which have become quite important in several sectors of society. People demand risk mitigation at a level which has not been foreseen by experts and politicians. If a kind of technological and economic development has reached a certain level, there may be commitments which are hard to phase out or deal with politically. Important examples are nuclear power and genetically modified food. In some cases, political conflicts may not yet have evolved although they may do so in a near future. It is therefore important to be prepared in good time. Mobile telephones are an example. Radiation both from receivers and transmitters is frequently discussed as a possible hazard (Siegrist et al., 2005).

In this chapter, we deal with risk perception. We first consider the question of the importance of real or objective risks, the factors that have been found to be of importance in risk perception, and models of risk perception. We then discuss demographic variables which are correlated with risk perception and which are important for understanding perception. Finally, we reflect on risk perception in a societal and political perspective, and give a brief discussion of its importance for risk communication. Many of the examples in this chapter have been taken from our own research on nuclear power and the ongoing process of choosing a site for a final repository for spent nuclear fuel in Sweden. For a comprehensive review of risk perception, the reader is referred to a volume edited by Paul Slovic (Slovic, 2000) and to a book by Breakwell (2007).

16.2 Objective Risk and Risk Perception

The first and simplest answer to the question of what determines perceived risk is of course the size of the "objective" or "real" risk. This is not a bad answer. People are affected by real risks and sometimes they have good ideas about how large they are. At the same time, there are deviations. There is a tendency to overestimate small risks and underestimate large risks. This tendency has frequently been discussed, but the bias is not particularly large. People can make reasonably good estimates of common and well-known risks, considering the average of risks judgments made by a group (Slovic et al., 1979).

However, many risk debates are about very small risks, and in such cases we are not greatly helped by intuition. When the risks are quite small, we usually have no direct personal experience of them. It is also in many cases quite problematic to determine the real or objective risk, especially in the case of very small risks. People have different opinions about risks. Experts, in particular, can make estimates very different from those of the public, especially with regard to risks within their special areas of expertise and responsibility.

Experts usually consider risks to be very much smaller than the public does, but there are exceptions. They may indeed sometimes consider risks to be larger than the public does, as in campaigns aiming to help people stop smoking or to test their homes for radon. Such campaigns are usually not very effective.

In the case of a controversial topic such as nuclear waste, the differences between experts and the public can be enormous. Figure 16.1 presents one result of a Swedish study of the opinions of experts and a representative sample of the public with regard to spent nuclear fuel (Sjöberg and Drottz-Sjöberg, in press). The difference between experts and the public is dramatic. Results of another study (Sjöberg et al., 1997) concerned with food risks are illustrated in Figure 16.2, which shows very large differences between experts and the public with regard to some food risks, but not all. It seems that experts make lower risk assessments in the case of hazards within their own area of responsibility, but not otherwise.

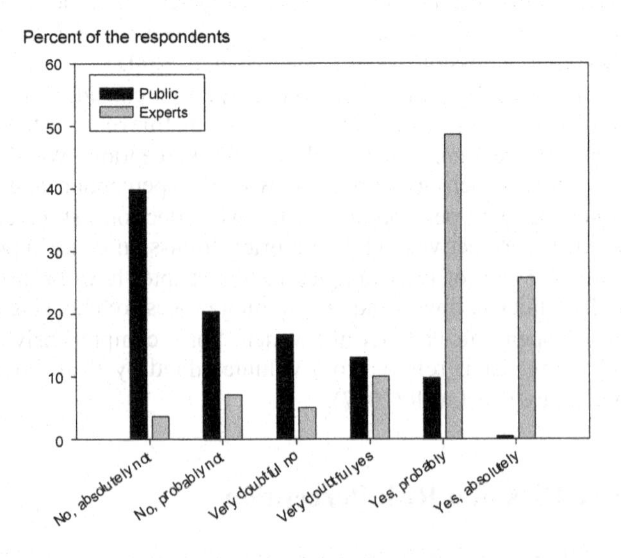

Figure 16.1. Response distribution to the question "Is it your opinion that the problems of a final repository for nuclear waste are solved today in a satisfactory manner?" Experts and members of the public (Sjöberg and Drottz-Sjöberg, in press)

It may seem reasonable to assume that experts know best, and that their risk assessment should therefore be the most correct one, but experts can also be wrong and some risk assessments are so difficult to make that even experts have to guess, more or less. In such cases, they are possibly no better than other people at assessing risks, and their conclusions should be received with a skeptical attitude.

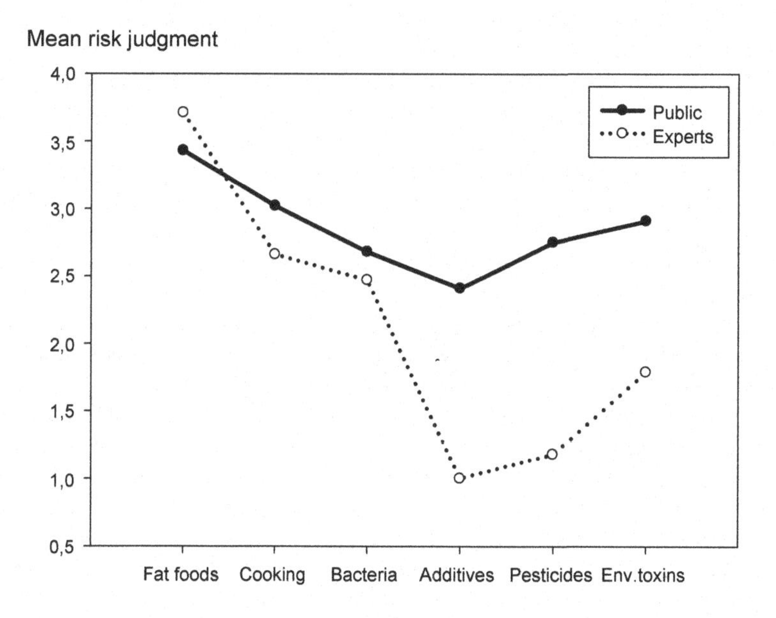

Figure 16.2. Mean risk ratings, by experts and members of the public, of six types of food risks (Sjöberg et al., 1997)

16.3 Factors in Risk Perception

The earliest attempts at explaining risk perception were based on research into subjective probability. Daniel Kahneman and Amos Tversky had shown that there were a number of biasing factors which make subjective probabilities deviate from calculated probabilities, based on probability theory, or based on empirical data (Kahneman and Tversky, 1984). The deviations could be drastic and in either direction: over- or underestimation. The most often discussed explanatory factor was availability (Tversky and Kahneman, 1973). A risk is available if we can easily think about it and can give examples of events such as accidents connected with the risk. Kahneman and Tversky argued that exposure to a risk in the media should make it more available and therefore increase the perceived risk. This seemed to be a reasonable argument and some data support it (Combs and Slovic, 1979). However, later research has led to some doubts (Sjöberg et al., 2000). There now seems to be no strong and pervasive relationship between media exposure and risk perception.

At the end of the 1960's, Chauncey Starr showed that risk acceptance seems to be dependent not only on the size of a risk but also on other factors, the most important, in his view, being voluntariness (Starr, 1969). It seemed that the public is willing to accept a risk level in private automobiles which is 10 times higher than that in professional road traffic (taxis and buses). At that time, at the end of the 1960s, nuclear power was the prime reason for interest in social science risk research. The risk of a major nuclear accident was very small, according to experts. Yet, the public was unwilling to accept nuclear power. Why? Maybe Starr was on the right track with his concept of voluntariness. People may have experienced nuclear technology as being imposed upon them without their prior consent.

Starr's article gave rise to much interest, but voluntariness was a far-fetched explanation for some of the risk perception phenomena. Riding on a bus is not less "voluntary" than driving your own car, for example. Several authors therefore developed his ideas further and suggested other concepts in order to explain risk acceptance or the lack of it, in addition to the size of the objective risk and voluntariness.

In 1978, an important paper was published by Baruch Fischhoff, Paul Slovic, and colleagues (Fischhoff et al., 1978). This work was based on a compilation of factors which had been suggested in the discussion following Starr's article. The authors showed that people could make meaningful judgments of risks according to a number of dimensions which could then be reduced to three underlying factors, namely "number of affected," "dread," and "new risk." This was the basis of the *psychometric model* of risk perception. Its dimensions could well explain average judgments of risks of a large number of hazards. Nuclear power was regarded as being both new and dreaded, and hence there was support for explaining opposition to it as based on emotional and irrational factors. This was probably also in line with how supporters of nuclear power regarded opposition to it. Together with the seemingly high level of explanatory power of the model, it is easy to understand how this model became very popular and widely spread.

The psychometric model must be taken with a large pinch of salt. It is easy to explain average values of perceived risk, but it is much harder to explain the risk perception of individuals. The model has nevertheless been seen by many as a final answer to the question of the structure of risk perception. It is the basis of much risk communication. It is unclear what practical value it has, however, since its explanatory power is limited.

The model can be improved by the incorporation of new dimensions. One such dimension is "interfering with nature." This factor also includes moral aspects. Adding it to the psychometric model yields the *extended psychometric model* (EPM) and empirical studies have shown that EPM is about twice as powerful as the traditional psychometric model (Sjöberg, 2000a). At the same time, when "interfering with nature" is added, the traditional explanatory factors lose much of their explanatory power.

This is an interesting result. It seems that too few explanatory dimensions were included in the original formulation of the psychometric model. This was not noticed since data analysis was done at the level of means and a very high explanatory power was therefore attained. Opposition to nuclear power was explained in a way that seemed plausible to many. However, when the set of explanatory factors is extended, it appears that reactions to technology have more to do with ideology and morality and ideas about "nature" than anything else.

Table 16.1. Percentage of the respondents who classified a number of concepts as "natural" or "unnatural." "Don't know" responses are not shown in the table (Drottz-Sjöberg, 1994)

Concept	"Natural" [%]	"Unnatural" [%]
Primeval forest	92	1
Cornfields	91	2
Human intelligence	86	1
Earthquakes	82	7
Illnesses	73	9
Technical development	60	9
Insulin	47	17
Uranium	43	25
Pesticides	36	29
Vaccines	32	25
X-ray diagnostics	30	30
Wolf reintroduction	22	27
Radioactivity	26	47
HIV virus	20	47
Human violence	18	57
New dog breeds	2	77
New types of fruit through genetic modification	2	82
Radiation of vegetables to increase durability	1	87

What is regarded as "natural" or "unnatural"? One study investigated how people conceived certain phenomena to be natural or unnatural (Drottz-Sjöberg, 1994). Table 16.1 shows that a majority felt that "primeval forest" and "cornfields" were natural phenomena, while a large majority responded that "new dog breeds" and "radiation of vegetables" were unnatural. Phenomena which were less clearly classified were "insulin" and "uranium." The results suggest that people sometimes have peculiar notions about what is natural and unnatural, and that there are variations among people with regard to such notions. What lies behind these judgments? Why should, for example, a cornfield be so much more natural than radiation of vegetables or the reintroduction of wolves? How can one see the HIV virus as unnatural when the concept of illness is seen as natural?

These are questions which have not been discussed in the literature on risk and the psychometric model, but the model was perhaps not intended to deal with all aspects of risk.

In the 1980s another, and very different, model was introduced: *cultural theory*. Cultural theory was developed within the field of social anthropology and is based on studies of primitive cultures, and notions of taboo and purity. The theory claims that there are four types of people who have "chosen" to be worried about different types of risk, depending on properties of the groups to which they belong. Most of the empirical basis of the theory consists of case studies. It is impossible to know whether the results give substantial support to the theory. An attempt was made to construct scales for measuring the four dimensions of central importance to the theory and to relate them to risk perception data (Wildavsky and Dake, 1990). The initial results were promising but they could not be replicated. The dimensions of cultural theory seem to be very weakly related to perceived risk (Sjöberg, 1997).

Other factors in risk perception such as trust, knowledge, morality, and risk definitions have also been introduced. Some suggestions are close to common-sense thinking, such as existential anxiety, radiophobia (Drottz-Sjöberg and Persson, 1993), "New Age" beliefs (Sjöberg and af Wåhlberg, 2002), and risk sensitivity (Sjöberg, 2004a).

In some fields of social science, there is currently much interest in social constructivist approaches (Hacking, 2000). According to social constructionist theory, everything, or almost everything, is a social construct. Are risks social constructs – and if so – what does it mean that they are? Our answer is "yes, of course they are social constructs." Risks are abstract, theoretical notions and nothing "out there;" they are expectations of future negative events. But it is pointless to argue that there is nothing harmful "out there." The risk that my neighbor's dog will bite me is surely a "construction," but the dog is not a construction and a bite would actually hurt. Thus, although thoughts and expectations are subjective and may be socially constructed, the conceptualizations are often related to real-world phenomena. There is enormous confusion if this distinction is not upheld.

Another factor, which has been discussed in the risk perception literature, is trust. Some studies have found relatively strong correlations between trust and perceived risk, e.g., between trust and the attitude to a local nuclear waste repository. However, we have found that epistemic trust is a more important dimension than the traditional trust measures, which are all concerned with social aspects: trust in people or organizations (Sjöberg, 2001, 2008a). People sometimes have a skeptical attitude. They do not believe that science has the final answers.

Even the extended psychometric model discussed above leaves a lot of the variance of perceived risk unexplained. It would seem that a new approach is needed in order to obtain a more efficient model. We have attempted such an approach and found that about 70 % of the perceived risk of nuclear waste can be explained by the following factors:

- *Attitude to the technology* that has created the risk, in this case nuclear power (Sjöberg, 1992). The attitude can be measured on a simple scale rating how good or bad nuclear power is seen to be in the view of the respondent.
- *Risk sensitivity*, a measure of how large or small the respondent judges all the involved risks to be (Sjöberg, 2004a). It can be measured by the average risk judgment. It is psychometrically sound to do so, because risk judgments tend to be correlated regardless of the nature of the risk.
- *Specific risk*, which is a component in the larger risk context under study. In the case of nuclear power, specific risk is the risk of ionizing radiation, regardless of the technology responsible for it (Sjöberg, 2000b). It can be measured by the risk associated with background radiation.

16.4 Emotions and Risk Perception

Does risk perception have anything to do with emotions? The question tends to give rise to acrimonious debates because people criticize each other for being irrational and emotional (Sjöberg, 1980). But there is nothing wrong with being emotional, nor is it wrong to show one's emotions. Many people who are deeply emotionally engaged have a neutral facade, which can be very misleading. Secondly, studies have shown that "dread" in the sense of emotion does not correlate at all, or only weakly, with perceived risk. The matter is complicated because dread has often been defined much more broadly than just as an emotion, and it has involved a number of aspects measuring the severity of consequences as well (Sjöberg, 2003b).

The crucial question is whether emotions are attributed to others or to oneself. People's own emotions are consistently correlated with perceived risk – those attributed to others are not (Sjöberg, 2007). The problem with the traditional definition of "dread" is that it refers to the emotional reactions of others.

The link between emotions and perceived risk need not be strong. Table 16.2 gives data from a dramatic example: patients facing major surgery the next day (Nyth, 1979). Both anxiety and perceived risk were measured. The table shows that there was no consistent relationship between the two dimensions. Some people who saw a large risk did not feel much anxiety. Conversely, some people who saw only a small risk did feel a lot of anxiety.

Does risk have a relationship to morality and ethics? First it must be noted that morality is not the same as emotions, although there is often an association in the sense that people are upset if important moral values are threatened. There is a strong tendency to reject activities which are risky if they are also seen to be immoral. "Interfering with nature" is, in the eyes of many people, immoral and against God's will. To some people, it is deeply upsetting that human beings interfere with Nature. An important example is that of genetically modified crops (Sjöberg, 2008b).

Table 16.2. Relationship between risk and anxiety, number of patients, on the day before major surgery (Nyth, 1979)

Perceived risk	Anxiety	
	Low	High
Small	11	7
Large	9	11

Values is another factor which is possibly of some relevance for understanding risk perception. Several scales have been developed in order to measure values, but it has been found that there are only weak relations between values and risk perceptions. A related, and more promising, approach is the measurement of "New Age" beliefs. An example would be the statement "Humanity is developing a higher level of consciousness." Such beliefs have been found to correlate with the perceived risk of technology (Sjöberg and af Wåhlberg, 2002) at a moderately high level.

16.5 Background Variables of Importance for Risk Perception

This section is about background variables of importance for risk perception. It deals with characteristics of the people who judge the risks and with how the risks are defined.

Let us first make a distinction between personal risks and general risks. Personal risks are those that are judged as they pertain to one's own person. Sometimes judgments are also made of "the risk to you and your family." General risks denote risks to "people in general." Many researchers have ignored this distinction and simply asked for judgments of "risk." We have found that such judgments tend to be close to general risks.

It is usually the case that personal and general risks differ both with regard to level and rank order. Some typical results are given in Figure 16.3. Note that the personal risks were judged to be smaller than the general risks, often much lower. Most people think that others are more at risk than they are. This, cannot, of course, be true for most people. It is an illusion.

What is the cause of this strong and common illusion? There are some hints as to its causes in the nature of the hazards where the differences between personal and general risks are especially large. The difference is striking with regard to the use of alcohol. When data are also collected about the possibility of protecting oneself from a risk, it is seen that those cases where there is a large difference between general risk and personal risk are associated with a larger perceived possibility of protecting oneself, as shown in Figure 16.4.

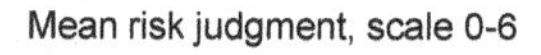

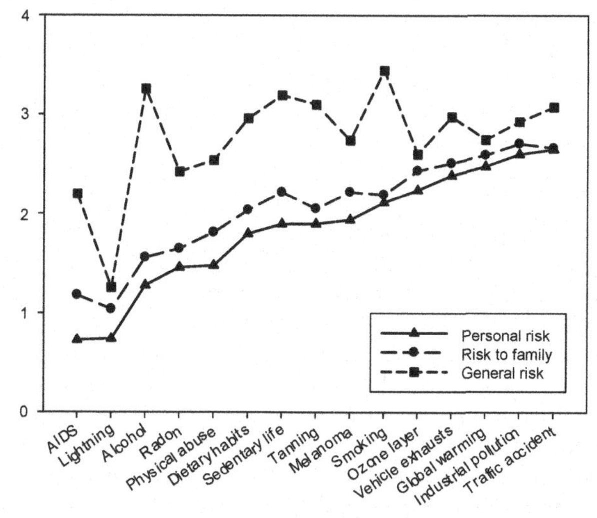

Figure 16.3. Mean judgments of risk to oneself, to family, and to people in general (Sjöberg, 1994)

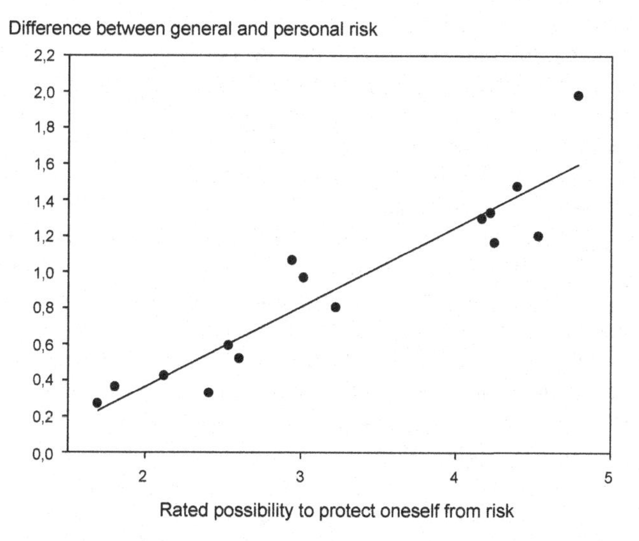

Figure 16.4. Difference between general and personal risk plotted against possibility of protecting oneself from the risk, mean values (Sjöberg, 1994)

The perceived work environment risk is a case of special interest. It can be described by a number of dimensions. For example, the risk of violence is quite important for some types of jobs. In Sweden, a bank employee runs a large risk of being the victim of a bank robbery at some time during his or her working life (Lind and Drottz-Sjöberg, 1994).

Even if there are considerable job risks, they are not necessarily perceived as such by employees. The reason can be that people feel that they have the situation under control. In addition, there is the familiarity factor; they have been working in the same environment for a number of years and have had no or few negative experiences – yet. There is a tendency towards denial of risks in the work environment.

Most studies of risk perception show that it is more common to deny risks than to be very worried about them (Sjöberg, 2006). However, worried people make themselves heard more frequently. They seem to have a greater tendency to be active. Maybe their opinions are also more interesting to the media.

Those who are active in a risk debate, e.g., those who are for or against a local nuclear waste repository, tend to have more extreme views than people in general. These groups belong to the so-called stakeholders (Sjöberg, 2003a). Corporations and authorities can easily get a biased view of public opinion if they listen mostly to active groups, which are those most likely to voice their anxieties and get in touch with organizations responsible for risk management. This is an important reason for making surveys of the risk perception of the general public, not only case studies of active persons or groups.

Risk denial is an important phenomenon. Lifestyle risks are perceived to be important for others, but not for oneself. Every doctor has encountered patients who come too late for help, because they have disregarded symptoms and told themselves, for much too long a time, that they do not have a serious illness.

Gender is an especially important factor in risk perception. Women tend to judge risks higher than men do. Ionizing radiation is a very clear example. Women tend to judge risks as larger especially if they concern risks to others. There is a gender difference also with regard to personal risk but it is only about half as large. Education and socio-economic status (SES) tend to be of importance as well. People with a higher education and a higher SES tend to judge risks to be smaller. The reason could be that such people are in fact exposed to smaller and fewer risks, or that they believe they have a better opportunity to protect themselves against risks. American studies have shown a "white male effect." Afro-Americans and white women judge risks to be greater than white men do (Finucane et al., 2000). This effect does not have an analog in the effects of SES in Sweden: a low SES is associated with larger perceived risks in both men and women.

There have been few studies of the risk perception of children and adolescents. The gender difference seems to emerge in puberty but is not pronounced before puberty (Sjöberg and Torell, 1993). However, it is known that younger boys take greater risks than girls do and that they have a higher accident rate. Age seems to bring with it more risk taking and more risk denial. Young men are a special group of risk deniers, and men in this group are more likely to be victims of accidents.

The concept of risk has different meanings for different people (see Chapter 2). Some people emphasize the probability of harm, others the magnitude of consequences. The latter group contains more women than men, and fewer experts than non-experts. The size of perceived risk is associated with how the concept of risk is defined. Those who emphasize the probability aspect tend to make lower risk judgments than those who emphasize the magnitude of the consequences of a negative event (Drottz-Sjöberg, 1991).

16.6 Risk Perception in a Social and Political Perspective

Psychologists have dominated social and behavioral risk research. They have worked on the perspective of individuals' risk reactions, and such work is important. Yet, they have been criticized for applying a limited individualistic approach. There is some truth in that argument. Psychologists have had a tendency to assume, more or less explicitly, that there is a simple and straightforward relationship between individual risk perceptions and societal reactions to, for example, siting proposals. It is easy to realize that societal reactions are affected by many factors besides individual reactions: NGO activities, the media, the involvement of politicians and the judiciary system, to mention a few. Some researchers have suggested a framework for these complex processes and called it "amplification of risk" (Kasperson et al., 1988). It builds on an analogy with technical systems. It is not a theory of societal risk management, but it is useful as a conceptual framework.

It is interesting to try to understand what leads to strong opposition to actions and suggestions in some cases and the lack of it in others. The question of management of nuclear waste is a question which is important in all countries which use nuclear power. In the USA, there have been several studies associated with perceptions and attitudes related to the Yucca Mountain project for a final repository of nuclear waste (Kunreuther et al., 1990). In Sweden, there have been discussions about repository siting in a large number of municipalities and in most cases, the result was a strong opposition. Local referenda were held in the 1990s (Storuman and Malå), and in both cases proposals to investigate siting feasibility were rejected. In Storuman, 71 % rejected the proposal, and in Malå there was a smaller majority against it, 54 %. The local referenda may have led to increased involvement in the issue and perhaps to increasingly polarized attitudes (Drottz-Sjöberg, 1996).

Figure 16.5 summarizes and structures the value dimensions which were found to be of importance after the local referendum in Storuman. It is based on interviews with people who had been active in the referendum process (Drottz-Sjöberg, 1996).

A prominent dimension in these discussions was that of "individualism-solidarity," represented by the vertical axis in the figure. The horizontal axis represents the Yes and No responses in the referendum. There are four sectors in the figure, corresponding to types of interest:

1. Small entrepreneurs, people who owned or wanted to start a small business
2. Established businesses on a larger scale
3. Municipal initiatives
4. Self-subsistent households

Persons from small enterprises and self-subsistent households usually placed themselves on the No side in the referendum. The established, larger businesses and central actors in the municipality were found on the Yes side. In each quadrant, we give the core arguments of each group with respect to the siting proposal. Hence, there were people who voted No in the referendum who preferred the development of small-scale business rather than a large industrial development. They wanted to have control over developments, and wanted to be employers rather than employees. They stressed that they had been born and raised in the community and had not moved there recently. They were very concerned about the environment and about retaining pristine Nature. The municipal initiative, which was the starting point of the siting discussions in the municipality, was explained and justified as supporting new job opportunities, badly needed, and development needed for the survival of the community. Those who supported the proposal argued that it was sufficient to have the promised influence over the process and that "guaranteed safety" beyond that was not necessary or achievable. Those who rejected the proposal were, on the other hand, greatly concerned with guaranteed safety – they wanted no risk at all.

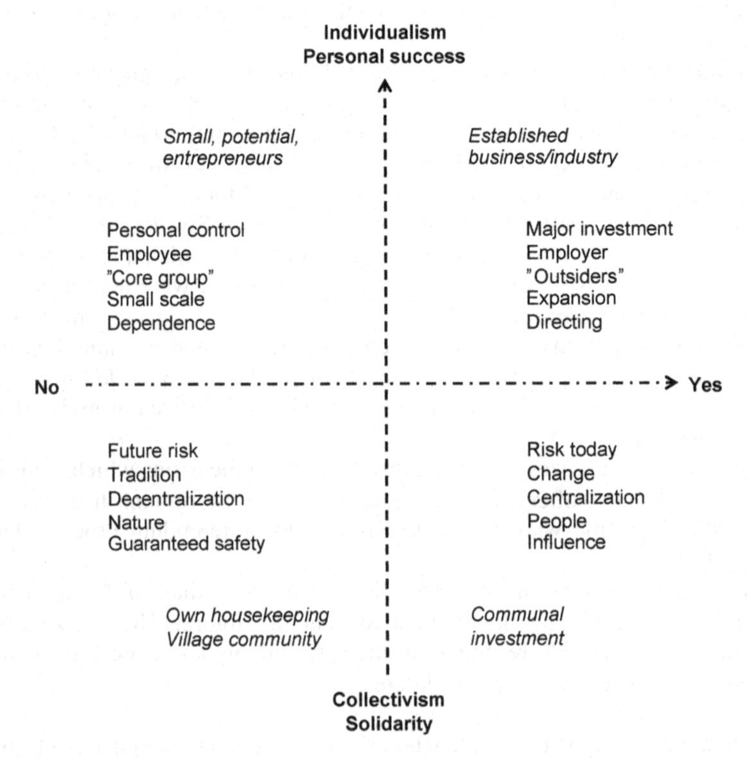

Figure 16.5. Schematic structure of the interview responses ordered in relation to the "No" and "Yes" groups' argumentation and in relation to a main "lifestyle" dimension of "individualism" versus "solidarity" (Drottz-Sjöberg, 1996)

Starting in 2001, two municipalities (Östhammar and Oskarshamn) accepted further feasibility studies (Sjöberg, 2004b). There are probably several reasons for their positive attitude: many years of experience with safe nuclear technology and the management of nuclear waste, information provided by the Swedish Nuclear Fuel and Waste Management Co (SKB), increasingly positive attitudes toward nuclear power, and local job opportunities.

Organizations and corporations also have risk opinions, and they may be on a collision course with the public or their own employees. In some cases, there is a very high level of safety, e.g., aviation. Overall, however, corporations tend not to be eager to listen to risk arguments. Employees who voice such concerns may be seen as "whistle blowers" and ostracized, however correct their arguments may be (Alford, 1999). There is a deep value antagonism between regarding industrial production as positive and profitable on the one hand, and seeing it as inherently very risky on the other. Economists and engineers have probably, in most cases, chosen their professions because they perceived them to be interesting and positive. It is improbable that the risk perspective was on their minds. The perspective of some people who may fear being affected by pollution or accidents can quite naturally be very different.

16.7 Consequences of Risk Perception

What is the importance of risk perception? It is often assumed, more or less implicitly, that perceived risk is a major factor behind demands for risk mitigation: The greater the perceived risk, the greater the requirements to mitigate it. This should be true both for insurance decisions made by consumers and for voters' requirements to phase out a technology which they see as too risky. This is a simple and self-evident approach, and much of risk debates seem to support it. But the matter is not quite that simple.

Some time ago, everyday and banal risks were included in the list of hazards investigated (Sjöberg, 1999). This was an unusual approach. Almost all earlier work had been concerned with fatal and catastrophic risks. The results of the study were that some trivial risks were judged as larger than some of the serious mortal risks. The risk of catching a common cold was seen to be greater than the risk of being infected with the HIV virus! But, at the same time, we require protection from the HIV risk, rather than from the common cold. This result calls for some new thinking about risk and policy.

If you ask people to judge the probability of injury, the size of the consequences, and the size of the "risk," it is found that risk and probability correlate very strongly and seem almost to represent the same concept. The demand for risk mitigation has little to do with risk and probability, and much more to do with the perceived consequences.

It is possible that these results are related to the fact that "probability" is a concept which it is difficult to understand, and that very small probabilities in particular are hard to grasp intuitively. Small probabilities are also difficult to determine because there is little or no empirical evidence to build upon and they have to be estimated with the help of models. Models can be questioned because they are based on assumptions. It is difficult to retain credibility in such cases, if risk assessments are challenged.

Demand for mitigation has economic consequences. If a risk can affect children or handicapped people, we are willing to allocate more money towards mitigating it. For adults who are capable of dealing with risks, or should be able to do so, we are less willing to mitigate risks at the expense of society. The sum of money spent on "saving a life" in different sectors of society has been calculated in several countries, among them Sweden (Ramsberg and Sjöberg, 1997; Tengs et al., 1995). When children are concerned, it can amount to hundreds of millions of SEK per life, and likewise for some especially dreaded types of illnesses and accidents (see Chapter 18). It is very hard to understand the large variation in implicit life values calculated in this way, even if some of the dynamics can be imagined. Is suicide a type of risk? Society has a low level of interest in suicides and many people would argue that it is a matter for the individual. It could even be argued that it is a human right. It is a very different matter with children's suicide, or preventing accidents where children may be involved. In these cases, society is willing to allocate large resources to "save a life."

Many questions about risk perception remain. The dynamics are different in different areas of application. Much of the work so far has been concerned with nuclear technology and emerging technologies. Much of the early applications and research on risk communication has been based on findings in risk perception work related to nuclear technology. The models which have been formulated for risk perception usually explain only a minor part of the variance of perception – with the exception noted above. The application to risk communication should be seen in this context.

16.8 Risk Communication: The Case of Nuclear Waste

Risk information is often focused on dangers related to substances or behaviors which should be avoided, performed with caution or treated in a cautious manner. There are lifestyle risks (alcohol, smoking, driving a car, etc.), and environmental risks such as indoor radon gas. In other cases, the risk information is needed to show that the risk is small and the behavior or technology in question need not be avoided and can be utilized. Examples are being close to a person infected with HIV or the use of a mobile telephone.

In the old "engineering" model of risk communication, an attempt was made to provide correct information in a form which was attractive. Communication and providing information were seen as being virtually the same thing. However, it was soon noted that simply "informing" about a risk had little effect on people's risk perceptions. There were problematic assumptions behind this model of risk communication. It was for example assumed that a neutral non-partisan informer could be used, that the message was easy to understand and that the effects of a message were those intended. Although this approach often resulted in failure it was hard to understand why. Several reasons for the failure were suggested. Perhaps the informer was the wrong type of person or had some personal characteristics which were counter-productive. Perhaps the timing of the message was wrong, or the medium. Perhaps the media had distorted the message, or people had not understood it, or alternatively they were uninterested or too emotional and frightened to listen.

It has become clear that the process is much more complex than that just described. There are several interacting parties involved in any attempt at risk communication. The reliance on one-way "communication" led to failure, and interactive communication models were developed in the 1990s. The key concepts became process, integration, and problem solving. It was realized that a continuous process involving feed-back channels was necessary and that it was essential to listen carefully to the public's or other parties' concerns, including the concerns of special groups. Furthermore, communication had not always been understood in the intended manner in the "engineering model" of risk communication. Therefore, in later development, problem-solving became much more important than an attractive "packaging" of messages. The models used in advertising failed, probably because serious risk matters need to be approached along a central route of cognitive processing, not a peripheral one (Petty and Cacioppo, 1986). The problem-solving approach goes hand in hand with the two-way interactive information exchange of the later risk communication models.

At this point, problems became differentiated. What perspectives exist with regard to a risk topic? Whose perspectives should be considered? What do people want to know? The concept of stakeholder came into the foreground, but it caused some confusion. Who "owns" a given risk topic? Who is a "stakeholder"? This is a complex issue seldom clearly defined in practice, and is by no means always legally clear. For example, there is a high level of self-determination of municipal authorities in Sweden. Municipal authorities are stakeholders in the national process of selecting a site for a nuclear waste repository. However, within a municipality several other types of organizations and groups may claim the status of stakeholder as well.

Representatives of different political parties are members of the municipal councils, elected by people locally. They may have different options with regard to the siting of nuclear waste. They could cooperate or go for separate political profiles, and what is defined as a stakeholder could take on a new meaning. There are also other organizations such as NGOs outside the politically elected bodies (Figure 16.6). They can act according to the same philosophies as political parties or they can develop schemes of their own. They are groups of people who claim to have special perspectives, which must be taken into account. More-or-less well-organized local groups of this kind can consist of people with a special interest in Nature or the environment, employers, religious groups, youth organizations, and owners of summer houses, land, or water. In addition there may be national or international organizations with an interest in risk management issues at a certain location or in relation to a certain issue. In discussions of siting of a repository for spent nuclear fuel, there are also neighboring municipalities, as well as neighboring countries, with an interest in the outcome of the siting planning and decisions.

To accommodate many interests and to promote full information on an issue, research on risk communication has emphasized the importance of transparency in the decision-making process. Transparency is related to several dimensions of the decision process: what is to be done and what are the goals, who are the actors, and what are their roles, and how can they contribute. The RISCOM model of transparency, developed in an EU project, considers three basic elements: technical and scientific questions, normative questions, and credibility (Andersson et al., 1998). Transparency must pertain to all these elements. Research on local risk decision processes has shown that transparency contributes in an important way to achieving success in risk communication and in the continuing work process.

Figure 16.6. Greenpeace activists plant apple trees on the rail line used to transport nuclear waste, Gleisbett, Germany, May 11, 2001. (Photo: Kay Nietfeld/EPA PHOTO/DPA/ Scanpix)

In this chapter, we have emphasized that the experience, or perception, of risk varies with personal background and perspectives, and that such perceptions can be captured based on empirical data. The experiences and reactions of individuals are but a part of extensive decision processes related to, e.g., large technological projects. The risk communication field involves a number of techniques and methods of data collection, and we have given an example of results obtained by an interview approach. We have underlined that a number of interests, or stakeholders, are usually involved and interact in risk communication processes, and that openness or transparency is essential for developing a fruitful approach to the management of complex projects of large societal concern. Such an approach incorporates different perspectives at an early stage and facilitates a broadening of the basis for later decision-making. Research related to risk perception of nuclear energy and waste management, and the extensive and multi-organizational risk communication process that surrounds the Swedish plans for a final repository for spent nuclear fuel reflect both individual and societal concerns. Experiences from these fields will also hopefully prove useful is other areas related to social controversies and individual reactions to large industrial projects.

References

Alford CF (1999) Whistle-blowers – How much we can learn from them depends on how much we can give up. American Behavioral Scientist 43:264–277

Andersson K, Espejo R,Wene C-O (1998) Building channels for transparent risk assessment. RISCOM Final Report. Swedish Nuclear Power Inspectorate, Stockholm

Breakwell GM (2007) The Psychology of Risk: An Introduction. Cambridge University Press, Cambridge

Combs B, Slovic, P (1979) Newspaper coverage of causes of death. Journalism Quarterly 56:837–843,849

Drottz-Sjöberg B-M (1991) Non-expert definitions of risk and risk perception. Center for Risk Research, Stockholm

Drottz-Sjöberg B-M (1994) Experiences of nature and the environment. Rhizikon: Risk Research Report, No. 21. Center for Risk Research, Stockholm School of Economics, Stockholm

Drottz-Sjöberg B-M (1996) Stämningar i Storuman efter folkomröstningen om ett djupförvar. (Moods in Storuman after the repository referendum). SKB, Stockholm

Drottz-Sjöberg B-M, Persson L (1993) Public reaction to radiation: Fear, anxiety or phobia? Health Physics 64:223-231

Finucane ML, Slovic P, Mertz CK, Flynn J, Satterfield T (2000) Gender, race, and perceived risk: The 'white male' effect. Health, Risk & Society 2:159–172

Fischhoff B, Slovic P, Lichtenstein S, Read S, Combs B (1978) How safe is safe enough? A psychometric study of attitudes towards technological risks and benefits. Policy Sciences 9:127–152

Hacking I (2000) The social construction of what? Harvard University Press, New York.

Kahneman D, Tversky A (1984) Choices, values, and frames. American Psychologist 39:341–350

Kasperson RE, Renn O, Slovic P, Brown HS, Emel J, Goble R, et al. (1988) The social amplification of risk. Risk Analysis 8:177–187

Kunreuther H, Easterling D, Desvousges W, Slovic P (1990) Public attitudes toward siting a high-level nuclear waste repository in Nevada. Risk Analysis 10:469-484

Lind F, Drottz-Sjöberg B-M (1994) Risk, säkerhet och arbetsmiljö: Yrkesgrupper som handhar kontanta medel. (Risk, safety and work environment: Occupational groups handling cash) RHIZIKON: Rapport från Centrum för Riskforskning, Handelshögskolan i Stockholm No. 4. Centrum för Riskforskning, Handelshögskolan, Stockholm

Nyth A-L (1979) Ängslan och riskupplevelse inför operation. (Anxiety and risk perception on the eve of major surgery). Rapport från Psykologiska Institutionen No. 1, 1979. Göteborg: Psykologiska institutionen, Göteborgs universitet

Petty RE, Cacioppo JT (1986) Communication and persuasion: Central and peripheral routes to attitude change. Springer-Verlag, New York

Ramsberg J, Sjöberg L (1997) The cost-effectiveness of life saving interventions in Sweden. Risk Analysis 17:467–478

Siegrist M, Earle TC, Gutscher H, Keller C (2005) Perception of mobile phone and base station risks. Risk Analysis 25:1253–1264

Sjöberg L (1980) The risks of risk analysis. Acta Psychologica 45:301–321

Sjöberg L (1992) Psychological reactions to a nuclear accident. In: Baarli J (ed.) Conference on The radiological and radiation protection problems in Nordic regions, Tromsö 21–22 November, 1991 (Paper 12). Nordic Society for Radiation Protection, Oslo.

Sjöber L (1994) Solstrålningens risker: Attityder, kunskaper och riskuppfattning. (The risks of sunrays: attitudes, knowledge and risk perception). RHIZIKON: Rapport från Centrum för Riskforskning, Handelshögskolan i Stockholm No. 3. Centrum för Riskforskning, Handelshögskolan, Stockholm

Sjöberg L (1997) Explaining risk perception: An empirical and quantitative evaluation of cultural theory. Risk Decision and Policy 2:113–130

Sjöberg L (1999) Consequences of perceived risk: Demand for mitigation. Journal of Risk Research 2:129-149

Sjöberg L (2000a) Perceived risk and tampering with nature. Journal of Risk Research 3:353-367

Sjöberg L (2000b) Specifying factors in radiation risk perception. Scandinavian Journal of Psychology 41:169–174

Sjöberg L (2001) Limits of knowledge and the limited importance of trust. Risk Analysis 21:189 –198

Sjöberg L (2003a) Attitudes and risk perceptions of stakeholders in a nuclear waste siting issue. Risk Analysis 23:739–749

Sjöberg L (2003b) Risk perception, emotion, and policy: The case of nuclear technology. European Review 11:109–128

Sjöberg L (2004a) Explaining individual risk perception: the case of nuclear waste. Risk Management: An International Journal 6:51–64

Sjöberg L (2004b) Local acceptance of a high-level nuclear waste repository. Risk Analysis 24:739–751

Sjöberg L (2006) Rational risk perception: Utopia or dystopia? Journal of Risk Research 9:683–696

Sjöberg L (2007) Emotions and risk perception. Risk Management: An International Journal 9:222–237

Sjöberg L (2008a) Antagonism, trust and perceived risk. Risk Management: An International Journal, 10:32–55

Sjöberg L (2008b) Genetically modified food in the eyes of the public and experts. Risk Management: An International Journal 10:168–193

Sjöberg L, af Wahlberg A (2002) Risk perception and new age beliefs. Risk Analysis 22:751–764

Sjöberg L, Drottz-Sjöberg B-M (in press). Attitudes towards nuclear waste and siting policy: Experts and the public. In: Lattefer AP (ed) Nuclear waste research: Siting, technology and treatment. Nova Publishers, New York

Sjöberg L, Torell G (1993) The development of risk acceptance and moral valuation. Scandinavian Journal of Psychology 34:223–236

Sjöberg L, Oskarsson A, Bruce Å, Darnerud PO (1997) Riskuppfattning hos experter inom området kost och hälsa. (Experts' risk perception with regard to food and health). Statens Livsmedelsverk, Uppsala

Sjöberg L, Jansson B, Brenot J, Frewer LJ, Prades A, Tönnesen A (2000) Radiation risk perception in commemoration of Chernobyl: A cross-national study in three waves. Center for Risk Research, Stockholm

Slovic P (ed.) (2000). The perception of risk. Earthscan, London

Slovic P, Fischhoff B, Lichtenstein S (1979) Rating the risks. Environment 21:14-20,36–39

Starr C (1969) Social benefit versus technological risk. Science 165:1232–1238

Tengs OT, Adams ME, Pliskin JS, Safran DG, Siegel JE, Weinstein MC, et al. (1995) Five-hundred life saving interventions and their cost effectiveness. Risk Analysis 15:369–390

Tversky A, Kahneman D (1973) Availability: A heuristic for judging frequency and probability. Cognitive Psychology 4:207–232
Wildavsky A, Dake K (1990) Theories of risk perception: Who fears what and why? Daedalus 119:41–60

[illegible]

[illegible]

17

The Value of a Life

Bengt Mattsson

17.1 Introduction

In this chapter we are primarily interested in how to assign a pecuniary value to personal injuries or deaths in connection with accidents. In all decisions where risks are included, one must in some way or another consider – i.e., assign a value to, the personal injuries and deaths. To simplify matters, only deaths will be considered here in detail. Other risks can be discussed in a similar way. The numbers of people who are killed or injured in different kinds of accidents have been discussed in other chapters of this book and will not therefore be discussed here.

Decisions affecting the risk of injury or death in accidents are taken in many parts of society, for instance concerning sea, air, and land transport and within companies, hospitals, nuclear power plants, and even in everyday life at home. The decision-maker can be a parliament, government, corporate CEO or head of security, employee, car driver, person working at home, etc.

The different decision-makers can adopt many kinds of measures to lower the risk. Parliament and government can legislate regarding speed limits and police surveillance on the roads. The road administration can rebuild road intersections or increase the proportion of roads with lighting. Municipalities can put resources into separating different types of road-users, e.g., by constructing bicycle lanes and underpasses for pedestrians. The road administration and the railway administration can build new intersections between roads and railways. Train companies can make it impossible to open the coach doors even at low speeds. Companies can abolish piecework rates in cases where a high work tempo involves a considerably increased risk of injury. Municipalities can decide about full-time fire and rescue services, instead of part-time, and thus reduce the risk of personal injuries and large material damage. As individuals, we can choose cars instead of motor cycles and also choose cars that are well equipped with safety features. We can also reduce the risk of traffic accidents, for instance by driving more slowly.

17.2 Risk Valuation and Socio-economic Efficiency

There is some knowledge about the effects of measures of the kind mentioned above. (The examples given below often refer to Sweden. Conversions to US dollars are made at the early 2008 exchange rate.) For example, when road lighting is introduced on a large road to a population center, several studies show that the night accident rate is reduced by about 30 %, which is equivalent to a reduction by 10 % of the total number of traffic accidents on that road.

According to one study (The Swedish Road Administration, 1997), the number of deaths per billion vehicle kilometers is approximately four for drivers and passengers in private cars, but 75–80 for road-users on mopeds or motor cycles. If part-time fire fighters have other occupations and leave for the fire station when the alarm is given, it may take approximately five minutes longer before they leave the fire station than when the station is manned by full-time employed fire-fighters. Short turn-out times are most important for traffic accidents, drowning cases and fires in buildings. For fires in buildings, it is foremost the material damage that is affected. With the distribution of alarms that exists in Sweden (including incorrect automatic alarms) it has been calculated that a five minutes shorter turn-out time on average would mean an accident cost reduction per alarm of approximately USD 8 000, at the 2007 price level (Mattsson and Juås, 1997; Mattsson, 2004).

In other cases, our knowledge of the effects of risk-reducing measures is more uncertain. Almost all measures, however, mean that resources are sacrificed, i.e., the measures have a monetary cost. With increasing, and ultimately severely increasing, expenditure, the risk of injury or death can be reduced. For instance, it would probably substantially reduce the number of accidents if pedestrians, cyclists, motor cyclists, private cars, and trucks were completely separated in our cities, but it would also mean huge reconstructions and costs. A full-time rescue service in each municipality, and several full-time services in all municipalities with more than, e.g., 50 000 inhabitants, would shorten the turn-out time and reduce the costs of the accidents. This would, however, result in significant increases in costs for the municipal authorities. It can hardly be a norm in our society to do everything that is technically possible. Of course we want safe traffic, a safe working life, and a safe home environment, but we also want to live well, perhaps travel a lot, enjoy ourselves, and so on. This means that we are all willing to take risks provided that the "cost" is right. We may rush across a street with heavy traffic to catch the bus, if we know that the next bus will not depart for another 30 minutes. We may refrain from using breathing protection even in a dusty working environment, if we feel that it restrains us too much. We may choose not to buy a fire extinguisher for our house. It is expected that society, i.e., all people taken together, will act in the same way – trading safety for other utilities. When resources (work force, machines, energy, raw materials) are to be divided between risk-reducing measures on the one hand and the production of other utilities (food, clothes, pleasure, etc.) on the other, we need to know how much people are willing to pay in order to increase safety.

On the basis of pure efficiency, we can demand that resources are divided between risk-reducing activities in such a way that the marginal costs are the same for each saved life or case of injury reduction. (This statement is modified later in

Section 17.7 headed "A More Balanced Approach to the Valuation of Life.") If the price is USD 5 million to reduce by one, the number of probable deaths per year in road traffic, and only USD 2 million to achieve the same reduction in the working life, we can probably increase efficiency by transferring resources from road traffic to the working life.

Socio-economic efficiency, according to the so-called *Hicks–Kaldor criterion*, means that in an optimal welfare situation there is no measure for which the benefits exceed the costs. One way of deciding if this is the case is through cost–benefit analyses for different possible measures. The main idea behind these calculations is simple. One can use the concept of "the socio-economic balance," where all the advantages (benefits) of a given measure are put into one scale and all the costs into the other, regardless of whether they affect enterprises, the state, municipalities, households, etc. If the benefits weigh more, then society benefits from the measure. If the cost is heavier, society should refrain from the measure, since the current situation is better than that achieved by the proposed change. The value of increased safety can be a component in the benefit scale of the balance.

When one refers to society, all the individuals in a nation are usually included. All the advantages and costs will in principle be measured by these individuals' willingness to pay. Socio-economic efficiency is thus based on the idea that it is the individuals, not the government, municipal commissioner, head of the Road Administration, etc., who assess the value of, for instance, a change in safety.

17.3 Some Important Concepts

Let us for the moment concentrate on a discussion of deaths, even though what we are about to say could also be valid in the case of injury. The risk we discuss concerns the probability of having an accident. Assume that the probability of dying in road traffic in a municipality with 100 000 inhabitants is 0.000 10 per year, and let us assume that by implementing some kind of measure, e.g., by constructing a tunnel for pedestrians in the city center, that probability can be reduced to 0.000 09 per year. The expected number of deaths per annum has thus decreased from 10 to 9, after the measure has been implemented. This "saved life" is usually called a *statistical life*. It is such statistical lives, or by analogy "statistically severely or mildly injured" that we consider here when we discuss valuation issues.

The question is thus how much a municipality should invest to achieve this effect, i.e., to save a statistical life. (We here assume that other effects, for example material damage, do not exist or that we have excluded them.) Thus, our problem refers to an assessment for a fairly anonymous group of people, where we perhaps know the distribution of age and gender. It is not about how much John Smith is worth, or about you or me. It is, however, very possible that the measure that is taken will affect us, since we may belong to the group of 100 000 individuals. The assessment discussion, however, applies only to a change in the probability of dying for a large group of people. We do not know who will benefit from the measure. It may be you, it may be me (Figure 17.1).

Figure 17.1. Arial view of pedestrians at Shibuya Crossing, Tokyo, Japan. The risk of an individual pedestrian being run over is unpredictable, while we may be able to predict how many fatal pedestrian accidents will happen on an average for different crossings or crossing designs. (Photo: Jeremy Woodhouse/Scanpix)

A recent recommendation to the Swedish Road Administration (SIKA, 2007) is to give a statistical life a value of SEK 21 million (about USD 3.4 million) at the 2007 price level. This means that if the cost of a measure that saves one statistical life is SEK 15 million, it should be carried out. If, on the other hand, it costs SEK 25 million it should not be done. We can call this way of determining a price an *explicit value*. The Swedish Road Administration also uses explicit values for severely injured (SEK 3.5 million, or about USD 0.6 million, per person) and mildly injured (SEK 135 000, or about USD 22 000) at the 2007 price level (SIKA, 2007). How these values have been obtained is described in Section 17.6 headed "Economic Assessment Methods."

Suppose that we know of an authority or municipality that has decided not to carry out a certain measure whose single effect would be to decrease the number of statistical lives lost per annum by four, at a cost of USD 10 million. In this case, we can say that the authority or municipality has set an *implicit value* of less than USD 2.5 million per statistical life. Instead, if it had been decided to carry out the measure, we would draw the conclusion that the implicit value was at least USD 2.5 million.

Implicit values can be calculated for a variety of decisions in different sectors. Accounts of explicit and implicit values of a statistical life are found in, e.g., Miller (2000) and in Viscusi and Aldy (2003). Explicit values have been used in many countries primarily with respect to road traffic safety, radiation protection and air safety. As a further example, the Swedish Radiation Protection Authority (SSI) has decided that the cost per "saved" statistical life should be in the range of SEK 5 to 25 million, or about USD 1–4 million (Valentin, 1997).

By studying decisions relating to measures, or rejections of proposals of measures, where the probability of death or injury is affected, it is thus possible to estimate implicit values. Before we can say that these values also correspond to the assessments (preferences) of the decision-makers, the following conditions have to be fulfilled:

- The decision-maker must have had an opportunity to choose, i.e., there must have been at least one more option.
- The decision-maker must have been informed of all the relevant facts; how many more or fewer killed and injured people are to be expected, what other effects may arise, and what the proposed measure will cost.
- The decision-maker must act rationally. Among other things this means that he considers risks and always reaches the same decision when the circumstances are the same.

17.4 The Economic Assessment Problem

Our problem is about decisions when the resources that influence the number of deaths or injuries in accidents are scarce. We want to study how much individuals, groups, authorities, etc., are willing to pay to reduce the number of deaths or injuries. As an example, we choose to study road traffic, but examples from working life, sea traffic, fires, etc., are not difficult to find.

There are several measures that can reduce a risk. One way is to influence *the exposure time*. Substantial petrol price increases make us drive less and therefore influence the time during which we are exposed to traffic in our own car. If the alternative to car driving is a less dangerous activity, e.g., reading in our favorite chair, the increase in the price of petrol can lead to a reduction in the number of accidents. Another way is to influence *the probability of injury* or death for a given exposure time. Experience has shown that stationary road lighting at a distance leads to a reduction in the number of deaths and injuries per million vehicle kilometers. A third way is to influence *the consequences of an accident.* Legislation requiring that everybody in a car uses seat belts when traveling, or that the car is equipped with air bags, hardly reduces the probability that an accident will occur, but it does relieve its consequences. Our discussion applies to all these types of measures.

To specify the problem, "to assess the risk of personal injuries and deaths," we can make use of a lottery analogy. Let us consider the risk of death. Suppose that there are two lotteries. Lottery 1 involves some probability that we shall die, and in lottery 2, the probability of death is lower. Better industrial safety, traffic safety, or fire protection are examples of measures that go from the worse lottery to the better. The main problem is how much the individual, the group, the county council or society wants to pay, how much one is willing to sacrifice of other utilities, to change a ticket in lottery 1 for a ticket in lottery 2.

Figure 17.2. People are generally more willing to take a greater risk, if they get sufficient compensation for this risk increase. The picture shows Egyptian fire-fighters trying to extinguish a fire at the nineteenth-century upper house of Egypt's parliament in Cairo, Egypt, August 20, 2008. (Photo: Amr Nabil/AP/Scanpix)

We can also look at it from another angle and say that we are interested in what compensation is demanded to change a ticket in the better lottery for a ticket in the worse. People are generally more willing to take a greater risk, e.g., to work in a mine, at a high altitude, or with different tools or materials, if they get sufficient compensation for this risk increase (Figure 17.2). As early as 1776, Adam Smith observed that workers wanted higher payment for a job with a higher risk (Smith, 1776).

17.5 Ethical Aspects of Economic Risk Assessment

Many readers may react, for ethical reasons, to the fact that we consider economic aspects on measures that affect the risk of death or injury to humans. There may be some who, while reading this text, get associations with the death camps of Nazi Germany and their classification of people capable of working for a while and the rest who were promptly executed. In the light of this, it can be important to discuss the moral aspects of what this chapter is about. To be more specific about when my own values are the starting point for the discussions, I write "I," "me," and "mine" from now on.

It is sometimes claimed that human life has an infinitely high value and that economic aspects of safety work are therefore nonsense. In some extreme cases, we can imagine that we do approach an infinitely high value. This can be the case in situations where individuals are in evident danger, for example when confined in a cave, sitting squeezed and severely injured in a car after an accident, or when they have lost their way in the mountains in a snow storm. In such cases, all available resources are usually mobilized and no cost limit is specified, but the question of how much society should invest in increased security hardly ever concerns situations like that.

Instead, the discussion usually concerns how much we shall invest in situations where there is a relatively small risk of dying in working life, in traffic, in fires etc. in order to make this risk even smaller. In such cases, several studies show (see Section 17.6 below on assessment methods) that individuals set a positive but far from an infinitely high value on increased safety. If adult, well-informed, sensible, relatively well-educated individuals act in this way in free elections in countries with a high standard of living, it corresponds to my values that central decision-makers in those countries do not deviate strongly from these values in their decisions.

It is also sometimes argued that it is unethical to put a monetary value on a change in the risk of accident or death. Here I want to rebuff those who, perhaps unwillingly, admit that the value per statistical life is hardly infinitely high, but claim nevertheless that it would be unethical to discuss or try to determine such values. When we decide to do or not to do something that affects security, we set an implicit value on, e.g., a life. It is difficult for me to see it as immoral to bring up these values for discussion. Even if we were to close our eyes to the problem and let chance decide (by flipping a coin or rolling a die) what measures to take, we cannot get away from the fact that our decisions result in some implicit values. Coin-flipping and refusing to discuss the problem probably leads to a lower safety within a given budget than could otherwise be attainable. I find it unethical to refuse to use relevant basic data to make a rational choice possible.

Politicians and other decision-makers often assert, although rarely in public, that they admittedly understand the problem but the voters will not. To avoid being dragged through the mud by mass media and risking their careers, they "put the lid on." Considerable responsibility naturally lies with the journalists, who should promote a meaningful discussion. Even though it may be tempting for decision-makers to be excused from such a discussion, my opinion is that we should demand not only that they make decisions but also that they accept a "popular adult education task." For instance, in this case they could use the lottery analogy above, present the problem, and start an assessment discussion with the voters.

17.6 Economic Assessment Methods

There are essentially two methods that have come into use to value statistical lives; the *loss of production approach* and the *willingness to pay approach*. In some cases other methods have been used, e.g., established by using compensations from traffic accidents as a starting-point, but these two former methods have predominated.

17.6.1 The Production-loss Approach

The idea in this approach is that society[1] saves human capital by preventing a death or a severe injury, at the expense of a mild injury, etc. What is human capital then worth? For an economist, the value of all capital is decided by the present value of future income from the capital. A narrow view can then be that human capital corresponds to the present value of future production, where the production value is measured as future income from work. The method has been diligently used since the 1950s, particularly regarding road traffic accidents and air safety, and in the area of health and medical care. In more recent times, calculations of the future decrease in production have been made to answer the question of how to value a change in the risk of death, e.g., Elvik (1988), Jones-Lee (1989). The values that have been established differ primarily because of different income levels in different countries, because different interests were used to obtain the present value, and because the assumptions of future increases in income can differ.

Comparisons between countries are complicated by the fact that devaluations or revaluations of currencies often mean that the comparison is dependent on the year or even the day of the exchange rates used for the calculations. We therefore choose here to show the size of the production loss in Sweden in the year 2007, rather than to refer to those values that emerged from the above and other investigations. Let us assume a 5 % (real) interest rate, an average income (including social fees) of SEK 25 000/month and a 1 % annual real growth per capita of this income in the future. Let us also assume a future unemployment rate of 10 %, i.e. that 90 % in each working age class will have these incomes.

Of course, the loss of production at the present value depends on the age of the individual. A 20-year-old may have 40 years of working left while a 50-year-old has on average about 10 years left. (The average retirement age is about 60.) The age distribution varies a lot between different types of injuries, as was pointed out earlier. In road traffic, the 15–22 year age group accounts for a large share of the total number of deaths. In working life, the average age is higher, and for accidents at home much higher. Table 17.1 gives the present value of future income of work for four different remaining times in professional life.

In other words, with these assumptions, we can say that with a remaining time in professional life of 20–30 years, the human value (the present value of future losses in production) is SEK 4–5 million.

But can the present value of future working income really give the value of a statistical life? Those who have used, and still use, this approach have not always specified what this value is intended to measure or what it is to be used for. If it is to support decisions affecting safety to be made by a road administration, a

[1] The term "society" is used with different meanings. Politicians sometimes use the term as a synonym for the state or public sector. Expressions like "the strong society," i.e., a public sector that is extensive and inclined to intervene, or "here society must intervene," which often means state contributions, exemplify this use. However, here "society" is used to describe something larger, often the same as a nation. If we identify society with a nation state, it means that municipalities, and companies as well as households, are included in this concept.

radiation protection authority, a municipality, the government or parliament, etc., the relevant question can be formulated as follows. "How much are we, individually or collectively, willing to spend in order to save a (statistical) life?" At the same time, society is also interested in using resources to reduce pain, suffering, etc. Many recent human capital (production loss) calculations have been characterized by the attitude that the calculated value should be a minimal amount, on top of which a politically established value to reduce pain, suffering, etc., should be added.

Table 17.1. The present value of future losses in production in SEK for different remaining times in professional life

Remaining time in professional life [Years]	10	20	30	40
Present value of future production losses [Million SEK]	2.19	3.68	4.66	5.34

Assumptions: SEK 25 000 per month in production value (price level of 2007), 1 % real future growth, 5 % (real) interest, and 10 % unemployment.
SEK 4–5 million is USD 600 000–700 000

The great popularity of the loss of production approach is probably explained by the fact that the method is operationally easy to handle. One could argue that the method shows the extent to which it is justified for a cynical slave owner to put resources into increased security (with deduction of the slaves' consumption) in order to maximize profit.

Since society includes us all, it is hardly likely that we in a democracy will arrive at the same conclusions as the cynical profit-maximizing slave owner. In our utility functions, future incomes or future consumption options are admittedly significant factors, but so are health, well-being, etc.

A high quality of life is not decided solely, and perhaps not even mainly, by consumption options but by a large number of other factors. We can imagine an exchange between different components in the utility function, for example increasing the risk of death or injury by taking a risky job provided we receive an adequate income compensation. (Compare the lottery analogy earlier.) However, we are not willing to say, like the slave owner, that it is the maximum present value of future incomes that determines our choice.

17.6.2 The Willingness-to-pay Approach

Two main methods have been used in studies of the willingness-to-pay (WTP) or the willingness-to-accept (WTA) approach. They measure it either directly by interviews (the contingent valuation method is often used), or indirectly by studying what decisions individuals or groups have made and what implicit values these studies indicate. For a presentation and discussion of the methodological problems, see for example Svensson (2007).

The interview surveys consistently include individual decisions, while studies of decisions made include the individual as well as the collective willingness to pay. Collective willingness to pay refers to the process whereby a group, e.g., a parliament, road administration committee, or municipal council, makes a decision that is equivalent to a certain willingness to pay per "saved" statistical life.

There are early examples of interviews in the US from the early 1970's concerning the willingness to pay for the reduction in risk related to the availability of special heart-infarction ambulances, and also concerning how much individuals were willing to pay for increased fire safety in the home. More recently, large surveys have been carried out in, e.g., England and Sweden concerning the willingness to pay for a change in traffic security, see for instance Jones-Lee et al. (1995) and Persson et al. (1999).

Indirect studies of WTP or WTA are based on the idea that the actions of individuals or groups reveal their preferences with regard to increased security; compare the discussion of implicit values above. One of the earliest studies on the demands of individuals concerned differences in salaries as a compensation for risk-filled jobs. Miller (2000) gives some examples of risk-compensating differences in salaries and the purchase of safety equipment and another study of collective decisions deals with traffic security (Mattsson, 1991).

The following is a brief summary of some well-performed surveys. More detailed descriptions are given by Miller (2000) and Viscusi and Aldy (2003).

Let us first consider *direct estimates* (interviews) of individual decisions. Three extensive interview surveys using modern methods, Jones-Lee (1989) concerning traffic safety, Gegax et al. (1985) concerning working life safety, and Persson (2004) concerning traffic safety, lead to an average of approximately SEK 32 million (about USD 5 million) at the 2007 price level. This can be compared with the value of 20 SEK million obtained in an extensive interview survey (approximately 6 000 persons) by Persson et al. (1999) concerning the willingness of individuals to pay per statistical life. See also Persson (2004).

In the context of *indirect estimates* and *collective decisions*, a decision by Swedish authorities was formulated as "the requirement of an annual reduction of the number of deaths by three per cent and the number of injuries by two per cent from 1990 to 2000" (TSV, 1989). A group of traffic safety experts calculated what measures were the most cost-efficient when trying to reach the goals. According to these calculations, it would cost about SEK 24 million at the 2007 price level (about USD 4 million) per saved statistical life to achieve the goal (Mattsson, 1991).

Finally, let us consider *indirect estimates* of *individual decisions*. There are numerous such studies, mainly regarding differences in salary. Based on 15 such studies that were considered to be of high quality (Mattsson, 1990), the value of a statistical life was on average approximately SEK 53 million (about USD 9 million) at the 2007 price level. Most of these studies concerned the US and salary compensation for an increase in risk.

Thus there is a spread from SEK 20 million to about 53 million (about USD 3–9 million) in WTP/WTA for a statistical life. This may appear to be a large spread, but it probably indicates a surprising consistency. The value interval also agrees well with the statement (Viscusi, 1992) that "although the estimates of the risk-dollar trade-off vary considerably depending on the population exposed to the risk, the nature of the risk, and similar factors, most of the reasonable estimates of the value of life are clustered in the $3 to $7 million range." Translating this interval to the 2007 level using the Consumer Price Index, gives SEK 24–55 million or almost the same as above.

The main advantage of using both direct studies (interviews) and indirect studies of the WTP/WTA is that this procedure helps to measure what is relevant. In the interviews, one can choose the group whose values one wants to examine and can arrange it exactly like the decision situation one is interested in. In the study of decisions already made, one is to a greater extent limited to what is actually available. The advantage of the indirect studies lies, however, in the fact that a real sacrifice has been made. One has actually bought the airbag, one has taken the risky job, etc. There is an uncertainty in interviews in that one does not know whether the people's answers as to how much they are prepared to stake agrees with what they would actually stake in a real decision. Thus both methods have their advantages and disadvantages. Instead of choosing one of them, it may thus be wise to use results of well-performed studies of both kinds.

Let us now compare the loss-of-production approach and the WTP/WTA-approach. A decision as to which of them is the "correct "approach when it comes to the value of increased safety naturally depends on what values provide the basis. In the slave-owner example a pure loss-of-(net)production approach can be justified, but for me, and surely for many others, the following values are important starting points.

1. *All individuals – both today and in the future – must be considered.* By achieving safer transport, lower risks in working life, lower risk of fires, etc. we also affect unborn generations. A new road or bridge has an economic life–time of at least 50–60 years. Our decisions taken today as to whether or not to build a bridge thus affect our unborn grandchildren and great grandchildren. We do not know of course what they will think about, e.g., safer roads, and this may be a problem. For example, we may assume that an increase in income levels will make them value safety somewhat more highly than we do. The problem of establishing the preferences of unborn generations cannot, however, conceal what is in principle important; that all individuals who will be affected shall also be included when we make a decision.

2. *Individuals are usually the best judges of their own welfare.* If an individual is prepared to pay USD 2 to get home 20 minutes faster from work, we should accept this valuation. If the individual thinks that USD 500 is too much to pay for an airbag on the driver's seat we should also accept this. It is, however, essential that the individuals are well-informed, i.e., that they know roughly what effect an airbag has, how long it lasts, what it costs, etc. On the other hand, we do not need to accept the basis for the decision of an individual. We can argue that an individual driving at a high speed may disregard consequences such as a worse environment, an increase in accident risk for others, etc. Our starting point means only that we accept the individual's assessment, provided that the individual has

basic data that in some parts include effects, which are relevant for the individual as well as for all other individuals (society). The word "usually" in the first sentence above allows for exceptions. Children may have problems assessing and valuing risks, and the same may apply to mentally disabled people.

3. *The preferences of the individuals can be measured by willingness to pay (WTP/WTA).* A factor is relevant for a decision in society if someone is willing to make a sacrifice, to "pay" something, e.g., to have a safer home, avoid noise, and enjoy nature. The fact that such aspects may be difficult to quantify cannot be decisive. Whether or not there happens to be a market price is of no consequence in this context.

This method of counting "votes" differs from that employed in reaching political decisions, partly because we are here trying to measure the strength in the preferences (not merely for or against, not merely yes or no) and partly because we have no minimum age for voting, and also because the preferences of future generations are taken into consideration.

The great popularity of the loss-of-production approach, not only earlier but also more recently as shown by Lindell et al. (1997), is probably because it makes it possible to easily establish a value with high precision, although the method does not answer the relevant question, i.e., it has a very low validity.

17.7 A More Balanced Approach to the Valuation of Life

It has so far been assumed that people's willingness to pay per statistical life should be the same in all possible situations, for an efficient use of resources. There are, however, some important factors that can motivate that statistical lives should be valued differently in different situations.

17.7.1 The Size of the Initial Risk

A reasonable hypothesis is that the willingness to pay for a given statistical life increases when the initial risk increases. One may be willing to invest all one's assets and as much as one can borrow to get an untested treatment for cancer, if one thinks that it provides the only chance of survival.

Let us consider the game "Russian roulette." The idea is to put a bullet in a revolver with six cartridge positions, spin the drum, aim the revolver at one's head and pull the trigger. Those who can consider playing this game probably demand greater compensation if the initial number of bullets is increased. The interview surveys that were mentioned earlier may also be used to show that the willingness to pay for a certain risk reduction increases with the size of the initial risk, as reported by Persson (2004) and Jenni and Loewenstein (1997).

The efficient use of resources requires approximately the same value per statistical life "for the same or approximately the same initial risk." The risk of being killed or injured in road traffic, air traffic, in large areas of working life, etc, is generally very small and about the same in each case. In these areas an equal value per statistical life should apply. On the other hand, tunnels for pedestrians at street crossings and lower speed limits cannot be directly compared to improved intensive care departments in hospitals.

17.7.2 The Risk Distribution

As pointed out above, socio-economical efficiency is increased if the benefits of a certain measure, for example greater protection in working life, better sea rescue services or more automatic fire alarms, are greater than the costs of the measures. In this context, one does not consider who bears the costs and who reaps the benefits. Decision-makers in society, however, are often concerned about the distribution of costs and benefits. The decision-makers in a municipal authority may also want those living in peripheral regions to have a good fire protection. Even if a part-time fire service is not justified from the viewpoint of strict efficiency, it may nevertheless be established because it is felt that benefits must be distributed even to remote areas.

Another example of a risk distribution consideration is the objective formulated by the Swedish parliament in 1988: "The risk of dying or being injured in traffic shall be reduced to a greater extent for unprotected than for protected road-users. Problems of children shall in particular be taken into consideration." This statement means that there is a greater willingness to pay per statistical life for a 10-year-old on a bicycle than for a 50-year-old in a car.

17.7.3 To Feel Safe – to Be Safe

The willingness to pay is determined not only by a change in the probability of injury or death but also by the anxiety or sense of insecurity that may be involved. Even though it can be shown statistically that the number of deaths or injuries per million passenger kilometers is lower when traveling by air than in a car, the fear of flying is a factor. Choosing between flying and car is decided not by the objectively established risk but by a subjective perception of risk, where feelings of discomfort or helplessness can be relevant factors. The perception of how awful different ways of dying may naturally affect the willingness to pay. One way to consider this is by using the measure quality adjusted life years (QALY) which is explained in Section 17.7.7.

17.7.4 Large Accidents (Catastrophes) or Small

Is the value per statistical life influenced by the number of people dying in an accident? In the shipwreck of MS Estonia in 1994 approximately as many Swedish persons were killed in a few hours as are killed in Swedish road accidents during a whole year. (A description of this large accident is given in Chapter 9 in this book.) For politicians and other central decision-makers, it is easy to understand that measures against a potential catastrophe have a higher priority than hundreds of measures against small accidents with the same number of "saved" statistical lives. Whether or not this is rational decision-making is a moot point. One conclusion can, however, be that it is important to distinguish between regular accidents and catastrophes, Zeckhauser (1996).

17.7.5 Risks for Oneself and Others

To see what society, i.e., parliament, the government, the road administration, etc., is willing to pay it is not sufficient to measure the individual's willingness to pay to reduce a personal risk. It can be said that society's willingness to pay for a certain reduction in risk corresponds to the individual's own willingness to pay for this reduction plus the willingness of others to pay for the same reduction. Cropper and Sussman (1988) established that married couples were more willing than unmarried couples to pay for the same reduction in risk and the same income. A reasonable addition for married couples is 25 %. Most of those who mention this in the literature only point out its existence and conclude that the individual values that have been derived can be considered to be somewhat low as approximations of the willingness to pay in society.

17.7.6 Statistical Lives or Remaining Lifetime

If the life of a teenager is saved, a further lifetime of about 65 years on the average is made possible. The corresponding value for a 65-year-old person is 15–20 more years of life. It is therefore important to distinguish between *statistical lives* or *statistical years of life* as the relevant quantity. If statistical lives are considered, the municipal authority should perhaps invest in a better street crossing outside the old people's home, whereas the crossing would perhaps be placed outside the kindergarten if years of life are considered in the calculations.

If only the probability of dying is considered, cardiovascular diseases take first place as the cause of death in the USA. If lost years of life are included in the calculation, they take 12th place. On the other hand, perinatal diseases take 11th place regarding probable risk of death but first place if lost years of life are considered (Viscusi et al. 1997).

There are strong arguments against giving the same value to a life with a short life-expectancy as to a life that is "saved" at an early stage. Strong arguments, of both a theoretical and an empirical nature, indicate, however, that a discount should be applied to future years of life. This means that a year of life in 30 years' time will be less in present value than one in 10 years' time, which in turn has a lower value than one in two years' time.

When a cost–benefit analysis is carried out, a discounting interest is normally used to make future benefits and costs comparable to today's values, and figures are expressed in the present-day values. This means that the present value (the value at year 0) of a benefit or cost in 28 years of USD 1 million becomes about USD 0.25 million, or ($1/(1.05)^{28} = 0.25$), if the rate of interest is 5 %. According to this principle, future years of life should also be recalculated to the present value.

It is, however, sometimes argued that, although discounting of resources (work force, energy, materials, etc.) used in the future is acceptable, it is unethical to do a discounting of future years of life. Suppose that 0 % is therefore used for future years of life but 5 % (a commonly used discounting interest) in other cases. This would mean that a tunnel for pedestrians that costs USD 1 million if built now corresponds to a present value of USD 0.5 million if built in 14 years time. If a "saved" life is not discounted, it is always worth the same, regardless of when it is saved. The present value is thus always equal to one, whereas the present value of future costs for a measure becomes lower with increasing distance in time. The consequence of this is that it is profitable to postpone most traffic safety investments and refrain from implementing them now. The decision-makers become trapped in a *paradox of paralyzation*. Those who believe that avoiding of discounting for future years of life leads to more safety-increasing measures are thus completely wrong. The effect would be the opposite. In other words, postponing safety measures becomes more profitable without discounting than if, for example, we had a 5 % interest even for future years of life. The conclusion is that we should also discount future "saved" lives.

There are empirical findings that people do indeed de facto discount future years of life. Viscusi et al. (1997) calculated implicit interest rates between 3 % and 12 % for different measures. If a discounting interest rate of 3 % is chosen, the present value of a "saved life" in the USA is 21 discounted years for road traffic accidents, 17 discounted years for all accidents, 6 discounted years for cardiovascular diseases and 31 discounted years for perinatal diseases.

17.7.7 Quality-adjusted Life Years (QALY)

Some measures, for example tunnels for pedestrians, fire walls, or an extended sea rescue service, mean that those who are affected often can live on as before. Other measures, such as seat belts or airbags, may result in disablement instead of death, and severe injury instead of disablement, and so on. It is thus important to consider into what kind of life the human being is "saved." In health care, work has been carried out since the 1970s to develop measures that consider the extended life time as well as the quality of the prolonged life. The term quality-adjusted life years (QALY) is often used.

A commonly used method developed by Bush et al. (1973) begins with an assessment of different levels of functioning. Bush considered that the description could be limited to mobility and physical and social activity. Complete mobility (ability to travel unhindered), complete physical activity (ability to walk unhindered), and good social contacts were given the weight 1, whereas death was given the weight 0. A panel of doctors and medical students ascribed weights between 0 and 1 to different conditions. In other studies, patients with the relevant affliction have been involved in the panel. This makes it possible to compare the likely development of the QALY for groups with and without certain measures. By relating the cost of the measure to the difference in QALY, a value per quality-adjusted year of life can be given, in the same way as in the discussion of value per statistical life above. Examples of values per QALY can be found in, for example, Torrance (1986).

17.8 Concluding Remarks

Many persons are injured and die every year in various kinds of accidents. We can reduce their number through different measures. What we *can do* and what we *should do* differ, however. It cannot be the norm to do everything that is technically possible, since this becomes extremely resource-demanding and we need resources in society to produce other utilities. The question is how much should it cost to save lives or reduce personal injuries. Our discussion does not concern *identified* lives but *statistical lives,* i.e., for the probability of injury or death) for large groups where we perhaps know the age and gender distribution but not much more. That does not mean that a certain measure will actually save your life or my life.

Two methods have dominated in the assessment of the value of a statistical life. One is the loss-of-production (human capital) approach. The other is based on the willingness to pay (WTP) or willingness to accept (WTA), determined either directly by interview or indirectly by studying decisions made. The loss-of-production method was criticized for its low validity, i.e., that it does not measure what is relevant, in contrast to the WTP/WTA calculations. Both the direct and the indirect ways of determining the WTP/WTA have their advantages and disadvantages. Results from well-performed studies of both kinds should be valuable.

Well-performed willingness-to-pay and compensation-demand surveys lead to a value of approximately USD 3–9 million for a statistical life, expressed in 2007 price levels.

For road authorities, radiation protection authorities, air traffic security authorities, etc., the value of a statistical life is commonly used, although there are instead strong theoretical and empirical reasons for considering the value “per year of life” or, to be more specific, per prolonged number of years of life, adjusted to present-day values. Finally it has been pointed out that it is desirable to seek even better measures than statistical lives and remaining life years. An interesting example of such an attempt is the quality-adjusted life years (QALY) parameter which seeks to consider both the number of years and the quality of life during these years. There is, however, considerable dissonance as to what aspects shall be assessed and what values they shall have. In practice, we must wait until a better consensus has been achieved.

References

Bush JW, Chen MM, Patric DL (1973) Health status index in cost effectiveness: Analysis of PKU program. In: Berg RL (ed) Health Status Indexes. Health Research and Educational Trust, Chicago

Cropper ML, Sussman FG (1988) Families and the economics of risks to life. American Economic Review 78:255–260

Elvik R (1988) Ulykkeskostnader og verdien av undgåtte ulykker. (The costs of road traffic accidents.) Notat 877, Transportøkonomisk institutt, Norway

Gegax D, Gerking S, Schulze W (1985) Perceived risk and the marginal value of safety. Environmental Protection Agency, USA

Jenni KE, Loewenstein G (1997) Explaining the 'identifiable victim effect'. Journal of Risk and Uncertainty 14:235–257

Jones-Lee MW (1989) The Economics of safety and physical risk. Basil Blackwell, Oxford

Jones-Lee MW, Hammerton M, Philips PR (1985) The value of safety: Results of a national sample survey. The Economic Journal 95:49–72

Lindell B, Malmfors T, Sundström-Frisk C, Swarén U, Wallinder G (1997) Att rädda liv. (To save lives.) Riskkollegiets skriftserie, nr 7. Stockholm, Sweden

Mattsson B (1990) Priset för vår säkerhet. (The price of our security.) Rapport till Riksrevisionsverket. Stockholm, Sweden

Mattsson B (1991) Samhällsekonomisk beräkningsmetod. (How to calculate benefits and costs for the society.) TFB & VTI forskning 7:1, bilaga 1. Sweden

Mattsson B, Juås B (1997) The Importance of the time factor in fire and rescue service operations in Sweden. Accident Analysis and Prevention No. 6

Mattsson B (2004) Vad är lagom säkerhet nu? (What is optimal safety now?) Räddningsverket. (Swedish Rescue Services Agency)

Miller TR (2000) Variations between countries in values of statsitical life. Journal of Transport Economics and Policy, 34:169–188

Persson U (2004) Valuing reductions in the risk of traffic accidents based on empirical studies in Sweden. Doctoral thesis, Lunds Institute of Technology, bulletin 222. Sweden

Persson U et al. (1999) Värdet av att minska risken för vägtrafikskador – beräkning av Vägverkets riskvärden. (An estimation of the accident-values of the Swedish Road Administration.) Arbetsrapport LTH och IHE. Lund, Sweden

Smith A (1776) The wealth of nations. In: Cannan E (ed). The Modern Library, New York (1937)

Svensson M (2007) What is a life worth? Methodological issues in estimating the value of a statistical life. Doctoral Dissertation. Örebro Studies in Economics 14. Örebro, Sweden

Torrance GW (1986) Measurement of health state utilities for economic analysis: A review. Journal of Health Economics, 5:1–30

SIKA (2007) (Swedish Institute for Transport and Communication Analysis), Report without title from the ASEK4-group

TSV (Traffic Safety Authority) (1989) Trafiksäkerhetsprogram. (A program for the traffic safety 1990–2000). Sweden

Valentin J (1997) Strålskyddets ambition för att rädda liv. (The value of life in radiation protection.) Riskkollegiets skriftserie, nr 7. Stockholm, Sweden

Viscusi WK (1992) Fatal tradeoffs. Oxford University Press, Oxford

Viscusi WK, Aldy J (2003). The Value of Statistical Life: A Critical Review of Market Estimates Throughout the World. Journal of Risk and Uncertainty, 27:5-76

Viscusi WK et al. (1997) Measures of mortality risks. Journal of Risk and Uncertainty 14:213–233

Vägverket (The Swedish Road Administration) (1997) According to an article in the newspaper "Bohusläningen" 13 August 1998

Zeckhauser RJ (1996). The economics of catastrophes. Journal of Risk and Uncertainty 12:113–140

18

Tacit Knowledge and Risks

Bo Göranzon

18.1 Problems of Knowledge and Risks

How are risks and disasters to be prevented in a high-technology environment? This is a question that has many facets. In this chapter, I shall discuss the aspects related to the history of knowledge and to tacit knowledge in particular.

Many companies claim to be learning organizations, without explaining what this means in practice. Here, we argue that reflection on experience – case studies – can be more effective than reliance on theories. In all the rhetoric about "knowledge society" there has been little discussion of what is meant by "knowledge." It is often assumed that a single model of knowledge will cover all the different fields of study, and indeed such a view underpins positivist social science, suggesting consistency with the natural sciences.

It has become increasingly evident that conventional approaches to business and technology have failed to come to terms with the fundamental problems of knowledge. It had been imagined by many that knowledge could be commodified, and made available for commercial exploitation, without any dependence on the continuing presence of those experts whose knowledge had been elicited for use in expert systems. How can companies address this challenge in practice?

Knowledge has been seen to be increasingly important as a driver for economic development, but research has exposed the limits of what can be achieved by conventional means. Starting by working closely with companies, a new foundation has been developed based on a concern for epistemology. Learning is seen as arising from encounters with differences. Management is reinvented as the orchestration of reflection.

Only a fraction of expert knowledge and professional skills can be codified and expressed in an explicit form, as facts and rules. No other layer of implicit knowledge is usually represented, but it may exist in the form of accepted procedures which can be elicited and formalized using available methods. This leaves the submerged iceberg of tacit knowledge, which is not reliably accessible by traditional analytical approaches: deductive, inductive, and abductive methods. We can explore the significance of tacit knowledge and consider how access to it can be gained through analogous methods. For a further treatment of this subject, see also Göranzon (1993) and Göranzon et al. (2006).

This chapter combines three examples based on practical experience: a railway example from 1915, an example from the oil industry, and an example of learning to drive a car. From the literature and philosophy of science we have six examples: Joseph Conrad *The Nigger of the Narcissus*, Rousseau and his novel *Émile*, the Swedish author Harry Martinsson and his novel *The Path*, Peter Weiss and his last play *The New Trial*, Denis Diderot and his dialog novel *The Neveu of Rameau*, and Michelle de Montaigne's *Essays*. To transform practical experience into an aspect of knowledge, we use these examples from literature and philosophy as "theory" supporting reflection on risks and tacit knowledge.

18.2 Confronting the Unforeseen

The first example of how disasters can be prevented in high-technology environments is a description of the action taken by a railway employee, Chief Conductor Sterner, from the Swedish town of Linköping. The report is from 1915 and was published in the local newspaper *Östgöten* (translation from the Swedish by Struan Robertson):

> *It is not only the theatres of war that see heroic actions. From time to time, everyday civilian life provides opportunities for exploits that are far more praiseworthy than the exploits on the "field of honour", in as much as they save lives and property instead of wreaking destruction. Such a feat was performed at Kimstad last Tuesday evening, when Chief Conductor G J Sterner, with great presence of mind and no thought for his own life, averted an imminent rail disaster that could have cost many lives. /.../ One of my colleagues in Östgöten called on Mr. Sterner to obtain some more details about this sensational incident.*
>
> *"I was only doing my duty," he explained, when we congratulated him on his creditable action. But for heaven's sake don't make any fuss about it. /.../*
>
> *... as the train was entering Kimstad station a fault was discovered in the train's vacuum system – it was passenger train number 411 – and the train was stopped. The driver got down to locate the fault and opened some valves on the side of the carriage chassis, releasing air from the vacuum system. After the brake shoes had been freed on a couple of carriages, the driver climbed back up into his cab, but found he could still not move the train – other than lurching a couple of centimetres backwards – so he got down from the train again to take another look at the vacuum system. It is likely that he had moved the control lever to the forward position to set the train in motion, and the regulator was set to full speed, i.e., 70 to 80 kilometres an hour. The driver continued to check the train all the way back*

to the last carriage. As soon as he opened a valve on the last carriage, the brakes released and the train suddenly began to move.

"Don't let it move!" I shouted to the driver, who evidently thought the stoker was still on board. And that is what I thought too," explained Mr. Sterner. I expected he would bring the train into the station, where we had orders to meet the goods train number 1352 (foodstuffs). The driver rushed forwards along the train in an attempt to catch up with the locomotive, but the train picked up speed alarmingly. I jumped up on to the second to last carriage, and saw the stoker climb up on the other side. I was both surprised and alarmed, and asked him, "Who is in the locomotive?" "The engine-driver" he answered. "No, he was out here" I replied.

I immediately understood the situation: no one was at the controls of the locomotive. I rushed forward through the carriages. I opened an emergency brake valve, but because there was a fault in the vacuum system it did not work. If the system had been in working order I could have stopped the train. But it picked up speed instead, and there was nothing else to do but to try to reach the locomotive and help the engine-driver bring the train to a halt. I thought that the engine-driver had managed to get on board one of the front carriages, but I went through carriage after carriage without finding him. I later found out that he had tripped over the points on the line and hurt himself. I had to open several locked doors and put down gangways between the carriages, which caused delay. Finally I reached the tender. There is no proper gangway between the tender and the locomotive, just a couple of small footplates. Fortunately I had the signal lamp with me so I could find my way forward by its light. I managed to find a foothold, reached for the door of the driver's cabin and pulled myself in. It only took a moment to shut off the steam and apply the brakes, although the vacuum brakes were still not working. At the same time, Engmark, the conductor, who was at the back of the train, applied the brakes there.

This saved the situation. The entire episode had lasted about three minutes, during which time the train had covered about the same number of kilometres. We had been accelerating all this time, and reached a speed of between 70 and 80 kilometres an hour. I could feel the locomotive lurching and swaying as it does at top speed. We had hurtled through Kimstad station at about 60 kilometres an hour, and then negotiated some s-bends. At each bend I had looked anxiously ahead to see if the goods train was approaching. When we finally came to a stretch of straight line and the other train was not in sight, I felt immense relief. And I felt even greater relief when I heard the stop signals in Norsholm. It was the most beautiful music I had ever heard, as it told me that the goods train had been stopped there.

When the train had come to a halt, the stoker ran forward and I told him to back the train up to Kimstad. Shortly afterwards the engine-driver arrived, with blood on his face.

There were about 30 people on board the passenger train. None of them had the slightest idea of the danger they had been in. And that was just as well, for they would certainly have panicked, and who can tell what that might have led to?

This story of Chief Conductor Sterner contains some elements of the way leadership applies sound and reliable judgment in an encounter with the unforeseen: the way the struggle, together with familiarity with the practice, pervades the way responsibility is assumed for action.

Here we see a quartet working together to prevent panic spreading among the passengers: they were the Chief Conductor, the stoker, the engine-driver and the conductor. This incident in Kimstad attracted a great deal of attention, and Chief Conductor Sterner had honours heaped upon him, among them the Carnegie medal, for his "resolution and disregard of the risk to his own life".

18.3 Art as a Source of Knowledge

The story of Chief Conductor Sterner could have been taken from one of Joseph Conrad's novels. Conrad, who was a ship's captain at the end of the nineteenth century, wrote artistic accounts of his personal experiences, with leadership on board a vessel as a recurring theme. Conrad wrote that there is no room for "bluffers" on board a ship. Navigating a ship through a storm required a high level of skill.

In the Preface to *The Nigger of the Narcissus* Conrad writes about the purpose of art (Conrad, 1897):

—My task which I am trying to achieve is, by the power of the written word to make you hear, to make you feel—it is, before all, to make you see. That—and no more, and it is everything.

The artist creates a more intense image of the reality he describes. He takes hold of reality and recreates it so that everyone who reads what he has written says to himself, "That's the way it is. I recognise what I have always known but have had neither the peace nor the gravity to dwell upon." Or perhaps the opposite: "For me, this is a new experience which I will take with me. It gives me the chance to cast new light on well-worn clichés and to develop my imagination."

Reflection is not always a process that is in harmony and balance. It may occur after "touching bottom" as Joseph Conrad describes in the novel *Heart of Darkness* (Conrad, 1899):

Imagine a blindfolded man set to drive a van over a bad road. I sweated and shivered over that business considerably, I can tell you. After all, for a seaman, to scrape the bottom of the thing that's supposed to float all the time under his care is the unpardonable sin. No one may know of it, but you never forget the thump – eh? A blow on the very heart. You remember it, you dream of it, you wake up at night and think of it – years after – and go hot and cold all over.

This is an excellent description of the way the struggle with the practical builds up skill.

18.4 Three Kinds of Knowledge in Practice

Our knowledge, in an occupation for example, can be divided into three parts: the knowledge we acquire by practicing the occupation (knowledge expressed in skill), the knowledge we gain by exchanging experience with colleagues and fellow-workers (the knowledge of familiarity), and finally the knowledge we can gain by studying the subject (propositional knowledge).

There is a clear tendency to give too much emphasis to theoretical knowledge (propositional knowledge) at the expense of practical knowledge (experimental knowledge and the knowledge of familiarity); the last two kinds of knowledge are often overlooked in discussions of the nature of knowledge.

Nevertheless, these three kinds of knowledge are, in fact, interrelated. The relationship between the three kinds of knowledge may be expressed like this: we interpret theories, methods, and rules by means of the familiarity and experience we have acquired through our participation in practice. The dialogue between people who participate in a practice contains an element of friction between different perceptions, friction that arises from differences in experience and in examples of familiarity and experience. The more in-depth development of occupational skills requires the introduction of a constant dialog. To be professional means expanding one's perspective to achieve a broader overall view than one's own familiarity with the practice in question permits. We may conclude from this argument on the relationship between the different kinds of knowledge that, if we remove from an activity the experiential knowledge and the knowledge of familiarity, we are also emptying it of its propositional knowledge.

Practical knowledge is not susceptible to systematization in the same way as theoretical knowledge, but it plays an equally important part, and it is best maintained and applied by means of reflection on examples taken from places of work and from art.

18.5 Philosophy and Engineering

In the period of the French Enlightenment there was a lively debate about the relationships between different aspects of knowledge. To give an example, in *Émile* (1762), his book about education and apprenticeship, Rousseau maintains that the concepts of theoretical knowledge must be reconciled with a reality that the senses can experience (Rousseau, 1982).

It is only by walking, feeling, counting, measuring the dimensions of things that we learn to judge them rightly. But, also, if we were always measuring, our senses would trust to the instrument and would never gain confidence.

To Rousseau it was not only true that theoretical knowledge was enriched by experience. Without experience, it creates chaos. To borrow a term from physics, the entropy in the system increases. In his dissertation *Experience as Tacit Knowledge,* Terje Sørensen, chief engineer at Statoil, quotes a slurry engineer of many years' experience (Sørensen, 1996):

We have lost skill in the oil industry of today. Young people come straight from university with no experience. And there is something about their attitude; they do not want to learn from older employees, and they become aggressive. They seem to be more interested in computers than in getting a feeling for the job. Instead of using their "feelings" to gain experience, they sit at the computer. They should walk around the platform, be there all the time, look at the work, feel and smell the drilling slurry and learn what goes on.

Sørensen quotes the same slurry engineer again:

People have no real grasp of what they are doing when they just key raw data into the computer. And later, when they read out the result, it is nothing more than new figures. They have no feeling for them.

The discussion of skill development in engineers on Norwegian oil platforms is like an echo from the discussion of epistemology during the French Age of Enlightenment.

18.6 "The Tacit Knowledge of the Working Hand"

There is an epistemological dimension in the work of the Swedish author and Nobel Laureate, Harry Martinson. He used the expression "approximation" for what he called the "fresh grass" of the human spirit. He thought he could see how our culture had become a seedbed for what he called "the skilled commonplace."

In his novel *Vägen till Klockrike* (*The Path*), Martinson writes of Bolle, a lifelong vagabond (Martinson, 1998; translation from Swedish by Struan Robertson):

But sometimes the wood folk came out of the forests and crafted long channels on frames and trestles, which rose above the streams, sometimes to the height of a man and more, and then lower, depending on the slope.

Bolle never learned the half of all that. No, the more he travelled the roads the greater the respect he had for artless, ingenuous work; for occupations that were rarely spoken of, but that were, in fact, complex disciplines. The tacit knowledge of the working hand.

When Harry Martinson discusses the way concept formation develops to reinforce confidence in actions, he expresses this philosophy of language through his character, Sandemar:

Sandemar loved the unlikely, that is to say reality as it is for the most part, and outside the world of the entrepreneurs of probability. He would rather accept fragmentation in the truth than the dishonest, sensible attraction to symbols that were supposed to express everything /.../ but that in reality expressed nothing other than lies about themselves, the false trust-worthiness.

And he has Sandemar expand on the nature of all certainty and uncertainty:

The truth is that we do not know anything, and so we are tired of one thing and another. This comes from uncertainty. We visit with our senses and thought all kinds of things, but we are never certain what they are, or what they really signify.

I write here on the blackboard a lot of thoughts that come and go. But why do they come and go so much? Well, because we have found nothing.

We only find that almost everything can be expressed in words. But as time goes by we cross out everything that can be expressed. At least, that is what I do here on the blackboard. I crossed it out because other things that can be said want to have a place on the board for a few minutes. Before the sponge arrives.

In his work, Martinson takes the sponge on the blackboard as a metaphor for work in progress. In a dialog with himself and others, there is always something to add. There are no absolute syntheses. The tip of the iceberg in his work is the language that comes to be expressed in words. To develop certainty in action and judgement means developing a process of concept formation that contains a tacit dimension.

18.7 Personal Responsibility

In his last play, *The New Trial*, Peter Weiss depicts personal responsibility in the encounter with the clients in an insurance company. The play had its premiere at the Royal Dramatic Theatre, Stockholm, in the autumn of 1981. K, the lead in Weiss's play, has been promoted from his earlier position as a clerk in an insurance company, where he had had direct contact with the customers (Weiss, 1981; translation from Swedish by Struan Robertson):

K: I stopped by Kaminer's office. Wanted to explain a few cases to him. He said he knew about everything.

Rabensteiner: Dealt with everything yesterday. He had been in on everything for a long time anyway.

K: How is he to understand things that have cost me years of work?

Rabensteiner: But you know that we've introduced the new system. The simplified procedures mean that what used to require a sizeable team can now be accomplished by just a few workers. You yourself had become superfluous. You with your card files –

K: It's those card files that matter to me. They bring their human beings behind the particular case close to me –

Rabensteiner: Everything can be accessed by the new system. My dear fellow, you were our museum piece!

K: One must listen to everyone, Mr. Rabensteiner, one must talk with everyone –

K wants to take personal responsibility in the eyes of the company's customers. He feels that personal responsibility is gradually becoming thinned out.

Figure 18.1. To learn to drive is an example of experience transfer between instructor (master) and the learner (apprentice). A learner driver in the UK must display L plates in a conspicuous position on the vehicle. (Photo: Topfoto/Scanpix)

18.8 The Transfer of Experience of Driving a Car

Without a foundation in the knowledge of experience and familiarity, theoretical knowledge cannot work. The example of driving a car may illustrate the point of experience transfer between master and apprentice (Figure 18.1). This example is taken from a 1986 conference on road safety and was published in the newspaper *Dagens Nyheter* (translation from the Swedish by Struan Robertson):

> *Beginners run an 8–10 times greater risk than experienced drivers of being involved in accidents. There is not much we can do about their age or the psychological and social factors that affect the way they drive. But we can help them get more experience. /.../ The way inexperienced drivers observe the road is passive; they have a narrow field of vision, and this means that they are late in noticing obstacles, and they have an unsystematic way of watching the road ahead. This means there are gaps in their observation or they miss things completely. /.../ New drivers believe that they have very fast reactions, although in reality their reactions are slower than those of more experienced drivers. /.../ Novices behind the wheel spend more time looking at the edge of the road. They concentrate on the position of the vehicle on the road. Experienced drivers focus their gaze farther ahead. They manage to watch the road and look out for risks at the same time.*
>
> *The young drivers' accident curve peaks two years after they pass the driving test. This is because their ability to handle the car improves, while their ability to keep an eye on the road does not improve at the same pace. The method for transferring experience to young drivers is that they must drive with an experienced teacher by their side and constantly describe what they see and what they are looking at while they are driving. The teacher assesses the new driver, and after the session goes through the important things the driver missed and the unimportant things he allowed to distract him. The result is a real improvement in both attention and observation.*

18.9 Reflection – When Different Instruments Play Together

In his book *Kunskapsbegreppet i praktisk filosofi* (*The Concept of Knowledge in Practical Philosophy*), the philosopher Allan Janik writes that reflection is a process of restoring balance (Janik, 1996). The unexpressed knowledge of an occupational group is articulated by reflecting on situations in which one's judgment is very severely tested. Stories must be brought out, made visible, stories that are the group's collective interpretation of their practice. This collective aspect is important, since all knowledge originates in experience, but not only in one's own experience.

Figure 18.2. The French philosopher and writer Denis Diderot, 1713–84. (Photo: Topham Picturepoint/Scanpix)

Reflection requires peace of mind: "It is a sickness that must run its course," writes Denis Diderot (Figure 18.2), the French encyclopedia author (Diderot, 1951). It begins with a surprise, which may be pleasant, but is most often unpleasant. It is the unexpected that is the decisive factor. Tried and tested routines no longer work. Put briefly, we no longer have an instinctive knowledge of what to do. Where there used to be order, the opposite now prevails. In such a situation, reflection offers us an opportunity to restore the balance, which creates the conditions we need to get our bearings in a situation where incomprehensibility dominates. It is in the error that we find the greatest need for reflection. Among other things, leadership involves creating meeting places for collective reflection on "scraping the bottom of the hull," to refer to Joseph Conrad's example – to benefit from mistakes. Joseph Conrad made an ironic comment on the tradition of knowledge that perceives words as identical with the reality they are intended to portray (Conrad, 1911):

> *Words, as is well known, are the great foes of reality. I have been for many years a teacher of languages. It is an occupation which at length becomes fatal to whatever share of imagination, observation, and insight an ordinary person may be heir to. To a teacher of languages there comes a time when the world is but a place of many words and man appears to be a mere talking animal not much more wonderful than a parrot.*

A quotation from Michel de Montaigne to which Allan Janik refers in his book may illustrate this perspective on reflection. Montaigne – who lived in the latter part of the sixteenth century – is the philosopher of experience-based knowledge (Montaigne, 1990):

I should willingly tell them, that the fruit of a surgeon's experience is not the history of his practice and his remembering that he has cured four people of the plague and three of the gout, unless he knows how thence to extract something whereon to form his judgement, and to make us sensible that he has thence become more skilful in his art. As in a concert of instruments, we do not hear a lute, a harpsichord, or a flute alone, but one entire harmony, the result of all together. If travel and offices have improved them, 'tis a product of their understanding to make it appear. 'Tis not enough to reckon experiences, they must weigh, sort and distil them, to extract the reasons and conclusions they carry along with them.

18.10 Dealing with the Unexpected

We are used to regarding disasters as critical events that take place in a brief period of time. The course of a disaster describes, event for event, the swift development of that critical happening, precisely as in the report from the runaway train in Kimstad and the confident actions of Chief Conductor Sterner.

The example from Kimstad draws attention to the phrase "safety culture." To focus on the culture of safety, we must first develop a perspective that also accepts that disasters may be cumulative trends – a tidy, well-ordered deterioration over a long period of time whose results would be described as disastrous if this deterioration took place at a specific time.

We must not expect to avoid crises. The fact is that some crises are to be regarded as beneficial – they test the strength and flexibility of the organization. A learning organization learns from its mistakes by coping with the crises which these mistakes generate. The essential point in this essay is the question of how we can prevent crises developing into disasters.

With the examples presented here, I have attempted to give a picture of the tacit dimension that exists in all practical work. To nurture and develop what I have here called the knowledge of familiarity and knowledge expressed in skill are vital if we are to prevent disasters in the long term. Professional people who, by accepting responsibility in their actions, develop their tacit knowledge are well prepared to deal with the unexpected, and thereby reduce the level of risk. Creating the conditions this requires is a central task of leadership.

References

Conrad J (1897) The nigger of the 'Narcissus': a tale of the forecastle. Originally published by Doubleday, Page & Company. http://www.gutenberg.org/files/17731/17731-h/17731-h.htm

Conrad J (1899) Heart of darkness. Originally published as a three-part series in Blackwood's Magazine. http://www.americanliterature.com/Conrad/HeartofDarkness/HeartofDarkness.html

Conrad J (1911) Under western eyes. Originally published by Methuen. http://www.gutenberg.org/files/2480/2480-h/2480-h.htm

Diderot D (1951) Rameaus brorson (The neveu of Rameau). Tiden, Stockholm (original 1762)
Göranzon B (1993) The practical intellect: computers and skills (artificial intelligence and society). Springer-Verlag and EUNESCO
Göranzon B, Hammarén M, Ennals R (eds) (2006) Dialogue, skill and tacit knowledge. John Wiley & Sons, Chicester
Janik A (1996) Kunskapsbegreppet i praktisk filosofi (The concept of knowledge in practical philosophy). Symposion, Stockholm
Martinson H (1998) Vägen till Klockrike (The path). Bonniers, Stockholm (original 1948)
Montaigne M (1990) Essäer (Essays). Atlantis, Stockholm (original 1580, 1588, 1595)
Rousseau J (1982) Émile. J.M. Dent & Sons, New York (original 1762)
Sørensen T (1996) Experience as tacit knowledge: a method for learning in organisations. Established by investigating practice in Statoil's drilling department from a knowledge perspective, Doctoral thesis, NTNU, Trondheim
Weiss P (1981) The new trial. Royal Dramatic Theatre, Stockholm

19

How to Worry in Moderation

Inge Jonsson

19.1 The Golden Mean of Knowledge

"And the Lord God took the man, and put him into the Garden of Eden to dress it and to keep it. And the Lord God commanded the man, saying, Of every tree of the garden thou mayest freely eat; But of the tree of the knowledge of good and evil, thou shalt not eat of it: for in the day that thou eatest thereof thou shalt surely die." This divine prohibition issued in the second chapter of the first book of Moses in King James' authorized version of the Bible had already been broken in the midst of its third chapter by the first humans, and they were consequently driven out from the earthly paradise. In the Grecian myth, Prometheus was cruelly punished because he had provided humans with the power to create fire, thereby providing them with the means to build up a technologically based civilization (Figure 19.1). Juxtaposed, these old accounts explain why it is necessary to assess potential risks within the context of technological systems. The very act of living is literally dangerous, and anyone who seeks new and untested knowledge must be prepared to take grave risks.

Such risks can naturally be dealt with in different ways. From a purely scholarly standpoint it may be tempting to cite a manifesto of a philosopher from the Romantic Age: "Seek the truth! Even if it brings you to the very gates of hell, just knock on them!" However, that commandment was formulated long before technological developments had made it possible to blow off the hinges of the gates with nuclear weapons, and in all likelihood there are very few who are prepared to underwrite it today. But that is not to say that a proclaimed research prohibition would in any way be acceptable. In an open society, an enlightened debate shall instead take place, and the researcher's ethical concerns should serve as a guarantee against the risk of technology running amok and of the risks becoming too high. However, does it work, as it should?

There is a Swedish word *lagom*, which is so widely regarded as being characteristic of the temperament of its indigenous practitioners that it is close to being untranslatable into other languages; the English interpretation that has perhaps best mediated the term is in "moderation." Encompassing an association with compromise, neutrality, caution, and a generally healthy lifestyle, it hardly creates a particularly inspirational impression, despite the fact that the wise Horatio used the epithet "golden" when he praised the concept of striking a happy medium: that is to say a true middle way. Despite its lack of luster it seems to be an appropriate departure point for some reflections concerning how risks can be assessed, or indeed, experienced.

Figure 19.1. Satyrs lighting their torches from the fire stolen by Prometheus. (Photo: Topham Picturepoint/Scanpix)

19.2 The Benefit of Knowledge

The history of invention dates back to ancient times. People have throughout time exposed themselves to great dangers gripped with the ambition of personal gain in terms of profit or greater knowledge: more often than not, the one in tandem with the other. The progression of shipbuilding and the compass opened up the world's oceans for European expansion some 500 years ago, but sea traffic and navigation along the coasts and on inland waterways had already been taking place for several thousand years. In grand utopias of the sixteenth and the seventeenth centuries like Sir Francis Bacon's *The New Atlantis,* dreams of a happy society emerged, in which the systematic search for knowledge would reveal all natural secrets for the general benefit of all its citizens. However, modern risk assessment is most closely associated with "the modern project," which has been the focus of the Western World since the Enlightenment. It was at this point that the consumption of the fruits of the Tree of Knowledge began to gather pace and leading thinkers like Immanuel Kant proclaimed man to have come of age and therefore to be able, as well as obliged, to use his reason freely. The findings of science and the experience of professionals were to be distributed in the vernacular so that society at large could draw benefit from them.

This became a main task for several of the new academies of science, with the Royal Society at the vanguard since the latter part of the 1660s, which in this capacity as well as in many other ways profiled itself in explicit contrast to the Latin-speaking universities. Practical experience had often to be collected from societal institutions other than the traditional seats of learning. It is characteristic that the first public evaluation in Sweden of a technological innovation was conducted by the College of Mines around 1720: it concerned a steam engine fabricated by Newcomen and its possible application in mining production. The person who conducted the evaluation was Emanuel Swedenborg, later to become world famous as interpreter of the Bible and founder of the New Church, but until 1747 working as a zealous civil servant within the most technologically orientated agency of the realm. His report hardly represented a risk evaluation in its truest sense but rather amounted to a balanced assessment of the advantages and disadvantages inherent to the engine expressed in general terms: no concrete dangers arising from its operation were noted by Swedenborg. On the other hand he found the task an extraordinarily interesting one, because other principles of physics than the well-known mechanical ones had now found a new technological application.

In this respect, he gave voice both to his own and the general enthusiasm of the time for inventions of potential value, but at the same time he underestimated the practical difficulties in an equally characteristic way. Both attitudes are widely represented even today, and they affect both the way in which risks arise and how they are assessed in advance. It is only natural that enthusiasts are more inclined to disregard risks than those who try to convert their visions into reality. Ever since Antiquity therefore, admiration for genius has been intimately related to a parallel fear of it. The Roman philosopher Seneca asserted, for example, that there has never existed a genius who has not simultaneously displayed a certain degree of madness. As the importance and prestige of science grew, this theme became increasingly attractive, and it is no accident that the period around 1800 witnessed two of its most prominent representatives, Goethe's Faust and Mary Shelley's Frankenstein.

19.3 The Expansion of Knowledge

The romantic poets reacted strongly against the Enlightenment's cult of reason and utilitarian values. In their view, opposite ideals should be the defining mark of honor of mankind. Poets and thinkers competed with one another in rejecting a Zeitgeist, which had threatened to fetter man to the material world and its machines and factories. However, no matter how beautifully these young bards molded their dreams of what is godly in nature and in the spirit of man, it nonetheless had little or no impact on natural science and its intrinsic utilitarianism. Parallel with the Romantic Movement and in sharp contrast to it, a technical evolution started which would come to fundamentally transform human societies. Nature was forced to open up many of its most secret spaces for scientists and engineers, and undoubtedly to many the utopian dreams of Bacon and others were now fulfilled. Through the progression of steam engine technology, communications were improved radically already during the first half of the 1800s, and by the end of the century, applications of electro-physical discoveries by brilliant inventors banished darkness and silence from the dwellings of the industrialized countries. As a consequence of unrivalled advances in medicine, disease could be combated and pain and suffering be effectively alleviated for the first time in history. No wonder then that the physician was often given the role of representing reason and modern progressiveness in the radical literature of the time.

However, this progress was found to have a price just like everything else. Certainly it is easy to laugh at the plethora of proposed threats that were thrown up in connection with the construction of the railways for example, such as the suggestion that farmers would steal the rail-tracks during the night or that the human frame could not bear the pressure of traveling at speeds in excess of 35 kilometers per hour. All of these negative responses were not, however, equally easy to dismiss. In fact, many serious train accidents took place, and still do, and the risk of injury from unprotected machine parts was great for industrial workers during the early stages. In his profound distrust of modernity, Leo Tolstoy chose the locomotive as a symbol of life-threatening powers, most effectively in the introduction and final scene of his great novel *Anna Karenina*, and many other writers forcefully described the effects of a technological civilization which was built up without the slightest regard for any risks to human beings or the environment. In naturalist novels, the charcoal smoke hangs like an impenetrable dome over industrial society's conglomeration of slum cities in the countries which regarded themselves as being most progressive. The Promethean gift appeared to unleash significant, environmental damage; and soon the Garden of Eden could neither be dressed nor kept.

19.4 The Explosion of Knowledge

This is of course because mankind has not proven itself to be mature enough in its possession of its newly won knowledge. On the contrary, mankind had been seduced by it into hubris, the deadly sin, which the Gods of Olympus always punished. Both in detail and in structure it is almost too easy to find confirmation of the Greek concept of hubris in the modern history of technology. The sinking of the Titanic in 1912 was a classic example of this. Thus, the fate of the largest and best equipped passenger ship of its time, which was said to have been constructed to be unsinkable, clearly showed that some fatal mistakes had been made during the potential risk assessment process. The likelihood that an iceberg would breach so many watertight points in its bulkhead as actually became the case had more or less been dismissed as non-existant, and the vessel thus continued at full speed into an area where warnings had been issued regarding the risk of collision with icebergs. Few catastrophes have awoken such strong and long-lasting emotions as this one, and the primary reason is probably because it became a symbol for a kind of technological arrogance that remains a prevailing feature to this day.

A few years later "the lengthy 1800s" were replaced by the outbreak of WWI and the commencement of an epoch which many historians following Eric Hobsbawm categorize in a corresponding way as the "short 1900s", that is to say the 75 years which passed between 1914 and the collapse of the Soviet Empire in 1989. The Great War between 1914 and 1918 was the first industrialized war and, even if prior warning was not absent, for many people its outbreak engendered the same kind of shock as the Titanic catastrophe did, albeit on an incomparably greater scale. Belief in the continuous path of progression suddenly appeared to be something of an illusion, a gigantic form of self-deception. The general public was hardly aware of the risk that a war of these dimensions would break out at all, and not even the military experts had counted on the drawn out and costly losses incurred by all of the combatants due to the progress in weapons technology. The contrast between mankind's technical capacity and his inherent social incompetence seemed to threaten the very survival of civilization.

Despite this, there is little support for the not uncommonly repeated thesis that belief in progress died with the First World War. It is said that hope is the last to leave a person, and this is perhaps the primary reason why so many retain a belief in the future. For those who, like Voltaire's Candide, refrain from erecting any universal perspectives but instead limit their focus to their own backyard, the short 1900s have in addition offered an almost overwhelming quantity of experience, which can only be interpreted as evidence that the only conceivable route is forward. One of the pioneers of aviation is said to have changed underclothes prior to each flight, because he calculated the chances of needing to look like a respectable corpse as being quite high. From the absolutely maximum level of risk, aviation has developed to a means of mass transportation, which today hardly requires anyone within the flight cabin to change his or her apparel in a similar fashion. No one who has any memory of the period between the world wars and the early years subsequent to WWII with overheated and punctured cars along the highways can doubt that the development of vehicle technology and road construction has progressed considerably. Radio, television, electrical home appliances, personal computers, mobile telephones: even a fairly short list of the technological innovations and accomplishments which have been made available to the populations of the first world during the course of one or two generations gives an impression of incredible progress.

An even more outstanding list of merits can be compiled from the fields of medical science and technology. Wilhelm Röntgen received the first Nobel Prize in Physics in 1901 for his discovery of a new kind of electromagnetic radiation (X-rays or Röntgen rays), which was to provide the medical profession with an incredibly effective diagnostic instrument. A seemingly never-ending flood of new pharmaceuticals has freed mankind from a range of life-threatening illnesses. During an astonishingly brief period, a number of potentially very risky procedures such as, for example, coronary artery operations have been developed almost to the point of being standardized procedures. The detailed indexing of mankind's genetic code has opened up breathtaking horizons in terms of continuous progression at an ever more accelerating pace.

19.5 The Miraculous Fruits of Knowledge

All of this is well known, but for this reason it has become all the more necessary to reflect over this development. Partly because we who live today do so with a debt of gratitude to all those who dared to take great risks in pursuit of the creation of this enormous set of technical instruments to which few of us would prefer not to have access. Partly because of our historically unique range of options, we run the risk of losing a sense of proportion as well as an awareness of the limits of science and technology. When miracles look trivial, danger is en route. How many X-rayed patients would today be struck by the thoughts Thomas Mann developed in *Der Zauberberg*, one of the great novels of the 1900s, when he allows his hero to X-ray one of his own hands and thus look at something which had never been meant to be seen by a human being, namely down into his own grave?

Of course one cannot expect that people should display wide-eyed wonder in the face of these technological advancements, particularly when they have become a part of our daily routines. The insight that everyone's life is a constant miracle has to be communicated by other means: religion, music, art, and literature. However, there must be a difference between routine and rut. This relates particularly to those who are to manage complex, technical systems and who have the responsibility for ensuring that the requirements of security are met. Because all technical systems are the artifacts of human endeavor, they consequently are composed of the same limitations characteristic of their origins. In addition, during their practical management there arise a number of individual quirks, which no designer's imagination will be able to predict. Investigations following accidents often conclude that the incident must be classified as being due to the human factor. This may sound like a form of sweeping under the carpet or indeed a form of resignation in the face of the inexplicable, but perhaps it is no more remarkable than the fact that not even the most well-educated of technicians can always be expected to be completely hale and hearty. The risk of directly irrational actions can most often be minimized through sophisticated security systems, but it is probably more difficult to hinder that even high-level responsible work may one day be carried out in a rather careless way, when it has been done on a countless number of occasions without anything unexpected occurring whatsoever.

Therefore, one must absolutely insist that useful and necessary routines must never lapse into wilful carelessness. That the air-hostess's demonstration of a number of security aspects on board prior to take-off hardly captures the attention of passengers possibly does not matter much, because the vast majority almost certainly have experienced the same ritual on many occasions and are already aware of the routes to the emergency exits. However, if the pilot were to go through his checklist with the same level of low interest there would be reason for serious worry. This goes without saying also for those who are responsible for the security and maintenance of nuclear power stations. That these experts carefully respect their security routines means that they worry "in moderation" (again the Swedish word *lagom* comes to mind) about that which may occur despite all precautions, and this of course is a necessary requirement if we laypeople are to be able to do the same.

19.6 The Communication of Knowledge

Because our modern society has made itself so dependent on well-functioning and complex technological systems, we have no longer the option of taking flight from the commotion of the world, so as to simply till cabbage in one's own garden patch like Candide. One cannot use high garden walls to provide protection against airborne environmental threats, contaminated ground water, or radiation. Neither is it possible any longer to shut down national borders against the most dangerous effects of disastrous technological failures: the Chernobyl catastrophe provided undeniable evidence of this. Accordingly, it is no longer sufficient for us to have the responsibility for domestic risk-taking. Instead, in actual fact, developments have forced us all to think more universally. This has undoubtedly brought a lot of good with it, although it has not made the process any easier for the layman in terms of worrying or being concerned in moderation.

From several points of view, the political upheaval since 1989 has improved the prospects for coherent assessments compared with the conditions that dictated events during the many decades of the Cold War. Those who recall the chilling fear of a Third World War, which could have broken out at any time through a number of crises – the Berlin Blockade, Hungary, Cuba, the Berlin Wall, Prague, the deployment of medium-range missiles in Europe – can only feel relief and gratitude that the din of rattling weapons quietened. Nonetheless, in fact there still remains the threat of extinction in the form of a large number of nuclear weapon missile systems, which many seem to have forgotten. There has been incredible progress in that the large powers have ceased to use these terrible arsenals as part of their political discourse, as even those who argue that the balance in nuclear arms strength hindered an escalation of the Cold War from taking on the proportions of a global catastrophe, almost certainly agree (Figure 19.2). If they are right, the incredibly high risk which Einstein took when, in a letter to President Roosevelt in 1939, he suggested that further research be conducted with a view to producing an atomic bomb would be justified.

Not only have these direct threats ceased to hang over us today, but also previously hermetically sealed installations are now open to international inspection. This latter development is a result of risk management, which can be generalized: it is a matter of collecting the greatest conceivable amount of knowledge about the current situation so as to be able to draw the most likely conclusions about the future. This can be carried out in a meaningful way through forms that guarantee a high level of security, if it can be assumed that all of the affected actors conduct themselves rationally. However, because we know from overwhelming experience that by no means everyone does so, an evaluation of likely outcomes becomes in the final analysis a question of judgment. For my own part, I think I have learned during the passage of time that what one most fears is seldom that which actually occurs, while a great deal of totally unexpected misery is created.

This must not be interpreted as a dismissal of rational risk assessment whatsoever, nor indeed of the Enlightenment inheritance to make good use of our reason in each and every context. On the contrary, I am currently concerned about the many apparent signs of something of a renaissance in general superstition.

Perhaps most of it can be laughed at, as, e.g., when certain airlines refuse to use the number 13 to denote the thirteenth row: at least this has little or nothing to do with adequate risk assessment. I find myself becoming more hesitant given the current popularity of astrology and indeed even more concerned over the considerable assaults on a rational view of knowledge, which have become something of a fashion in post-modern discourse. My aim here is only to recall in simplified terms that a human being is much more than his or her common sense. The absolutely rationally calculating and profit-maximizing *homo oeconomicus* I dare say – perhaps for the best – is actually something of a rarity.

Figure 19.2. A slab of the Berlin Wall is lifted by an East German crane at Potsdamer Platz to make way for a border crossing, November 12, 1989. (Photo: Pat Benic/Reuters/Scanpix)

In practice this means that one must be conscious of the fact that researchers and technicians can also be misguided by their strong commitment toward their projects, even when they have the responsibility for assessing risks. One wishes intensively that the planning falls into place, so that potential dangers come to be underestimated or to be overlooked, as was the case with Swedenborg's evaluation of the steam engine in 1724. Naturally, in a corresponding fashion, the expectations of the purchasers influence their view of potential risks. There is a considerable amount of evidence to suggest that many of us do not permit ourselves to be influenced even by the most overwhelming of statistics relating to risky activities that are connected with pleasure; be it sufficient to name car driving and tobacco smoking. Perhaps it would be different if such activities were unique in terms of their potential dangers, but we know that this is not the case. It is recounted that the Nobel Prize winner, Albert Camus, who was a chain-smoker, had on one occasion decided to stop smoking. Subsequently, however, he read in the newspapers that the Soviet Union had exploded its first hydrogen bomb, and thus he could find no reason to implement his decision: a couple of years later he died in a car accident. Also in the welfare state, the individual is nonetheless a brittle and fragile vessel, and in fact it is not strange at all that many seem to live in a kind of ill-defined anguish despite all of their material trappings. I have even seen a medical term for the psychosomatic consequences stemming from this misery: trust-deficiency diseases.

How shall society resist this fear of poorly defined dangers? The obvious answer would be through more information, but in practice it cannot only be a question of quantity, because human history has never before provided mankind with so much information as is currently the case. In addition, every parent knows that the results of education depend as much upon the student's own willingness and ability to listen and learn as the very content of the message. Then what is the use of offering the public more of the same? It is often suggested that the welfare society has become trapped in a kind of fixation with material security which causes people to look at any change as a risk rather than as a possibility. Add on top of this a kind of collective bad conscience because of our well-being relative to the poor masses of the world, and a mental atmosphere arises in which evil portents appear to be received with sullen satisfaction: did I not think so? Surely things cannot remain this good for ever!

This is perhaps not the case at all, or in any event entirely unscientific speculation, but it is part of my subjective experience of a secularized, Western World undergoing radical transformation. My supposition is based upon a limited but regular consumption of the mass-media's news supply and a certain amount of reading of contemporary fiction, and in none of these sources will you meet many traces of ordinary, everyday happiness and confidence in the future. But how representative is this sombre outlook on life? A reasonably alert observer will immediately discover an entire row of opposing bright views, not least connected with the acronym of IT. Thus word opposes word just like two butting rams, to quote the great Swedish author Hjalmar Bergman.

19.7 The Lack of Knowledge and Information Overflow

This is the unavoidable result of most attempts to construct a widely generalized view, and indeed the IT phenomenon is itself no exception. The fact that almost all the information in the world will soon be available on PCs at home should more than adequately meet the requirements of the general public, even when it comes to decisions relating to risk-taking. Obviously IT opens up enormous possibilities, and to dismiss this new information technology would be just as stupid for a humanist today as it was when our predecessors in the 1400s complained that Gutenberg's invention of the printing press would destroy the market for beautiful manuscripts (Figure 19.3).

But it would also be irresponsible not to fulfill one's role as a slave to this triumphant bandwagon by not at least reminding ourselves that, so far, computers can only produce what they have been pre-programmed to do. Even this late product of the Promethean fire is threatened by hubris, and here there is reason to recall the hysteria that broke out in 1999 at the approach of the turn of the century. Many so-called experts expected an entirely apocalyptic outcome as a result of some lapses in computer programs. If the calamities issued by these prophets of doom had occurred, then the word "IT-revolution" would have taken on a completely different meaning in bank offices all over the world at the turn of the century. In fact, nothing of any substantial import happened, but the commotion surrounding it occupies a prominent place among the grotesquely overestimated threats in the mass media society.

Even if these programming mistakes did not present any difficulties, it is, however, comforting to think that most such problems can be solved, given time. It is easier to see greater difficulties arising from a future information society, because they are essentially already amongst us. TV and radio channels together with the Internet offer even today an overflow of information. Nonetheless, it is still extremely common that people claim that they have not been informed about decisions that affect them in one way or another. In most cases, this is probably not true. They have instead not had the capacity to separate relevant information from the constant, demanding din surrounding us. Thus it is not sufficient to have access to information. It must also be translated into structured knowledge. To have any reasonable opportunity of carrying out individual assessments of the risks related to complex systems, most people are probably dependent on such transformations.

In any event, each treatment of individual pieces of information implies that an interpretation takes place, and this immediately raises the additional question of public confidence in those who carry out this interpretation. If one is to believe current opinion polls, the general public's confidence in journalists and other information professionals is relatively low. It is somewhat higher than appears to be the case with regard to politicians, but that is of little or no comfort: indeed quite the opposite, since it amounts to a serious warning cry for democracy – this admittedly being far too big an issue to deal with here. Thus only the experts remain with all the accompanying complications attached to their roles – not only the familiar quotation that what is the truth in Berlin and Jena merely amounts to a poor joke in Heidelberg, but also the contrast between their modus operandi and that of the media. Scientists know that the truth can seldom be formulated unambiguously in everyday language – regardless of whether it takes place in one university or another – and therefore must be tested through a plethora of differing nuances. The inquisitive reporter, on the other hand, has column inches or the space of seconds in which to present sharp, as well as eye-catching statements, which can make a breakthrough amidst the general din of the news-flow. In addition, he or she is more often than not inclined to find researchers opposing established opinions. It may well be a healthy and even a refreshing trend on many occasions, but at the same time it risks making the general public and their political representatives more confused than informed. It is therefore a necessary requirement that the established view is presented equitably and preferably with the same aplomb as that which often characterizes the conceptions of the dissidents.

Figure 19.3. Johannes Gutenberg, circa 1400–1468, the inventor of book-printing with movable type. He is generally accredited with the world's first book printed with movable type, known as the Gutenberg Bible. (Photo: DPA/Scanpix)

The process of communicating knowledge about research and technology also entails some fundamental difficulties. It is true that the following famous couplets from 1711 were aimed at poets and their critics, but their message may be relevant for other writers as well:

A little learning is a dangerous thing;
Drink deep, or taste not the Pierian spring:
There shallow draughts intoxicate the brain,
And drinking largely sobers us again.

The precocious Alexander Pope expressed at 23 years of age his skepticism concerning easily accessible knowledge in this dynamic fashion. His words were sometimes presented as an argument against obligatory schools and have thus benefited opponents of the Enlightenment. But, yet again, even respectable contributors to today's debate share his skepticism to the extent that they find it simply not possible to translate scientific accounts into popular language without creating inappropriate distortion.

Naturally enough, this is also a question encompassing judgment and confidence. In my opinion, the citizens of a democracy have the right to demand an open and understandable account of what scientists and technicians discover, which often means that researchers must put up with fairly extensive over-simplifications and must attempt to adapt themselves to the demands of today's media. But, at the same time, one should not underestimate the difficulties arising particularly in an era when the TV medium dominates. The medium itself is a shining example of how modern technology has fundamentally changed the conditions of life – it is also probably fair to say that it has become something else than what the pioneers had earlier predicted and hoped for. The rapid news-flow in pictures has self-evidently broadened peoples' horizons to an incomparable degree, but this does not necessarily imply a larger supply of intellectually elaborate knowledge. What is in fact communicated may not be silence, as a poet has formulated it, but is first and foremost emotional experience. The strength of the medium lies in its ability to relay dramatic events, and the selection of news as well as the way of presenting it is dictated by this particular quality. Pictures of outrageous events and anguished faces awaken strong feelings but seldom stimulate the viewers to reflect and assess what they see.

The basis for rigorous, intellectual examination, which nonetheless is offered, often consists of figures: they have the advantage of both saving time and providing an impression of objectivity. However, figures are seldom as completely unambiguous as they appear to be: on the contrary they can contribute to serious misunderstanding. The controversy surrounding Swedish school pupils' lack of knowledge about the Holocaust some years ago made the scholar responsible for the study dispute the media coverage of his percentage calculation as being incorrect; but at that stage the damage was done and political initiatives taken which should have been much better founded. One can also wonder whether the frequent use of Becquerel figures following the Chernobyl catastrophe did not represent a greater threat to public health than the actual fall out; quite a few people appeared to believe that the radiation measure was itself a lethal infectious agent in meat and vegetables. The problem does not limit itself to the presentation of figures but also encompasses the application of scientific terminology in general. For example, alarming reports concerning the dangers of genetic modification techniques have resulted in some consumers demanding guarantees that groceries do not contain any genetic material whatsoever.

The only reasonable conclusion would be that today's, and even more tomorrow's, society has to strengthen both the production and the communication of well-grounded knowledge. It may well be expressed as a paradox: knowledge must serve as a buffer against the information overflow. The educational system has to be able to provide a knowledge base from which citizens can orient themselves in a world where the only permanent phenomenon is change itself. Because human life has become entirely dependent upon science and technology, regardless of whether we like it or not, a good general education in these areas is a necessity. This is really a self-evident demand, but it deserves nonetheless to be formulated with a certain emphasis, particularly by a humanist, although it will by no means be easy to accomplish. In all likelihood one must begin already in pre-school by strengthening the effective competence of the teachers, so that the children can be exposed to real knowledge and thereby be protected from ideologically motivated, environmental infatuations, however well meant these may be.

19.8 Knowledge and Reflection

Insight concerning fundamental facts is a necessary prerequisite if a person is to worry in moderation – although this is not sufficient in itself. Risk assessment must obviously have as its starting point well-established knowledge regarding how people react to different situations, and there is a considerable degree of empirical material to be gleaned from. However, the contributions of psychological research are not sufficient in this case either. From the very inception of the ethical dimension, the responsibility toward our environment and our fellow man today and in the future, should dominate all assessments of threats as well as possibilities. Consequently one cannot satisfy oneself with ever so complicated probability calculations or other formalized forms of risk assessment; there must also be room for reflection.

The question is whether this room for reflection has not become the greatest commodity in short supply during the times in which we find ourselves. The enormous increase in speed through our systems of communication has led to an increasingly shorter time for personal reflection, prior to decisions being made. Once more, television comes to mind. We witness on an ongoing basis how politicians and other decision-makers when placed in front of completely merciless cameras supply answers to difficult questions without having had time to reflect upon them. The highly predictable result of this is that it produces either completely meaningless phrases or also ill-considered statements, which must subsequently be reversed on the very next day. However, we can also note that men and women in public life carefully select *fora* in which to announce important decisions, in order to maximize their impact on the media. Thus we see a repeat of the same paradox: access to quick and correct information via well-functioning technological systems has resulted in more uncertain and short-term knowledge than the designers of the system had actually hoped for – indeed the exact opposite.

There is perhaps not a great deal to indicate that this tempo will taper off in the near future; rather the contrary, we await the risk of a further increase in this speed-up. This makes it all the more important to promote balancing counter-forces, and for my own part I am convinced that historical knowledge is needed more than ever before. Definitely not an infatuation with the past, nor indeed any form of reactionary nostalgia: history supplies no completely dependable analogies; it hardly offers us either comfort or edification. Nevertheless, it does offer us an overwhelmingly rich collection of examples, which among other things show how people may use and misuse their technological knowledge. That very insight alone should serve to strengthen the ability to carry out risk assessment with both common sense and imagination.

Whatever conclusions one can ultimately draw depend to a high degree on personal experience and societal conditions. If history seems to give me a strong argument for worrying in moderation (again *lagom*), others may regard such a risk assessment as being naïve or even cynical. Counter-arguments are almost always unavoidable, whenever one tries to draw some lesson from history, and there remains in the final analysis not much more than to try and find some wise words of comfort and direction. Here is a final attempt at reflection. Even if the speaker in Walt Whitman's poem has engaged himself in more abstract phenomena than those who create risks, he reminds us of some of the fundamentals upon which science and technology ultimately rest:

The base of all metaphysics

And now gentlemen,
A word I give to remain in your memories and minds,
As base and finalé too for all metaphysics.

(So to the students the old professor,
At the close of his crowded course.)

Having studied the new and antique, the Greek and Germanic systems,
Kant having studied and stated, Fichte and Schelling and Hegel,
Stated the lore of Plato, and Socrates greater than Plato,
And greater than Socrates sought and stated, Christ divine having studied long,
I see reminiscent to-day those Greek and Germanic systems,
See the philosophies all, Christian churches and tenets see,
Yet underneath Socrates clearly see, and underneath Christ the divine I see,
The dear love of man for his comrade, the attraction of friend to friend,
Of the well-married husband and wife, of children and parents,
Of city for city and land for land.

Appendix A

Some Basic Concepts of Statistics

Per Näsman

Introduction

In risk analysis it is essential to know the probability of one or more events. This appendix gives an introduction to some aspects of probability and explains some important concepts of probability theory. Instead of presenting strict mathematical expressions, the concepts are illustrated with calculations for a number of specific cases. More information on the subject can, for example, be found in Montgomery and Runger (1999), Triola (2002), and Weiss (2001).

Models

The concept of a model is of pivotal importance to risk analysis. A model is an idealized description of a certain occurrence, representing its essential properties without including all the details. A difference is made between *analog models* (e.g., maps, construction drawings or ball-and-stick models of matter where the balls represent atoms), *physical models* (e.g., a 1:75 scale cardboard model of a house, or the use of crash test dummies instead of real drivers in a car crash test), and *abstract models*. Abstract models are often used within technology and natural science. They can be deterministic or stochastic. Another name for a stochastic model is a random model.

In a *deterministic model*, occurrences are approximated with mathematical functions. One example of a deterministic model is Ohm's law, $V = IR$. Another example is if we want to determine the area of a circular ice rink. We picture the ice rink as a circle with the area πr^2, where r is the radius of the circle. This is obviously a model, as there are in reality no ice rinks with the shape of an exact circle; they exist only in an abstract world. The use of classical mechanics to describe how an object is falling due to gravity is yet another example of a deterministic model. A common property of all deterministic models is that occurrences are approximated and expressed with mathematical functions.

In probability theory, *random models* (stochastic models) are used when describing random experiments. By random experiment we mean an experiment that is repeatable under similar conditions and for which the results are not predictable, even when the experiment has been conducted many times before.

A typical example of a random experiment is to roll a die. Beforehand we do not know what value we will get. Another example is a lottery. We do not know in advance if the lottery ticket will win a prize. Radioactive decay is yet another example. The exact number of particles that will decay during a certain time can not be predicted. Before probability theory can be applied to these cases, a random model describing the unpredictable variation must be formulated.

Probability

One definition of the concept of probability can be based on an important property of random trials. For example, consider an unbiased die being rolled a great number of times. After each throw, we calculate the relative frequency of sixes, i.e., the ratio of the number of rolled sixes to the total number of throws. This relative frequency will become more and more stable and approach $1/6 = 0.166666...$ as the number of throws increases. If the die is not unbiased, but is crooked, showing some kind of distortion, the frequency will approach another value, which it is impossible to guess beforehand.

We can now define the probability of a rolling a six as the *relative frequency* when the number of throws approaches infinity. This definition has the advantage of being in accordance with what most people mean by probability, but it can lead to problems in cases where it is not possible to make repeated independent trials. If you wish, for example, to determine the probability that a new-born baby girl will live for at least 90 years, you can collect empirical material. The relative frequency in the collected material of women older than 90 years can be used as an estimate of the probability. With this kind of estimation of probability, the accuracy of the model will be highly dependent on the size of the collected material.

Another definition of something called probability, but which should be called risk measurement, is used for instance in some parts of the transport sector. The risk measurement is calculated by relating the number of occurred events to some sort of traffic workload. In railroad traffic, the number of killed and injured passengers per billion passenger kilometers is used as a risk measurement. For motor traffic, one measure of the risk of being killed as a car passenger is calculated by relating the number of dead passengers to the number of passenger kilometers. One must carefully consider what to use as denominator and numerator when calculating these types of risk measures. It is not obvious what should be the denominator when, for instance, a risk measurement is calculated for flight safety. Should it be the number of flights or the number of flying hours? The choice of denominator and numerator will of course determine the value of the risk measurement.

The concept of probability can also be given a *subjective* content. The word is used in everyday language about events that it is not possible to characterize as random trials. It is not possible to claim that a Swedish curling team will have a probability of 90 % of winning the gold medal in a future Olympic Games. Yet another example of subjective probability is this kind of statement: "The risk that a certain share will fall during the next week is at the most 10 %". Neither of these individual cases allows any interpretation in terms of frequency, as they are based on unique situations and conditions which are not repeatable.

At present there is a considerable interest in subjective probabilities and much has been written on this topic in scientific literature. Subjective probabilities can play a significant role in risk analysis, as the input values in large complex models. *Bayesian methodology* can be used to continuously revise the probabilities used as input values in the model as more data are obtained. In Bayesian methodology, the basis is a probability distribution estimated beforehand, the *a priori distribution*. This distribution is then recalculated, using obtained data, to give an *a posteriori distribution*.

The advantage of Bayesian methodology is that it is applicable within areas with small numbers of occurred events, i.e., where few data are available. The *a priori* distribution in these cases can be obtained by expert judgment. The few data on events or incidents that may be available can later be utilized to determine the *a posteriori* distribution.

Independent Events

The concept of "independence" is essential to random models. Two events, A and B, are said to be independent if and only if the probability of B happening is the same no matter whether it is known that A has happened or that A has not happened.

If the events A and B are independent of each other, the probabilities of the events can be multiplied to obtain the probability that both these events will take place. Suppose that we have a lottery ticket in a lottery, with 5000 tickets and one first prize. Then, the probability of winning first prize in the draw the next day is 1 in 5 000, i.e., 0.0002. Suppose also that, according to the weather forecast, the probability of sunshine the next day is 1 in 2, i.e., 0.5. Apparently these two events, to win first prize and that the sun will shine, are independent of each other. The probability of the sun shining and of winning first prize is then 0.0002 multiplied by 0.5, e.g., 1 in 10 000.

The concept of independence makes a clean sweep of some common misconceptions. If one tosses a *fair* coin and ends up getting heads ten times in a row, it is common to believe that the probability of getting a tail in the eleventh toss is greater than 0.5. However, this is totally wrong! The probability of getting heads is the same in every toss, and does not change due to what happens in previous trials.

However, one always has to be prepared that apparently independent events may actually not be independent. The probability of an airplane engine failing during flight is very small. The probability of all three engines of a plane failing is negligible, if these events are regarded as independent. However, in an accident where all the engines failed, it turned out that the same improper service and maintenance procedures had been used for all three engines when the plane was given a complete overhaul. In this case, the probabilities of an engine stopping were not independent.

Conditional Probabilities and Bayes' Theorem

When we have two random events A and B we can introduce the concept of *conditional probability*. Conditional probability means the probability that an event B will occur, given that an event A has already occurred. Hence, if we assume that the event A has happened when we estimate the probability that B will happen. The formula for this conditional probability is $P(B/A) = P(A \cap B) / P(A)$, and in the same manner, if we assume that the event B has happened when we estimate the probability that A will happen $P(A/B) = P(A \cap B) / P(B)$. Now we can combine these two formulas to the form $P(B/A) = P(B)\,P(A \cap B) / P(A)$ which is called *Bayes' theorem.* The theorem holds also when subjective probabilities are used and if one wants to estimate the probabilities of events being dependent on other events, Bayes' theorem can be very useful. We illustrate the theorem with an example.

Suppose that one type of surveillance instrument is manufactured at three factories, A, B and C, in numbers proportional to 10:20:70. Of the manufactured instruments 5 %, 4 % and 3 % respectively are defective. We mix the surveillance instruments and send them to the customers. What is the probability of a random instrument being defective? The answer is $0.10 \times 0.05 + 0.20 \times 0.04 + 0.70 \times 0.03 = 0.034$. If we wish to know the probability of an obtained defective surveillance instrument being manufactured in factory B, we need to pick a defective instrument and this instrument has to be manufactured in factory B. According to Bayes' theorem, the probability we are trying to ascertain is the probability of a defective surveillance instrument being manufactured in factory B divided by the probability of obtaining a defective instrument, i.e., $0.20 \times 0.04 / 0.034 = 0.235$.

Bayes' theorem makes it possible to apply quantitative reasoning. When several alternative hypotheses are competing for belief we can test them by deducing the consequences of each and then conducting experimental tests to observe whether or not these consequences actually occur. If a hypothesis predicts that something should occur, and it does happen, this strengthens our belief in the truthfulness of the hypothesis. Conversely, an observation that contradicts the prediction would weaken our confidence in the hypothesis.

In the terminology of Bayes' theorem, we first construct a set of mutually exclusive and all-inclusive hypotheses and spread our degree of belief among them by assigning a prior probability to each hypothesis. If we have no prior basis for assigning probabilities, we spread our belief probability evenly among the hypotheses. Then we construct a list of possible observable outcomes. This list should also be mutually exclusive and all inclusive. For each hypothesis we calculate the conditional probability of each possible outcome. This will be the probability of observing each outcome, if that particular hypothesis is true. For each hypothesis, the sum of the conditional probabilities for all the outcomes must add up to one. We then note which outcome actually occurred. Using Bayes' theorem, we can then compute revised *post priori* probabilities for the hypotheses.

Discrete Random Variables

The result of a random trial is often a number determined by the outcome of the trial. The number is thus not known before the trial but is determined by chance. For example, we can take the number of girls in a randomly picked family with two children and denote it X. What is the probability of X taking the value 0, 1, or 2 respectively?

X is an example of a *random variable* (stochastic variable). There are two kinds of random variables; *discrete* and *continuous*. The discrete variables can only take certain values (often integers) whereas continuous variables can take all values within an interval.

We define the probability distribution of a *discrete variable* as $P(X=x)$, which means the probability of the variable X taking the value x. In our example with the number of girls in a family with two children, if we assume that it is equally probable to give birth to a boy as to a girl and that all pregnancies are independent of each other (i.e., we do not have any deliveries with more than one child), then the probability distribution is as follows:

X	0	1	2
$P(X=x)$	1/4	1/2	1/4
$P(X \leq x)$	1/4	3/4	4/4

$P(X \leq x)$ denotes the distribution function of the variable which determines the probability of obtaining a certain value, or a lower one, as the outcome of the variable.

The *expected value* of the variable X is often denoted $E(X)$ or μ and is defined by $E(X) = \sum xP(X=x)$. The summation should include all values that the variable can take. Thus, in our example the expected value will be $0 \times 1/4 + 1 \times 1/2 + 2 \times 1/4 = 1$.

The interpretation of this is that, when studying a number of families with two children, we expect to find on average one girl per family.

The *variance* of the variable X is defined as the expected value of the function $(X-\mu)^2$, i.e., $E(X-\mu)^2 = \sum(x-\mu)^2 P(X=x)$. The variance is often denoted σ^2. The square root of the variance is the *standard deviation*. The standard deviation is denoted σ and is a measure of the variation in the material. A small standard deviation means a small dispersion around the expected value of the material and a large standard deviation means a large dispersion around the expected value of the material. Hence, the standard deviation is a measure of precision.

One discrete distribution of considerable interest is the *binomial distribution*. This distribution occurs when a number, n, of independent repeated trials of an experiment, which can result in two different outcomes, are conducted. Either event A will happen or it will not happen. If we let p denote the probability of A occurring, the probability of A not occurring is $1-p$.

If X denotes the number of times A occurs during the n independent trials, X follows a binomial distribution with the parameters n and p. A variable following the binomial distribution with the parameters n and p has the following frequency distribution:

$$P(X = x) = \binom{n}{x} p^x (1 - p)^{n-x}$$

where $x = 0, 1, 2, 3, \ldots, n$.

That a variable follows a binomial distribution is denoted the following way: $X \in \text{Bin}(n, p)$. The binomial distribution is tabulated in most books of statistics, but one can use the frequency distribution directly when calculating the probabilities. The expected value of an event occurring if the frequency is described by a binomial distribution is always np and the variance is always $np(1 - p)$. Thus, the standard deviation of a binomial distribution is

$$\sqrt{np(1-p)} \, .$$

Here follows an example of a problem that can be solved with the use of the binomial distribution. In an industrial plant there are eight machines, which work independently of each other. The probability of a shutdown during one day, for each machine, is 0.08. What is the probability of three of the machines failing during a certain day? Calculate the expected number of machines failing during a given day.

If the number of machines that will fail during a given day is denoted X, then $X \in \text{Bin}(8, 0.08)$. We now seek the probability that three machines fail during a given day, i.e., $P(X = 3)$. According to the frequency distribution this is given by

$$P(X = 3) = \binom{8}{3} 0.08^3 (1 - 0.08)^{8-3} = 56 \times 0.000512 \times 0.78688 \approx 0.022.$$

The expected number of machines failing during any given day is $8 \times 0.08 = 0.64$.

The *Poisson distribution* occurs in the study of phenomena happening independently of each other in time or space. The Poisson distribution also occurs as an approximation of the binomial distribution. If n is large and p is small (less than approximately 0.1) for a binomial distribution, it can be approximated by a Poisson distribution.

The Poisson distribution is tabulated in most books of statistics. The notation of a variable X having a Poisson distribution is $X \in \text{Po}(m)$ where m denotes the expected value, the mean intensity. The variance of a Poisson distribution is equal to the expected value m, and the standard deviation is thus $\sqrt{m}$.

An important application where the Poisson distribution, or rather the *Poisson process*, is of considerable interest is in time processes. Suppose that events of type A can happen at any time, and independently of each other, in a given time interval. For example, A can be the number of calls received by a switchboard operator, or customers entering a shop, or emergency cases arriving at an emergency ward. With the use of Poisson processes, it is possible to calculate the probability of the number of events in a given time interval and to estimate the expected number per time unit.

Continuous Random Variables

A continuous variable can take all values in an interval, or possibly in several separate intervals. The interval can have an infinite extension and the outcomes are infinitely close to each other, which means that no outcome can be assumed with a positive probability. Hence, there is no probability function. The distribution function, $F_x(x)$, is the integral of another function, $f_X(x)$, which is called the frequency function, or probability density function, of X.

The significance of outcomes being infinitely close to each other where no outcome can be assumed with a positive probability can be exemplified with the measure of a person's height. An individual person has no exactly measurable height. If we had access to a better measurement instrument, we could have added another decimal and then yet another decimal, to infinity. The number that we in reality give the height of the individual, e.g., 180 cm, is rounded off, whereas in a theoretical view there is no limit to the accuracy with which we could determine the height of the person.

The frequency function for a continuous variable is the derivative of the distribution function. The distribution function is the integral of the frequency function.

The probabilities of different events can thus be interpreted as the area under the frequency function in question for each of the events, i.e. the probability of X falling within the range between the values a and b is described by the area beneath the density function enclosed by the values a and b. This area can be determined by integration.

The Normal Distribution

One of the most important continuous distributions is the *normal distribution*. If a random variable has the frequency function

$$\frac{1}{\sigma\sqrt{2\pi}}e^{-(x-\mu)^2/2\sigma^2} \quad (-\infty < x < \infty)$$

where μ and σ are given quantities, the expected value and the standard deviation respectively, then X has a normal distribution, $X \in N(\mu, \sigma)$.

The normal distribution is often used to describe the variation of occurrences. A large body of statistical theory is based on the assumption that the data follow the normal distribution. Other terms used for the normal distribution are the bell curve, the Gaussian distribution, or the error distribution.

As there are an infinite number of normal distributions, one for each combination of μ and σ, the normal distribution is often standardized by transformation of the original distribution to give a new standard distribution with the expected value 0 ($\mu = 0$) and the standard deviation 1 ($\sigma = 1$). This distribution is called the *standard normal distribution* or the *z-distribution.*

The transformation used to standardize the original normal distribution is as follows. If $X \in N(\mu, \sigma)$ and we form $Z = (X - \mu)/\sigma$, then $Z \in N(0, 1)$. The standardized normal distribution is tabulated in most books of statistics.

The normal distribution has many important properties. Sums of two or more independent variables with normal distributions, the differences between two such variables, and the arithmetic mean of several variables, also have a normal distribution. This is the case even if the input variables have different expected values and different variances.

The normal distribution also occurs in a more general context, which makes the distribution very useful. It can be shown that the sum of independent, equally distributed random variables with arbitrary distributions in general has an approximate normal distribution when the number of components in the sum is sufficiently large.

Statistical Inference

In a statistical analysis, a *population* is studied in some respect. The population consists of a set of elements having one or more properties in common.

A population is a set of data, or in other words, a set of observations. In a *census* the whole population is studied. Censuses are of great significance in the production of official statistics, where information regarding for instance the number of babies born and the number of deaths are collected. Censuses are used by industry as well, for instance in the inspection of valuable or potentially dangerous products.

However, a census is often too costly and time-consuming. It can even be unrealistic to carry out. This is so in the case of destructive testing. If a census of crash tests of cars were required, all cars would be destroyed during the tests.

Instead of a census, a *sample survey* is often conducted, where only a part of the population is studied. The result for the sample is used to draw an inference about the population.

The quality of the inference depends, among other things, on the sampling method and on the size of the sample. One example is the pre-election polls continuously being conducted with the aim of estimating the proportion of people voting for different political parties in order to predict the final outcome of an election.

Questions concerning relationships among variables are also well suited for study using sampling methods. Do men have a more positive attitude than women to nuclear power? Is there a tendency that persons with higher incomes prefer a certain political party? Does the age of an individual influence his or her attitude to refugees? If there is a relationship among variables, this is called a correlation. However, one has to be cautious concerning relationships, a high correlation does not imply causality. The relationships found are relationships of association but not necessarily of cause and effect. It is not certain that a high correlation shows a causal relationship; it could be a nonsensical relationship.

When drawing inferences about the whole population from the results for a sample, there is a risk of making *systematic errors* as well as *random errors.*

The systematic error (bias) is constant and may be due to the use of a faulty measuring instrument, for instance a scale that has been wrongly calibrated and constantly shows 1 kg too much. Misleading and vague questions in a questionnaire can also lead to systematic errors.

The random error is a random variable with the expected value zero. Therefore, we can estimate the size of the random error by statistical methods. This is done and is presented as an *uncertainty interval* (*confidence interval*) for the quantity in question. It is also popularly called *the margin of error*.

One important and interesting observation is that the more times a trial is repeated, the narrower is the confidence interval for an estimation of a property. This is due to the *law of large numbers*, which says that the average of several independent random variables with the same expected value will be close to the expected value, if the number of observations is sufficiently large.

References

Montgomery D, Runger G (1999) Applied statistics and probability for engineers. John Wiley & Sons, New York

Triola M (2002) Essentials of statistics. Addison Wesley, Boston

Weiss N (2001) Introductory statistics. Addison Wesley, Boston

References

Appendix B

Study Help

Göran Grimvall, Åke J. Holmgren, Per Jacobsson, and Torbjörn Thedéen

Introduction

As an aid for those who are using this book as a course book, a number of exercises follow here, grouped according to the nature of the tasks.

Exercises

1. A risk is usually defined in relation to one of the following:

- The probability of an event with negative consequences.
- The negative consequences.
- The product of the probabilities and the consequences.
- The perceived risk.

Assume that you are to discuss technical risks with laymen, e.g., politicians, at a local or a national risk level. What definitions do you judge to be suitable if you are discussing traffic risks, nuclear power, a new distribution network for gas to households, or the risks associated with genetically modified organisms? Give an example of a risk for which the third definition above is unsuitable.

2. In connection with the risks in technical systems, one usually considers three groups of persons: decision-makers, users, and risk-carriers. Who are included in these groups with respect to nuclear power, medical radiation treatment, ski-lifts, petrol stations, genetically modified organisms, and the closing-down of a petrol station?

3. Why do people often react much more strongly if 500 persons are killed on a single occasion, e.g., in a ferry disaster, than if 500 persons are killed in road accidents in a year?

4. What can be done from a general point of view to increase people's understanding of messages which include risk information? What such messages have you yourself received during the last few days, and which of them have you ignored?

5. Give examples from your own everyday life of voluntary and involuntary risks involving bodily injury and property damage. Are you exposed to any involuntary risk as a result of a democratic majority decision which you do not support?

6. Voluntariness, personal control, and influence in decision-making processes are among the central factors behind the acceptance of, e.g., new technology. Discuss the extent to which this can influence the possible introduction of a fuel system for automobiles based on hydrogen gas (fuel cells).

7. Do you believe that, in 25 years time, people will look back on our time and consider it to be a more, or a less, risk-filled era with respect to technology?

8. Give examples of large accidents where a change, whether technical or organizational, has been a strong contributory factor to the accident.

9. Give an example of the way in which a change intended to reduce a technical risk can instead temporarily increase the risk.

10. Give examples of large accidents where organizational deficiencies have been a strongly contributing cause. How could a change in the organization in question reduce the risk of this type of accident?

11. A teacher approves a professionally active person who has completed a technical continuation course, although the examination results show great and obvious gaps in the person's knowledge within the sphere. Soon after, due to these gaps, the person makes several serious misjudgments with an accident as the result. What responsibility does the teacher have for what has occurred?

12. Study for a week the reporting of a daily newspaper with regard to accidents with a technical connection, and the reason for these. Would you yourself draw the same conclusions as the newspaper regarding the causes?

13. What differences or similarities are there in principle between large accidents which have occurred within the building sector (e.g., the Tjörn Bridge, the Hyatt Regency), the transformation of energy (e.g., Chernobyl, Three Mile Island) and scheduled airline services (e.g., Tenerife, Gottröra)?

14. Give examples from your own experience as a private person of limit values, redundancy, and inherent security in technical systems.

15. Discuss whether or not the risks in the transport of oil can be reduced by employing several small tankers instead of one large vessel.

16. Describe possible threats against the European railway system. Are there supplementary infrastructures?

17. The critical infrastructure for financial transactions is largely electronic in character. Does this mean that the system is more vulnerable than yesterday's system which was based on the post, telephone, and telefax?

18: Choose a simple IT system, e.g., a class list or the booking of a service shop. Indicate a probable event which may lead to one of the three principal types of damage – loss of availability, loss of correctness, and loss of secrecy. Then, give the worst scenario which you can imagine for these three principal types. Are these real risks? Do you know whether anybody has tried to provide protection against them?

19. Give examples from your own everyday experience of recurring inspections or the like which are intended to reduce technical risks.

20. Compare international rules, etc., regarding the risks within electric distribution, nuclear energy, shipping, aviation, and transport of goods with a truck. Who has formulated the rules and who supervises that they are being followed? What are the sanctions in the case of an offence against the rules?

21. In what way can the rules regarding compensation to a person, organization, etc., influence the damage statistics?

22. In what way can the possible penalties for a person, organization, etc., after a damage event has occurred influence the damage statistics?

23. Limit values are one of several ways of specifying the requirement that a plant shall not give rise to harmful exposure levels. What other ways are there of reaching the same target?

24. Assume that a person is exposed to noise with a noise level of 96 dB(A) for one hour. What exposure time gives the same risk of hearing impairment if the noise level is 99 dB(A) according to the ISO Standard for the calculation of equivalent noise levels?

25. In a population, 7 % suffer from a certain type of lung disease. Of these, 90 % are smokers. Of those who do not suffer from the disease in question, 25 % are smokers. What is the probability that a randomly chosen smoker from the population suffers from the lung disease?

26. Assume that there is a method which rapidly tests whether you have received so much cadmium in the cortex of the kidney that you have to take some remedial measure. However, the test method is not infallible. If the cortex of the kidney has a higher content than the limit value, the test will show this in 99 % of the cases. If the content is actually lower than the limit value, the test method shows this in

99 % of the tests. Now, assume that 0.1 % of a population has a cadmium content higher than the limit value, and that 100 000 persons are subjected to the test. How certain is it that the cadmium content is really higher than the limit value in those cases where the test shows this?

27. The braking systems of a car consist of a footbrake and a handbrake. The footbrake consists of two interacting subsystems, each of which acts on three wheels. The handbrake acts on the rear wheels. The braking effect remains when at least one subsystem or the handbrake still functions. Draw an error-tree, where the top event is the total loss of the braking effect. Is there any redundancy in the system?

28. Give examples of technical systems where the type of risk is obvious, but where data that could give a quantitative modeling of the risk situation are almost completely lacking.

29. How can a human life be valued in connection with a risk decision? Give examples of explicit and implicit valuations.

30. In this book, the word “risk” is found in many contexts. Define or discuss, with examples from concrete situations, the concepts of objective risk, risk experience (risk apprehension, risk perception), risk source, risk measure, risk analysis, risk handling.

Appendix C

Editors and Contributors

Editors

Göran Grimvall is Professor Emeritus of Physics at the Royal Institute of Technology (KTH), Stockholm, Sweden where he has served as Dean at the School of Engineering Physics. In addition to three monographs in materials science he has published extensively on everyday physics and written several textbooks on engineering physics. Professor Grimvall is a member of the Royal Swedish Academy of Engineering Sciences since 1986.

Åke J. Holmgren is Senior Analyst at the Swedish Civil Contingencies Agency (MSB), where his work concerns critical infrastructure protection (CIP), and information assurance, mainly SCADA (Supervisory, Control and Data Acquisition) and industrial control system security. Previously he has been employed as Senior Analyst at the Swedish Emergency Management Agency (SEMA), and as Scientist at the Swedish Defence Research Agency (FOI). He is a faculty member of the Department of Safety Research at the Royal Institute of Technology (KTH), Stockholm, Sweden, and his areas of research include quantitative risk and vulnerability analysis of large technical systems, as well as operational risk and crisis management. He holds a Ph.D. degree in Safety Analysis and a M.Sc. degree in civil engineering, both from KTH, and a B.Sc. degree in business administration and economics from Stockholm University.

Per Jacobsson holds university degrees in surveying, mathematics and philosophy, education, and sociology and has a training in environmental medicine. After a brief career in surveying engineering, he has served, inter alia, as secretary on several commissions, especially concerning the interplay between technology and society, sustainable development and the quality of university education, in which areas he has also been responsible for several courses and projects. He is former director of the Centre for Environmental Science at the Royal Institute of Technology (KTH), Stockholm, Sweden and is currently Senior Administrative Officer at its University Administration.

Torbjörn Thedéen is Professor Emeritus of Safety Analysis and a former director of the Center for Safety Research at the Royal Institute of Technology (KTH), Stockholm, Sweden. He holds a Ph.D. in Mathematical Statistics. His research has been on the theory of point processes, applications to road traffic, and risk analyses of transportation, energy, and infrastructure systems. He has been vice dean of the Faculty of Engineering Physics, a member of the research board of the Swedish Emergency Management Agency (SEMA) and of the editorial board of Transportation Research.

Contributors

Evert Andersson (Chapter 9) is Professor Emeritus of Railway Technology at the Royal Institute of Technology (KTH), Stockholm, Sweden.

Terje Aven (Chapter 12) is Professor of Risk Analysis and Management at University of Stavanger, Norway.

Britt-Marie Drottz-Sjöberg (Chapter 16) is Professor of Social Psychology at the Norwegian University of Science and Technology, Trondheim, Norway.

Mats Ericson (Chapter 15) was Professor of Work Science at the Royal Institute of Technology (KTH), Stockholm, Sweden and is currently President of the Swedish National Defence College in Stockholm.

Viiveke Fåk (Chapter 10) is Associate Professor of Computer Security and Cryptography at Linköping University, Sweden.

Håkan Frantzich (Chapter 14) is Associate Professor of Fire Safety Engineering at Lund University, Sweden.

Bo Göranzon (Chapter 18) is Professor Emeritus of Professional Skills and Technology at the Royal Institute of Technology (KTH), Stockholm, Sweden .

Göran Grimvall (Chapter 1), see Editors.

Monica Gullberg (Chapter 6) holds a Licentiate of Engineering from the Royal Institute of Technology (KTH), Stockholm, Sweden and is currently employed at AB ÅF, Stockholm, Sweden.

Sven Ove Hansson (Chapter 8) is Professor of Philosophy at the Royal Institute of Technology (KTH), Stockholm, Sweden.

Åke J. Holmgren (Chapters 1 and 13), see Editors.

Göran Holmstedt (Chapter 14) is Professor Emeritus of Fire Safety Engineering at Lund University, Sweden.

Jan Hult (Chapter 4) is Professor Emeritus of Solid Mechanics at Chalmers University of Technology, Gothenburg, Sweden.

Per Jacobsson (Chapter 1), see Editors.

Inge Jonsson (Chapter 19) is Professor Emeritus of Comparative Literature and former President of Stockholm University, Sweden.

Lena Mårtensson (Chapters 9 and 15) is Professor of Industrial Ergonomics at the Royal Institute of Technology (KTH), Stockholm, Sweden.

Bengt Mattsson (Chapter 17) is Professor Emeritus of Economics at Karlstad University, Sweden.

Per Näsman (Appendix) is Research Scientist at the division of Safety Research at the Royal Institute of Technology (KTH), Stockholm, Sweden.

Birgitta Odén (Chapter 3) is Professor Emerita of History at Lund University, Sweden.

Olle Rutgersson (Chapter 9) is Professor of Naval Architecture, and Head of Chalmers Lindholmen at Chalmers University of Technology in Gothenburg, Sweden.

Lennart Sjöberg (Chapters 2 and 16) is Professor Emeritus of Economic Psychology at Stockholm School of Economics, Sweden.

Håkan Sundquist (Chapter 5) is Professor of Structural Design and Bridges at the Royal Institute of Technology (KTH), Stockholm, Sweden.

Torbjörn Thedéen (Chapters 1, 2, 6, 9, 11, and 13), see Editors.

Ulf Ulfvarsson (Chapter 7) is Professor Emeritus of Industrial Ergonomics at the Royal Institute of Technology (KTH), Stockholm, Sweden.

Index